Tharwat F. Tadros
Suspension Concentrates
De Gruyter Graduate

Also of Interest

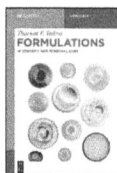

Formulations.
In Cosmetic and Personal Care
Tadros, 2016
ISBN 978-3-11-045236-5, e-ISBN 978-3-11-045238-9

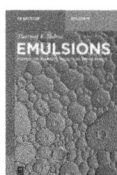

Emulsions
Formation, Stability, Industrial Applications
Tadros, 2016
ISBN 978-3-11-045217-4, e-ISBN 978-3-11-045224-2

Nanodispersions
Tadros, 2015
ISBN 978-3-11-029033-2, e-ISBN 978-3-11-029034-9

Interfacial Phenomena and Colloid Stability:
Volume 1 Basic Principles
Tadros, 2015
ISBN 978-3-11-028340-2, e-ISBN 978-3-11-028343-3

Interfacial Phenomena and Colloid Stability:
Volume 2 Industrial Applications
Tadros, 2015
ISBN 978-3-11-037107-9, e-ISBN 978-3-11-036647-1

An Introduction to Surfactants
Tadros, 2014
ISBN 978-3-11-031212-6, e-ISBN 978-3-11-031213-3

Tharwat F. Tadros

Suspension Concentrates

Preparation, Stability and Industrial Applications

DE GRUYTER

Author
Prof. Tharwat F. Tadros
89 Nash Grove Lane
Workingham RG40 4HE
Berkshire, UK
tharwat@tadros.fsnet.co.uk

ISBN 978-3-11-048678-0
e-ISBN (PDF) 978-3-11-048687-2
e-ISBN (EPUB) 978-3-11-048695-7

Library of Congress Cataloging-in-Publication Data
A CIP catalog record for this book has been applied for at the Library of Congress.

Bibliographic information published by the Deutsche Nationalbibliothek
The Deutsche Nationalbibliothek lists this publication in the Deutsche Nationalbibliografie;
detailed bibliographic data are available on the Internet at http://dnb.dnb.de.

© 2017 Walter de Gruyter GmbH, Berlin/Boston
Cover image: greenphotoKK/iStock/thinkstock
Typesetting: PTP-Berlin, Protago-TEX-Production GmbH, Berlin
Printing and binding: CPI books GmbH, Leck
♾ Printed on acid-free paper
Printed in Germany

www.degruyter.com

Preface

Suspension concentrates find application in almost every industrial preparation, e.g., paints, dyestuffs, paper coatings, printing inks, agrochemicals, pharmaceuticals, cosmetics, food products, detergents, ceramics, etc. The powder particles can be hydrophobic, e.g. organic pigments, agrochemicals, ceramics or hydrophilic, e.g. silica, titania, clays. The liquid can be aqueous or nonaqueous. The average particle size of suspensions can be within the colloid range (1 nm–1 µm) or outside the colloid range (> 1 µm). The formulation of suspension concentrates and maintenance of their physical stability over long periods of time under various conditions, e.g. temperature variation, transport, etc. still remains a challenging problem to the industrial chemist or chemical engineer. This requires an understanding of the various interfacial phenomena involved in their preparation and stabilization.

There are two main processes for the preparation of suspensions. The first depends on the "build-up" of particles from molecular units, i.e. the so-called "top-up" or condensation method, which involves two main processes, namely nucleation and growth. A particular case of the condensation process is the preparation of polymer latex particles by emulsion or suspension polymerization. The second procedure for preparation of suspension concentrates is usually referred to as "bottom-down" or dispersion process. Dispersion is a process whereby aggregates and agglomerates of powders are dispersed into "individual" units, usually followed by a wet milling process (to subdivide the particles into smaller units) and stabilization of the resulting dispersion against aggregation and sedimentation. The larger "lumps" of the insoluble substances are subdivided by mechanical or other means into smaller units. Once a suspension is prepared, it is necessary to control its properties on storage. Three main aspects must be considered. Firstly, control of its colloid stability which requires the presence of a repulsive energy that overcomes the everlasting van der Waals attraction. Two main types of stabilization may be considered: (i) electrostatic repulsion produced by the presence of electrical double layers surrounding the particles; (ii) steric repulsion produced by adsorption of nonionic surfactants or polymeric surfactants. The second property that must be controlled is the process of Ostwald ripening (crystal growth) that occurs with most suspensions of organic substances. The latter have a finite solubility in the medium that may reach several hundred ppm (parts per million). The smaller particles with higher radius of curvature have a higher solubility when compared to the larger particles. This difference in solubility between small and large particles is the driving force for Ostwald ripening. The third instability problem with suspensions is particle sedimentation, which occurs when the particle size is outside the colloid range and the density difference between the particles and the medium is significant. In this case, gravity force exceeds Brownian motion. It is important to distinguish between "colloid" and "physical" stability. Colloid stability implies absence of particle aggregation and this requires a high repulsive energy as discussed above.

DOI 10.1515/9783110486872-001

However, a colloidally stable suspension may undergo separation, e.g. as a result of particle sedimentation as discussed before. Physical stability, on the other hand, implies absence of any separation, ease of redispersion on gentle shaking or dilution. In many cases, "physical stability" may require weak flocculation and formation of a "gel-network structure". An important factor that controls physical stability is the bulk rheology of the suspension.

The present text addresses the above problems at a fundamental level and gives examples for the practical application of these principles. Chapter 1 gives a brief introduction to the main factors responsible for the stability/instability of suspensions concentrates. It also highlights the outlook of the book. Chapter 2 describes the process of preparation of suspensions by the "bottom-up" process. The advantage of this method over the "top-down" process is highlighted, namely the ability to control the particle size distribution. A section is devoted to the process of preparation of suspensions by precipitation. The effect of surface modification on precipitation kinetics is described, followed by other methods that can be applied for preparation of suspension particles. A section is devoted to the process of emulsion and suspension polymerization for preparation of latex suspensions. Chapter 3 describes the methods of preparation of suspensions using the "top-down" process. The process of wetting of powder aggregates and agglomerates (both external and internal surfaces) is described with special reference to the role of surfactants (wetting agents). The various methods that can be applied for measuring powder wetting are described. The different classes of surfactants for enhancement of wetting are described. This is followed by a section on dispersion of the aggregates and agglomerates using high speed stirrers. Finally, the methods that can be applied for size reduction are described with reference to bead milling. The main factors responsible for maintenance of colloid stability are described.

Chapter 4 deals with electrostatic stabilization of suspensions. The origin of charge in suspensions and the structure of the electrical double layer are discussed at a fundamental level. The interaction between particles containing double layers is described and a description is given of how the double layer repulsion changes with separation distance between the particle surfaces. The van der Waals attraction between the particles or droplets is described using the microscopic theory of Hamaker. This results in description of the variation of van der Waals attraction with separation distance. The electrostatic repulsion is combined with the van der Waals attraction at various separation distances in the well-known theory of colloid stability due to Deryaguin–Landau–Verwey–Overbeek (DLVO theory). The methods that can be applied for measuring the zeta potential are briefly described.

Chapter 5 describes the steric stabilization of suspensions. It starts with a section on the adsorption and conformation of nonionic surfactants and polymers at various interfaces. This is followed by describing the interaction between particle or droplets containing adsorbed surfactant or polymer layers. The steric interaction is described in terms of the unfavourable mixing of the adsorbed layers (when these are in good solvent conditions) and the unfavourable loss of entropy on significant overlap of the adsorbed layers. Combination of these two effects with van der Waals attraction results in the theory of steric stabilization. The energy-distance curves of sterically stabilized dispersions are presented with particular reference to the effect of the ratio of adsorbed layer thickness to particle radius.

Chapter 6 describes the various processes of flocculation of electrostatically and sterically stabilized suspensions. The theories of fast and slow flocculation of electrostatically stabilized suspensions are described with particular reference to the effect of electrolyte concentration and valency. Four different flocculation mechanisms are described for sterically stabilized suspension: weak (reversible) flocculation, incipient flocculation, depletion and bridging flocculation. The parameters that affect each type of flocculation are described.

Chapter 7 describes the Ostwald ripening of suspensions and its prevention. It starts with the Kelvin theory that describes the effect of curvature (particle size) on the solubility of the disperse phases. It shows the rapid increase in solubility when the particle size is reduced below 500 nm. The kinetics of Ostwald ripening is described showing the change of the cube of the mean radius with time and the effect of particle solubility on the rate. The effect of Brownian motion, phase volume fraction and surfactant micelles are described. This is followed by the methods that can be applied to reduce Ostwald ripening for suspensions.

Chapter 8 deals with sedimentation of suspensions and prevention of formation of dilatant sediments. It starts with the effect of particle size and its distribution on sedimentation. Sedimentation in non-Newtonian liquids and correlation of sedimentation rate with residual (zero shear) viscosity is illustrated. The role of thickeners (rheology modifiers) in preventing sedimentation is discussed at a fundamental level. A section is given over to the various methods that can be applied for preventing sedimentation.

Chapter 9 describes the rheological behaviour of suspension concentrates and the various parameters that affect the flow characteristics of suspensions, namely the balance between Brownian diffusion, hydrodynamic interaction and interparticle forces. It starts with a section describing the various rheological methods that can be applied, namely steady state shear stress-shear rate, constant stress (creep) and oscillatory measurements. The Einstein equation for very dilute suspensions ($\phi \leq 0.01$) and its modification by Batchelor for moderately concentrated suspensions (with $0.2 \geq \phi \geq 0.1$) is described. This is followed by a section on the rheology of concentrated suspensions ($\phi > 0.2$). It starts with hard-sphere suspensions where both repulsion and attraction are screened. The models for analysis of the viscosity versus volume fraction curves are described. This is followed by electrostatically stabilized suspensions where the rheology is determined by double layer repulsion. The rheology of sterically stabilized suspensions is described with particular reference to the effect of the adsorbed layer thickness. Finally, the rheology of flocculated suspensions is described and a distinction is made between weakly and strongly flocculated systems.

The various semi-empirical models that can be applied for analysis of the flow curves are described.

Chapter 10 describes nonaqueous (oil-based) suspensions where the dispersion medium can be a polar solvent (with relative permittivity > 10), such as alcohols, glycols, glycerol, etc. or a nonpolar solvent (with relative permittivity < 10), such as hydrocarbons which can have a relative permittivity as low as 2. In the first case, charge separation occurs and double layer repulsion plays an important role. In the second case, charge separation and double layer repulsion are not effective and one has to depend on the use of dispersants that produce steric stabilization. The colloid stability of suspensions in polar and nonpolar media is considered at a fundamental level. This is followed by a section on settling of suspensions in nonaqueous media and methods that can be applied for its prevention. Finally, a brief discussion is given of the rheological properties of these oil-based suspensions.

Chapter 11 describes the characterization, assessment and prediction of the physical stability of suspensions. For full characterization of the properties of suspensions, three main types of investigations are needed: (i) Fundamental investigation of the system at a molecular level. This requires investigations of the structure of the solid/liquid interface, namely the structure of the electrical double layer (for charge stabilized suspensions), adsorption of surfactants, polymers and polyelectrolytes and conformation of the adsorbed layers (e.g. the adsorbed layer thickness). (ii) Investigation of the state of suspension on standing, namely flocculation rates, flocculation points with sterically stabilized systems, spontaneity of dispersion on dilution and Ostwald ripening or crystal growth. (iii) Bulk properties of the suspension, that is particularly important for concentrated systems. This requires measurement of the rate of sedimentation and equilibrium sediment height. More quantitative techniques are based on assessment of the rheological properties of the suspension (without disturbing the system, i.e. without its dilution and measurements are carried out under conditions of low deformation) and how these are affected by long-term storage. These rheological methods can be applied for prediction of the physical stability of suspensions.

Chapter 12 summarizes the application of suspensions in pharmacy. Several examples are given to illustrate the various applications such as oral suspensions, topical suspensions, parenteral suspensions, antacid and clay suspensions and reconstitutable suspensions. Chapter 13 summarizes the application of suspensions in cosmetics, personal care and liquid detergent products. Three main applications in cosmetics are considered, namely the use of suspensions in sunscreen protection, colour cosmetics and make-up products. Chapter 14 discusses the application of suspensions in paints and coatings. Two main aspects are considered, namely the use of latexes in emulsion paints and the dispersion of various pigments for paint applications. Chapter 15 summarizes the application of suspension concentrates in formulation of water insoluble agrochemical products. The main advantages of the use of suspension concentrates are discussed with particular reference to the lack of hazard and possibility of incorporation of various adjuvants that are necessary for biological control.

This book gives a comprehensive text on the methods that can be used for preparation of suspension concentrates, the control of their colloid and physical stability as well as assessment of their stability and their applications. It will provide the formulation chemist, pharmacist, chemical engineer and research worker with a fundamental approach on how to arrive at a required state of a suspension and how to maintain the of ease of its application.

Tharwat Tadros
February 2017

Contents

1 General introduction

Suspensions are solid/liquid dispersions that find application in almost every industrial preparation, e.g. paints, dyestuffs, paper coatings, printing inks, agrochemicals, pharmaceuticals, cosmetics, food products, detergents, ceramics, etc. The powder particles can be hydrophobic, e.g. organic pigments, agrochemicals, ceramics or hydrophilic, e.g. silica, titania, clays [1–5]. The liquid can be aqueous or nonaqueous. The average particle size of suspensions can be within the colloid range (1 nm–1 μm) or outside the colloid range (> 1 μm). Solid-in-liquid dispersions within the colloid range are usually referred to as colloidal suspensions. Coarse suspension concentrates are to be distinguished from colloidal suspensions in the sense that the particles of the former settle to the bottom of the container (as a result of the gravitational field on the particles) whereas with colloidal suspensions, with particle density not significantly larger than that of the medium, the very mild mixing produced by ambient thermal fluctuations and/or Brownian motion can keep the particles uniformly dispersed in the continuous medium [1]. The formulation of suspension concentrates and maintenance of their physical stability over long periods of time under various conditions, e.g. temperature variation, transport, etc. still remains a challenging problem to the industrial chemist or chemical engineer. This requires an understanding of the various interfacial phenomena involved in their preparation and stabilization [6, 7].

The concentration of a suspension is described by its volume fraction ϕ, namely the ratio between the total volume of particles to the volume of the suspension. The value of ϕ above which a suspension may be considered "dilute", "concentrated" or "solid" can be defined from consideration of the balance between the particle translational motion (Brownian diffusion) and interparticle interaction [8–10]. If Brownian diffusion predominates over the imposed interparticle interaction, the suspension may be described as "dilute". In this case the particle translational motion is large and only occasional contacts may occur between the particles which are then separated by the Brownian force. The particle interactions can be represented by two-body collisions. This "dilute" suspension generally has time-independent properties and if the particle size is within the colloid range and the density difference between the particles is very small, no gravitational sedimentation of particles occurs and the system maintains its properties over long periods of time. These "dilute" suspensions show Newtonian flow, i.e. their viscosity is independent of the applied shear rate [11]. In contrast, if interparticle interaction predominates over Brownian diffusion, i.e. the interparticle distance h becomes much smaller than the particle radius, the system may be described as a "solid" suspension. In this case the particles may vibrate with a distance h that is much smaller than the particle radius. The interaction produces a specific order between the particles and a highly developed structure is reached. The resulting "solid" suspension becomes predominantly elastic with very little energy dissipation during flow [11]. Again the properties of these elastic systems are

DOI 10.1515/9783110486872-002

time independent. In between the two extremes one may define a volume fraction ϕ at which the suspension may be considered "concentrated". In this case the inter-particle distance h is comparable to the particle radius and the suspension shows time-dependent spatial and temporal correlations. The particle interactions occur with many body collisions and the translational motion of the particles is restricted. These "concentrated" suspensions show viscoelastic behaviour, i.e. a combination of viscous and elastic response [11]. At low stresses the suspension may show a predominantly elastic response, whereas at high stress a predominantly viscous response is obtained (see Chapter 9)

There are two main processes for the preparation of suspensions [12]. The first depends on the "build-up" of particles from molecular units, i.e. the so-called "top-up" or condensation method, which involves two main processes, namely nucleation and growth. In this case, it is necessary first to prepare a molecular (ionic, atomic or molecular) distribution of the insoluble substances; then, by changing the conditions, precipitation is caused leading to the formation of nuclei that grow into the particles in question. A particular case of the condensation process is the preparation of polymer latex particles by emulsion or suspension polymerization. In emulsion polymerization, the water insoluble monomer such as styrene or methyl acrylate is emulsified in water using a surfactant such as sodium dodecyl sulphate. An initiator such as potassium persulphate is added and the process of polymerization is carried out by heating the system. Initially, oligomers of styrene sulphate are produced which aggregate to form nuclei that grow by a process of coagulation and finally the polymer latex particles are produced. In suspension polymerization, the monomer is dissolved in an organic solvent in which the resulting polymer is insoluble. An oil soluble initiator is added together with a polymeric surfactant (protective colloid) to prevent aggregation of the resulting polymer particles. The second procedure for preparation of suspension concentrates is usually referred to as the "bottom-down" or dispersion process. Dispersion is a process whereby aggregates and agglomerates of powders are dispersed into "individual" units, usually followed by a wet milling process (to subdivide the particles into smaller units) and stabilization of the resulting dispersion against aggregation and sedimentation. The larger "lumps" of the insoluble substances are subdivided by mechanical or other means into smaller units. In the whole "bottom-down" process it is essential to wet both the external and internal surfaces of the aggregates and agglomerates. For hydrophobic solids dispersed in aqueous media, it is essential to use a wetting agent (surfactant) that lowers the surface tension of water (under dynamic conditions) and reduces the solid/liquid interfacial tension by adsorption on the surface of the particles. After wetting, the aggregates and agglomerates are dispersed into single particles by high speed mixing. The resulting suspension, referred to as "mill base" is then subjected to a milling process to reduce the particle size to the desired value.

Once a suspension is prepared, it is necessary to control its properties on storage. Three main aspects must be considered. Firstly, control of its colloid stability, which requires the presence of a repulsive energy that overcomes the everlasting van der Waals attraction. Three main types of stabilization may be considered: (i) Electrostatic repulsion produced by the presence of electrical double layers surrounding the particles [13]. These double layers are extended in solution, particularly at low electrolyte concentration and low valency of ions forming the double layers. When two particles with these extended double layers approach each other to a separation distance h that is smaller than twice the double layer extension ("thickness") strong repulsion occurs. This repulsion G_{elec} increases with decreasing electrolyte concentration and valency. Combining G_{elec} with the van der Waals attraction G_A results in an energy-distance curve that forms the basis of the theory of colloid stability due to Deryaguin–Landau–Verwey–Overbeek (DLVO theory) [14, 15]. This energy-distance curve shows a maximum (energy barrier) at intermediate separation distances (when the 1:1 electrolyte concentration is $< 10^{-2}$ mol dm^{-3}). This energy barrier prevents strong aggregation of the particles thus maintaining effective stability. (ii) Steric repulsion produced by adsorption of nonionic surfactants or polymeric surfactants. These surfactants consist of an "anchor" chain(s) that strongly adsorbs on the particle surface and a stabilizing chain(s) that remains in solution and becomes strongly solvated by the molecules of the medium [16]. One can define an adsorbed layer thickness δ which increases with increasing molar mass of the stabilizing chain(s). When two particles with adsorbed surfactant or polymer layers approach to a distance h that is lower than 2δ, the stabilizing chains may overlap or become compressed resulting in an increase in the segment concentration in the overlapped or compressed layers. Provided the latter are in good solvent conditions (highly solvated by the molecules of the medium), this effect results in an increase in the osmotic pressure in these overlapped or compressed layers. Solvent molecules will now diffuse to these layers, thus separating the particles. This repulsive effect is referred to as the mixing interaction energy G_{mix} (unfavourable mixing of the stabilizing chains). In the overlapped or compressed layers, the configurational entropy of the chains is significantly reduced resulting in another repulsive energy G_{el} (entropic, volume restriction or elastic interaction). The sum of G_{mix} and G_{el} is referred to as G_s (steric repulsive energy). Combining G_{mix} and G_{el} with G_A gives G_T–h curves and this forms the basis of the theory of steric stabilization [16]. In this case the G_T–h curve shows a shallow minimum at separation distance h $\approx 2\delta$, but when h $< 2\delta$, G_T increases sharply with further decreases in h. (iii) Electrosteric stabilization where G_{elec} and G_s are combined with G_A. In this case, the G_T–h curve shows a shallow minimum at large h values, a maximum at intermediate distances (DLVO type maximum) and a sharp increase at distances comparable to 2δ. Electrosteric stabilization is generally produced when using polyelectrolytes to stabilize the suspension or when using a mixture of nonionic or polymeric surfactant with an ionic one.

The second property that must be controlled is the process of Ostwald ripening (crystal growth) that occurs with most suspensions of organic substances [1–3]. The latter have a finite solubility in the medium that may reach several hundred ppm (parts per million). The smaller particles with higher radius of curvature have a higher solubility when compared with the larger particles. This difference in solubility between small and large particles is the driving force for Ostwald ripening. With time, molecular diffusion occurs from the small to the large particles. With the ultimate dissolution of these small particles their molecules become deposited on the larger particles. Thus, on storage of the suspension, the particle size distribution shifts to larger particles. This will result in instability of the suspension, e.g. enhanced sedimentation. Another mechanism of Ostwald ripening is due to polymorphic changes. The particles (e.g. a drug) may contain two or more polymorphs with different solubility. The more stable polymorph has a lower solubility when compared with the metastable polymorph. With time, the more soluble polymorph gradually changes to the less soluble stable one. This Ostwald ripening problem may result in reduction of bioavailability of the drug and hence it must be reduced or eliminated. Several methods can be applied to reduce Ostwald ripening in suspensions, e.g. incorporation of impurities that strongly adsorb on the particle surface, thus blocking the active sites for growth. Alternatively one can use strongly adsorbed polymeric surfactants which have the same effect as the added impurities.

The third instability problem with suspensions is particle sedimentation [1–4], which occurs when the particle size is outside the colloid range and the density difference between the particles and the medium is significant. In this case the gravity force $(4/3)\pi R^3 \Delta \rho g h$ (where R is the particle radius, $\Delta \rho$ is the density difference between the particle and the medium, g is the gravity force and h is the height of the container) exceeds the Brownian motion kT (where k is the Boltzmann constant and T is the absolute temperature). With most practical suspensions with a wide particle size distribution, the larger particles sediment at a higher rate than the smaller particles. A particle concentration gradient of the particles occurs across the container. Several methods may be applied to reduce sedimentation, e.g. balance of the density of the disperse phase with that of the medium, reduction of particle size (i.e. formation of nanosuspensions) and addition of thickeners. The latter can be high molecular weight polymers such as xanthan gum, or addition of "inert" fine particles such as silica or clays. In all cases these thickeners produce a "gel network" in the continuous phase which produces a very high viscosity at low shear rates that prevents particle sedimentation.

It is now important to distinguish between "colloid" and "physical" stability. Colloid stability implies absence of particle aggregation and this requires a high repulsive energy as discussed above. However, a colloidally stable suspension may undergo separation, e.g. as a result of particle sedimentation as discussed above. Physical stability, on the other hand, implies absence of any separation, ease of redispersion on gentle shaking or dilution. In many cases "physical stability" may require weak flocculation and formation of a "gel-network structure".

An important factor that controls physical stability is the bulk rheology of the suspension.

One way to distinguish between colloid and physical stability is to consider the various states of the suspension on standing as schematically illustrated in Fig. 1.1. These states are determined by the interaction energy between the particles, the effect of gravity and addition of other components such as surfactants, polymers and thickeners [4].

States (a)–(c) correspond to a suspension that is stable in the colloid sense. The stability is obtained as a result of net repulsion due to the presence of extended double layers (i.e. at low electrolyte concentration), as the result of steric repulsion producing adsorption of nonionic surfactants or polymers, or as the result of a combination of double layer and steric repulsion (electrosteric). State (a) represents the case of a suspension with small particle size (submicron) where the Brownian diffusion overcomes the gravity force producing uniform distribution of the particles in the suspension, i.e. $kT \gg (4/3)\pi R^3 \Delta \rho g h$.

A good example of the above case is a latex suspension with particle size well below $1\,\mu m$ that is stabilized by ionogenic groups, by an ionic surfactant or nonionic surfactant or polymer. This suspension will show no separation on storage for long periods of time.

States (b) and (c) represent the case of suspensions in which the particle size range is outside the colloid range ($> 1\,\mu m$). In this case, the gravity force exceeds the Brownian diffusion, i.e. $(4/3)\pi R^3 \Delta \rho g h \gg kT$.

With state (b), the particles are uniform and initially they are well dispersed, but with time and the influence of gravity they settle to form a hard sediment (technically referred to as "clay" or "cake"). In the sediment the particles are subjected to a hydrostatic pressure $h\rho g$, where h is the height of the container, ρ is the density of the particles and g is the acceleration due to gravity. Within the sediment each particle will be acting constantly with many others, and eventually an equilibrium is reached where the forces acting between the particles are balanced by the hydrostatic pressure on the system. The forces acting between the particles will depend on the mechanism used to stabilize the particles, for example electrostatic, steric or electrosteric, the size and shape of the particles, the medium permittivity (dielectric constant), electrolyte concentration, the density of the particles, etc. Many of these factors can be incorporated to give an interaction energy in the form of a pair potential for two particles in an infinite medium. The repulsive forces between the particles allow them to move past each other until they reach small distances of separation (that are determined by the location of the repulsive barrier). Due to the small distances between the particles in the sediment, it is very difficult to redisperse the suspension by simple shaking.

With case (c), consisting of a wide distribution of particle sizes, the sediment may contain larger proportions of the larger sized particles, but still a hard "clay" is produced. These "clays" are dilatant (i.e. shear thickening) and they can be easily detected by inserting a glass rod in the suspension. Penetration of the glass rod into these hard sediments is very difficult.

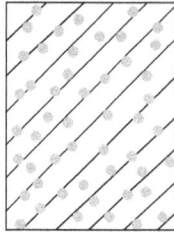
(a) Stable colloidal suspension

(b) Stable coarse suspension (uniform size)

(c) Stable coarse suspension (size distribution)

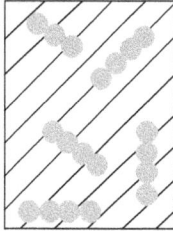
(d) Coagulated suspension (chain aggregates)

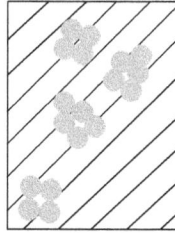
(e) Coagulated suspension (compact clusters)

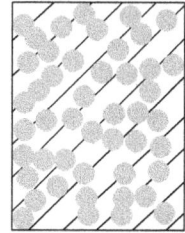
(f) Coagulated suspension (open structure)

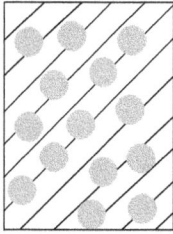
(g) Weakly flocculated structure

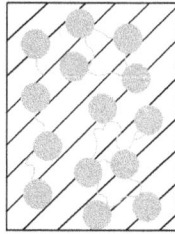
(h) Bridging flocculation

(i) Depletion flocculation

Fig. 1.1: States of the suspension.

States (d)–(f) represent the case for unstable, coagulated suspensions which either have a small repulsive energy barrier or it is completely absence. State (d) represents the case of coagulation under the condition of no stirring: chain aggregates are produced that will settle under gravity forming a relatively open structure. State (e) represents the case of coagulation under stirring conditions: compact aggregates are produced that will settle faster than the chain aggregates and the sediment produced is more compact. State (f) represents the case of coagulation at high volume fraction of the particles, ϕ. In this case whole particles will form a "one-floc" structure that is formed from chains and cross chains that extend from one wall to the other in the container. Such a coagulated structure may undergo some compression (consolidation) under gravity leaving a clear supernatant liquid layer at the top of the container. This phenomenon is referred to as syneresis.

State (g) represents the case of weak and reversible flocculation. This occurs when the secondary minimum in the energy-distance curve is deep enough to cause flocculation [4]. This can occur at moderate electrolyte concentrations, in particular with larger particles. The same occurs with sterically and electrosterically stabilized suspensions. This occurs when the adsorbed layer thickness is not very large, particularly with large particles. The minimum depth required for causing weak flocculation depends on the volume fraction of the suspension. The higher the volume fraction, the lower the minimum depth required for weak flocculation.

The above flocculation is weak and reversible, i.e. on shaking the container redispersion of the suspension occurs. On standing, the dispersed particles aggregate to form a weak "gel". This process (referred to as sol-gel transformation) leads to reversible time dependence of viscosity (thixotropy). On shearing the suspension, the viscosity decreases and when the shear is removed, the viscosity is recovered. This phenomenon is applied in paints. On application of the paint (by a brush or roller), the gel is fluidized, allowing uniform coating of the paint. When shearing is stopped, the paint film recovers its viscosity and this avoids any dripping.

State (h) represents the case in which the particles are not completely covered by the polymer chains. In this case, simultaneous adsorption of one polymer chain on more than one particle occurs, leading to bridging flocculation. If the polymer adsorption is weak (low adsorption energy per polymer segment), the flocculation could be weak and reversible. In contrast, if the adsorption of the polymer is strong, tough flocs are produced and the flocculation is irreversible. The latter phenomenon is used for solid/liquid separation, e.g. in water and effluent treatment.

Case (i) represents a phenomenon, referred to as depletion flocculation, produced by addition of "free", nonadsorbing polymer [17, 18]. In this case, the polymer coils cannot approach the particles to a distance Δ (that is determined by the radius of gyration of free polymer R_G), since the reduction of entropy on close approach of the polymer coils is not compensated by an adsorption energy. The suspension particles will be surrounded by a depletion zone with thickness Δ. Above a critical volume fraction of the free polymer, ϕ_p^+, the polymer coils are "squeezed out" from between the particles and the depletion zones begin to interact. The interstices between the particles are now free from polymer coils and hence an osmotic pressure is exerted outside the particle surface (the osmotic pressure outside is higher than in between the particles) resulting in weak flocculation. The magnitude of the depletion attraction free energy, G_{dep}, is proportional to the osmotic pressure of the polymer solution, which in turn is determined by ϕ_p and molecular weight M. The range of depletion attraction is proportional to the thickness of the depletion zone, Δ, which is roughly equal to the radius of gyration, R_G, of the free polymer.

Outline of the book

Chapter 2 describes the preparation of suspensions by the "bottom-up" process. The advantage of this method over the "top-down" process is highlighted, namely the ability to control the particle size distribution. A section is devoted to the process of preparation of suspensions by precipitation. The classical theory of nucleation and growth (Gibbs–Volmer theory) is described in terms of the free energy components of the process, namely the surface free energy and the bulk energy in producing a new phase. This results in definition of the critical nucleus size above which spontaneous particle growth occurs. The dependence of the critical nucleus size on supersaturation is described and particular reference is given to the effect of added surfactants. A section is devoted to the kinetics of precipitation and control of particle size distribution. The effect of surface modification on precipitation kinetics is described, followed by other methods that can be applied for preparation of suspension particles. A section is devoted to the process of emulsion and suspension polymerization for preparation of latex suspensions.

Chapter 3 describes the methods of preparation of suspensions using the top-down process. The process of wetting of powder aggregates and agglomerates (both external and internal surfaces) is described with special reference to the role of surfactants (wetting agents). Wetting of the external surface requires the presence of a surfactant that lowers the surface tension of the liquid and the interfacial tension of the solid/liquid interface. This results in lowering the contact angle at the three phase region of solid/liquid/vapour. For adequate wetting of the external surface, a contact angle approaching zero is required. Wetting of the internal surface (pores of aggregates and agglomerates) requires adequate penetration of the liquid inside the pores. Again a low contact angle is required, but good penetration of the liquid requires a high liquid surface tension. Thus, a compromise is needed for good wetting of the external and internal surfaces, namely a low contact angle but not too low surface tension. This shows the difficulty of choosing the right wetting agent for any particular powder. The various methods that can be applied for measuring powder wetting are described. The different classes of surfactants for enhancement of wetting are described. This is followed by a section on dispersion of the aggregates and agglomerates using high speed stirrers. Finally, the methods that can be applied for size reduction are described with reference to bead milling. The main factors responsible for the maintenance of colloid stability are described.

Chapter 4 deals with electrostatic stabilization of suspensions. The origin of charge in suspensions and the structure of the electrical double layer are discussed at a fundamental level. The charge and potential distribution at the solid/liquid interface are described with particular reference to the effect of electrolyte concentration and valency on the double layer extension. The interaction between particles containing double layers is described and a description is given of how the double layer repulsion changes with separation distance between the particle or droplet surfaces. The van der Waals attraction between the particles or droplets is described using the microscopic theory of Hamaker. This results in the description of the variation of van der Waals attraction with separation distance. The electrostatic repulsion is combined with the van der Waals attraction at various separation distances in the well-know theory of colloid stability due to Deryaguin–Landau–Verwey–Overbeek (DLVO theory) [14, 15]. Energy-distance curves are presented at various electrolyte concentrations to distinguish between stable and unstable systems. The methods that can be applied for measuring the zeta potential are briefly described.

Chapter 5 describes the steric stabilization of suspensions. It starts with a section on the adsorption and conformation of nonionic surfactants and polymers at various interfaces. This is followed by describing the interaction between particles containing adsorbed surfactant or polymer layers. Steric interaction is described in terms of the unfavourable mixing of the adsorbed layers (when these are in good solvent conditions) and the unfavourable loss of entropy on significant overlap of the adsorbed layers. Combining these two effects with van der Waals attraction results in the theory of steric stabilization. The energy-distance curves of sterically stabilized dispersions are presented with particular reference to the effect of the ratio of adsorbed layer thickness to particle or droplet radius.

Chapter 6 describes the various processes of flocculation of electrostatically and sterically stabilized suspensions. The theories of fast and slow flocculation of electrostatically stabilized suspensions are described with particular reference to the effect of electrolyte concentration and valency. This leads to the definition of the critical coagulation concentration. Four different flocculation mechanisms are described for sterically stabilized suspension: weak (reversible) flocculation, incipient flocculation, depletion and bridging flocculation. The parameters that affect each type of flocculation are described.

Chapter 7 describes the Ostwald ripening of suspensions and its prevention. It starts with the Kelvin theory that describes the effect of curvature (particle or droplet size) on the solubility of the disperse phases. It shows the rapid increase in solubility when the particle size is reduced below 500 nm. The kinetics of Ostwald ripening is described showing the change of the cube of the mean radius with time and the effect of particle solubility on the rate. The effects of Brownian motion, phase volume fraction and surfactant micelles are described. This is followed by the methods that can be applied to reduce Ostwald ripening for suspensions.

Chapter 8 deals with sedimentation of suspensions and prevention of formation of dilatant sediments. It starts with the effect of particle size and its distribution on sedimentation. The sedimentation of very dilute suspensions with a volume fraction $\phi \leq 0.01$ and application of Stokes law is described. This is followed by the description of sedimentation of moderately concentrated suspensions (with $0.2 \geq \phi \geq 0.1$) and the effect of hydrodynamic interaction. Sedimentation of concentrated suspensions ($\phi > 0.2$) and models for its description are described. Sedimentation in non-Newtonian liquids and correlation of sedimentation rate with residual (zero shear) viscosity is illustrated. The role of thickeners (rheology modifiers) in prevention of sedimentation is discussed at a fundamental level. One section is given over to ways of preventing sedimentation: balance of density, reduction of particle size and use of thickeners and finely divided inert particles. The application of depletion flocculation for reducing sedimentation is described. The use of liquid crystalline phases for reducing sedimentation is also described.

Chapter 9 describes the rheological behaviour of suspension concentrates and the various parameters that affect the flow characteristics of suspensions, namely the balance between Brownian diffusion, hydrodynamic interaction and interparticle forces. It starts with a section describing the various rheological methods that can be applied. The first method is steady state shear stress–shear rate measurements that can distinguish between Newtonian and non-Newtonian flow. The various rheological models that can be applied for analysis of the flow curves are described. This is followed by a description of constant stress (creep) measurements and the concept of residual (zero shear) viscosity and critical stress. This is followed by dynamic or oscillatory techniques and calculation of the complex, storage and loss moduli. The Einstein equation for very dilute suspensions ($\phi \leq 0.01$) and its modification by Batchelor for moderately concentrated suspensions (with $0.2 \geq \phi \geq 0.1$) is described. This is followed by a section on the rheology of concentrated suspensions ($\phi > 0.2$). It starts with hard sphere suspensions where both repulsion and attraction are screened. The models for the analysis of viscosity versus volume fraction curves are described. This is followed by electrostatically stabilized suspensions where the rheology is determined by double layer repulsion. The rheology of sterically stabilized suspensions is described with particular reference to the effect of the adsorbed layer thickness. Finally, the rheology of flocculated suspensions is described and a distinction is made between weakly and strongly flocculated systems. The various semi-empirical models that can be applied for analysis of the flow curves are described.

Chapter 10 describes nonaqueous (oil-based) suspensions where the dispersion medium can be a polar solvent (with relative permittivity > 10), such as alcohols, glycols, glycerol, etc., or a nonpolar solvent (with relative permittivity < 10), such as hydrocarbons, which can have a relative permittivity as low as 2. In the first case, charge separation occurs and double layer repulsion plays an important role. In the second case, charge separation and double layer repulsion are not effective and one has to depend on the use of dispersants that produce steric stabilization. The colloid stability of suspensions in polar and nonpolar media is considered at a fundamental level. This is followed by a section on settling of suspensions in nonaqueous media and methods that can be applied for its prevention. Finally, a brief discussion is given on the rheological properties of these oil-based suspensions.

Chapter 11 describes the characterization, assessment and prediction of the physical stability of suspensions. For full characterization of the properties of suspensions, three main types of investigations are needed: (i) The fundamental investigation of the system at a molecular level. This requires investigation of the structure of the solid/liquid interface, namely the structure of the electrical double layer (for charge stabilized suspensions), adsorption of surfactants, polymers and polyelectrolytes and conformation of the adsorbed layers (e.g. the adsorbed layer thickness). It is important to know how each of these parameters changes with conditions such as temperature, solvency of the medium for the adsorbed layers and effect of addition of electrolytes. (ii) Investigation of the state of suspension on standing, namely flocculation rates, flocculation points with sterically stabilized systems, spontaneity of dispersion on dilution and Ostwald ripening or crystal growth. All these phenomena require accurate determination of the particle size distribution as a function of storage time. (iii) Bulk properties of the suspension, that is particularly important for concentrated systems. This requires measurement of the rate of sedimentation and equilibrium sediment height. More quantitative techniques are based on assessment of the rheological properties of the suspension (without disturbing the system, i.e. without its dilution and measurement under conditions of low deformation), and how these are affected by long-term storage. These rheological methods can be applied for prediction of the physical stability of suspensions.

Chapter 12 summarizes the application of suspensions in pharmacy. Several examples are given to illustrate the various applications such as oral suspensions, topical suspensions, parenteral suspensions, antacid and clay suspensions and reconstitutable suspensions.

Chapter 13 summarizes the application of suspensions in cosmetics, personal care and liquid detergent products. Three main applications in cosmetics are considered, namely the use of suspensions in sunscreen protection, colour cosmetics and make-up products. One section is devoted to applications of suspensions in liquid detergents and laundry products.

Chapter 14 discusses the application of suspensions in paints and coatings. Two main aspects are considered, namely the use of latexes in emulsion paints and the dispersion of various pigments for paint applications.

Chapter 15 summarizes the application of suspension concentrates in formulation of water insoluble agrochemical products. The main advantages of the use of suspension concentrates are discussed with particular reference to the lack of hazard and possibility of incorporation of various adjuvants that are necessary for biological control.

References

[1]   Tadros, Th. F., Advances in Colloid and Interface Science, **12**, 141 (1980)
[2]   Tadros, Th. F., "Solid/Liquid Dispersions", Academic Press, London (1987).
[3]   Tadros, Th. F., "Applied Surfactants", Wiley-VCH, Germany (2005).
[4]   Tadros, Th. F., "Dispersion of Powders in Liquids and Stabilisation of Suspensions", Wiley-VCH, Germany (2012).

[5]   Tadros, Th. F., "Suspensions", in: 2010 Encyclopedia of Colloid and Interface Science", Th. F. Tadros (ed.), Springer, Germany (2013).

[6]   Tadros, Th. F., "Interfacial Phenomena and Colloid Stability", Volumes 1&2, De Gruyter, Germany (2015).

[7]   Tadros, Th. F., Advances in Colloid and Interface Science, **68**, 97 (1996).

[8]   Ottewill, R. H., in "Concentrated Dispersions", J. W. Goodwin (ed.), Royal Society of Chemistry Publication, N.43, London (1982), Chapter 9.

[9]   Ottewill, R. H., in: "Science and Technology of Polymer Colloids", G. W. Poehlein, R. H. Ottewill and J. W. Goodwin (eds.), Martinus Nishof Publishing, Boston, The Hague, Vol. II (1983), p. 503.

[10] Ottewill, R. H., "Properties of Concentrated Suspensions", in: "Solid/Liquid Dispersions", Th. F. Tadros (ed.), Academic Press, London (1987).

[11] Tadros, Th. F., "Rheology of Dispersions", Wiley-VCH, Germany (2010).

[12] Tadros, Th. F., "Nanodispersions", De Gruyter, Germany (2016).

[13] Bijesterbosch, B. H., in "Solid/Liquid Dispersions", Th. F. Tadros (ed.), Academic Press, London (1987).

[14] Deryaguin, B. V. and Landau, L., Acta Physicochem. USSR, 14, 633 (1941).

[15] Verwey, E. J. W. and Overbeek, J. Th. G., "Theory of Stability of Lyophobic Colloids", Elsevier, Amsterdam (1948).

[16] Napper, D. H., "Polymeric Stabilisation of Colloidal Dispersions", Academic Press, London (1983).

[17] Asakura, S. and Oosawa, F., J. Chem. Phys., **22**, 1235 (1954).

[18] Asakura, S. and Oosawa, F., J. Polymer Sci., **93**, 183 (1958).

2 Preparation of suspension concentrates by the bottom-up process

2.1 Introduction

As mentioned in Chapter 1, the bottom-up process for preparation of suspensions involves forming particles from molecular units. A good example is the preparation of particles of inorganic materials, such as silica, titania, ZnO, etc., i.e. a process of precipitation, nucleation and growth. This will form the first part of this chapter. Another example is the preparation of polymer colloids by emulsion or dispersion polymerization and this will form the second part of this chapter.

One of the main advantages of the bottom-up process over the top-down process is the possibility to control particle size and shape distribution as well as the morphology of the resulting particles [1, 2]. By controlling the process of nucleation and growth it is possible to obtain suspensions with a narrow size distribution. This is particularly important for many practical applications such as photonic materials and semiconductor colloids. However, for other processes, such as in paints and ceramic processes, a modest polydispersity can be beneficial to enhance the random packing density of the spheres and consequently the viscosity of the mixtures is generally below that for monodisperse spheres at the same volume fraction [1].

A very important aspect of suspensions is the maintenance of their colloid stability, i.e. absence of any flocculation. This can be achieved by three different mechanisms which will be discussed in detail in Chapters 4 and 5 and only a brief summary is given here: (i) Electrostatic stabilization by creation of a surface charge as a result of ionization of surface groups or adsorption of ionic surfactants and formation of electrical double layer. Stabilization is obtained by double layer overlap on the approach of the particles. This repulsion will overcome the van der Waals attraction, particularly at intermediate separation distances. As a result an energy barrier is produces that prevents any flocculation [3, 4]. (ii) Steric stabilization produced by adsorption of nonionic surfactants or polymeric surfactants [5]. These molecules consist of an "anchor" chain B that strongly adsorb on the particle surface and a stabilizing chain A that is strongly solvated by the molecules of the medium. When two particles approach to a separation distance h that is smaller than twice the adsorbed layer thickness (2δ), overlap of the layers occurs resulting in strong repulsion as a result of two main effects: unfavourable mixing of the A chains when these are in good solvent conditions and reduction of configurational entropy of the A chains on considerable overlap. As a result of these two effects, the energy-distance curve between the particles shows a very rapid increase in repulsion when $h < 2\delta$ and this prevents any flocculation. (iii) Electrosteric repulsion, which is a combination of electrostatic and steric repul-

DOI 10.1515/9783110486872-003

sion. This can be produced when using a mixture of ionic and nonionic surfactants or polymers or when using a polyelectrolyte.

2.2 Preparation of suspensions by precipitation

This method is usually applied for preparation of inorganic particles such as metal oxides and nanoparticles of metals. As an illustration, ferric oxide and aluminium oxide or hydroxide are prepared by hydrolysis of metal salts [1],

$$2FeCl_3 + 3H_2O \rightarrow F_2O_3\downarrow + 6HCl, \qquad (2.1)$$

$$AlCl_3 + 3H_2O \rightarrow AlOOOH + 3HCl. \qquad (2.2)$$

Another example of preparation of particles by precipitation is that of silica sols. These can be prepared by acidification of water glass, a strongly alkaline solution of glass (that essentially consists of amorphous silica). Acidification is necessary to achieve a highly supersaturated solution of dissolved silica. Another method for obtaining high supersaturation of silica is by changing the solvent, instead of changing pH. In this method a stock solution of sodium silica solution ($Na_2O\cdot SiO_2$, 27 wt% SiO_2) is diluted with double distilled water to 0.22 wt% SiO_2. Under vigorous stirring 0.2 ml of this water glass solution is rapidly pipetted into 10 ml of absolute ethanol. A sudden turbidity increase manifests the formation of small, smooth silica spheres with a diameter around 30 nm and a typical polydispersity of 20–30 % [1]. A third method for preparation of silica spheres is the well-known Stober method [3]. The precursor tetraethyl silicate (TES) is dissolved in an ethanol-ammonia mixture which is gently stirred in a closed vessel. Silica spheres with a radius of about 60 nm and typical polydispersity of 10–15 % are produced.

An example of nanometal particles is the reduction of metal salt [1],

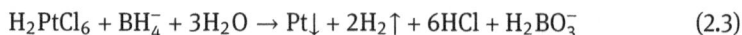

$$H_2PtCl_6 + BH_4^- + 3H_2O \rightarrow Pt\downarrow + 2H_2\uparrow + 6HCl + H_2BO_3^- \qquad (2.3)$$

To understand the process of formation of nanoparticles by the bottom-up process we must consider the process of homogeneous precipitation at a fundaments level. If a substance becomes less soluble by a change of some parameter, such as a temperature decrease, the solution may enter a metastable state by crossing the bimodal, as is illustrated in the phase diagram (Fig. 2.1) of a solution which becomes supersaturated upon cooling [1].

In the metastable region, the formation of small nuclei initially increases the Gibbs free energy. Thus, demixing by nucleation is an activation process, occurring at a rate which is extremely sensitive to the precipitation in this metastable region. In contrast, when the solution is quenched into the unstable region on crossing the spinodal (see Fig. 2.1) there is no activation barrier to form a new phase.

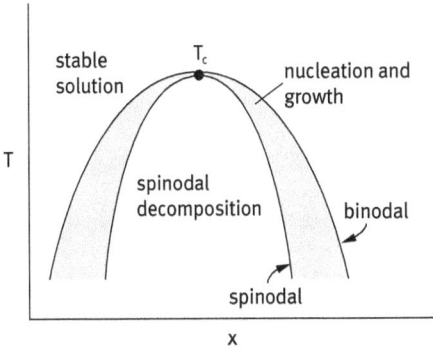

Fig. 2.1: Phase diagram of a solution which becomes supersaturated upon cooling; x is the solute mole fraction and T is the temperature.

2.2.1 Nucleation and growth

Classical nucleation theory considers a precipitating particle (referred to as a nucleus or cluster) to consist of a bulk phase containing N_i^s molecules and a shell with N_i^s molecules which have a higher free energy per molecule than the bulk. The particle is embedded in a solution containing dissolved molecules i. This is schematically represented in Fig. 2.2. The Gibbs free energy of the nucleus G^s is made of a bulk part and a surface part [6],

$$G^s = \mu_i^s N_i^s + \sigma A, \tag{2.4}$$

where μ_i^s is the chemical potential per molecule, s is the solid/liquid interfacial tension and A is the surface area of the nucleus.

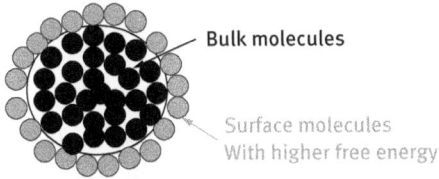

Bulk molecules

Surface molecules
With higher free energy

Fig. 2.2: Schematic representation of a nucleus.

In a supersaturated solution the activity a_i is higher than that of a saturated solution $a_i(\text{sat})$. As a result molecules are transferred from the solution to the nucleus surface. The free energy change ΔG^s upon the transfer of a small number N_i from the solution to the particle is made up of two contributions from the bulk and the surface,

$$\Delta G^s = \Delta G^s(\text{bulk}) + \Delta G^s(\text{surface}). \tag{2.5}$$

The first term on the right-hand side of equation (2.5) is negative (it is the driving force), whereas the second term is positive (work has to be carried out in expanding the interface). $\Delta G^s(\text{bulk})$ is determined by the relative supersaturation, whereas $\Delta G^s(\text{surface})$ is determined by the solid/liquid interfacial tension s and the interfacial area A which is proportional to $(N_i^s)^{2/3}$.

ΔG^s is given by equation (2.6),

$$\Delta G^s = -N_i kT \ln S + \beta \sigma N_i^{2/3} , \qquad (2.6)$$

where k is the Boltzmann constant, T is the absolute temperature and β is a proportionality constant that depends on the shape of the nucleus. S is the relative supersaturation that is equal to $a_i/a_i(\text{sat})$.

For small clusters, the surface area term dominates whereas ΔG^s only starts to decrease due to the bulk term beyond a critical value N^*.

N^* can be obtained by differentiating equation (2.6) with respect to N and equating the result to 0 $(dG^s/dN = 0)$

$$(N^*)^{1/3} = \frac{2\sigma\beta}{3kT \ln S} . \qquad (2.7)$$

The maximum in the Gibbs energy is given by,

$$\Delta G^* = \frac{1}{3}\beta(N^*)^{2/3} . \qquad (2.8)$$

Equation (2.7) shows that the critical cluster size decreases with increasing relative supersaturation S or a reduction of σ by addition of surfactants. This explains why a high supersaturation and/or addition of surfactants favours the formation of small particles. A large S pushes the critical cluster size N^* to smaller values and simultaneously lowers the activation barrier as illustrated in Fig. 2.3, which shows the variation of ΔG with radius at increasing S.

Assuming the nuclei to be spherical, equation (2.6) can be given in terms of the nucleus radius r

$$\Delta G = 4\pi r^2 \sigma - \left(\frac{4}{3}\right)\pi r^3 \left(\frac{kT}{V_m}\right)\ln S , \qquad (2.9)$$

where V_m is the molecular volume.

ΔG^* and r^* are given by,

$$\Delta G^* = \frac{4}{3}\pi(r^*)^2\sigma , \qquad (2.10)$$

$$r^* = \frac{2V_m\sigma}{kT \ln S} . \qquad (2.11)$$

When no precautions are taken, precipitation from a supersaturated solution produces polydisperse particles. This is because nucleation of new particles and further particle growth overlap in time. This overlap is the consequence of the statistical nature of the nucleation process; near the critical size particles may grow as well as dissolve. To narrow down the particle size distribution as much as possible, nucleation should take place over a short time, followed by equal growth of a constant number of particles. This can be achieved by rapidly creating the critical supersaturation required to initiate homogeneous nucleation after which particle growth lowers the saturation sufficiently to suppress new nucleation events. Another option is to add nuclei (seeds) to a solution with subcritical supersaturation. A fortunate consequence of particle growth is that in many cases the size distribution is self-sharpening.

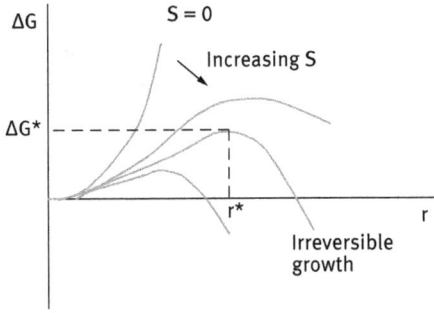

Fig. 2.3: Schematic representation of the effect of supersaturation on particle growth.

2.2.2 Precipitation kinetics

The kinetics of precipitation in a metastable solution can be considered to follow two regimes, according to Fig. 2.3. The initial regime is that where small particles struggle with their own solubility to pass the Gibbs energy barrier ΔG^*. This passage is called a nucleation event, which is simply defined as the capture of one molecule by a critical cluster [1], assuming that after this capture the cluster enters the irreversible growth regime in which a new colloid is born. In this case the number I of colloids that exist per second is proportional to c_m (the concentration of single unassociated molecules) and c^* (the concentration of critical clusters),

$$I = kc_m c^*,$$ (2.12)

where k is the rate constant. Equation (2.12) predicts second-order reaction kinetics.

To quantify I, one must evaluate the frequency at which molecules encounter a spherical nucleus of radius a by diffusion. This can be evaluated using Smolu-chowski's diffusion model for coagulation kinetics [7]. The diffusion flux J of mole-cules through any spherical envelop with radius a is given by Fick's first law,

$$J = 4\pi r^2 D \frac{dc(a)}{da},$$ (2.13)

where D is the molecular diffusion coefficient relative to the sphere positioned at the origin a = 0. Each molecule that reaches the sphere surface irreversibly attaches to the insoluble sphere and it is assumed that the concentration c_m of molecules in the liquid far away from the sphere radius remains constant [1],

$$c(a - r) = 0; \quad c(a \rightarrow \infty) = c_m.$$ (2.14)

For these boundary conditions equation (2.13) becomes,

$$c(a - r) = 0; \quad c(a \rightarrow \infty) = c_m,$$ (2.15)

if it assumed that J is independent of a, i.e. if the diffusion of the molecules towards the sphere has reached a steady state. Such a state is approached by the concentration gradient around a sphere over a time of the order of r^2/D needed by the molecules to diffuse over a sphere diameter. Assuming that sphere growth is a consequence of stationary states, one can identify the nucleation rate I as the flux J multiplied by the concentration c^* of spheres with critical radius r^*,

$$I = 4\pi D r^* c_m c^* \quad [m^{-3} s^{-1}]. \tag{2.16}$$

The concentration c^* can be evaluated by considering the reversible work to form a cluster out of N molecules using the Boltzmann distribution principle,

$$c(N) = c_m \exp\left(\frac{-\Delta G}{kT}\right). \tag{2.17}$$

$c(N)$ represents the equilibrium concentration of clusters composed of N molecules and ΔG is the free energy of formation of a cluster. Applying this result to clusters with a critical size r^*, we obtain on substitution in equation (2.16),

$$I = 4\pi D r^* c_m^2 \exp\left(\frac{-\Delta G^*}{kT}\right) \tag{2.18}$$

$$\Delta G^* = \left(\frac{4\pi}{3}\right), \tag{2.19}$$

where ΔG^* is the height of the nucleation barrier. The exponent may be identified as the probability (per particle) that a spontaneous fluctuation will produce a critical cluster. Equation (2.18) shows that the nucleation rate is very sensitive to the value of r^* and hence to supersaturation as given by equation (2.11). The maximum nucleation rate at very large supersaturation, i.e. the pre-exponential term in equation (2.18) can be obtained by substitution of D using the Stokes–Einstein equation,

$$D = \frac{kT}{6\pi\eta r^*}, \tag{2.20}$$

where η is the viscosity of the medium. This maximum nucleation rate is given by,

$$I \approx \frac{kT}{\eta} = c_m^2. \tag{2.21}$$

For an aqueous solution at room temperature with a molar concentration c_m, the maximum nucleation rate is of the order of 10^{29} m^{-3} s^{-1}. A decrease in supersaturation to values around S = 5 is sufficient to reduce this very high rate to practically zero. For silica precipitation in dilute, acidified water glass solutions, $S \approx 5$ and nucleation may take hours or days. As mentioned above, when no precautions are taken, precipitation from a supersaturated solution produces polydisperse particles. Fortunately, in many cases the size distribution is self-sharpening. This can be illustrated by considering the colloidal spheres with radius r, which irreversibly grow by the uptake of molecules from a solution according to the following rate law [1],

$$\frac{dr}{dt} = k_0 r^n, \tag{2.22}$$

where k_0 and n are constants. This growth equation leads either to spreading of the relative distribution, depending on the value of n. Consider at a given time t any pair of spheres with arbitrary size from the population of independently growing particles. Let $1 + \varepsilon$ be their size ratio such that $r(1 + \varepsilon)$ and r are the radius of the larger and smaller spheres, respectively. The former grows according to,

$$\frac{d}{dt} r(1 + \varepsilon) = k r^n (1 + \varepsilon)^n ,$$ (2.23)

which can be combined with the growth equation (2.22) for the smaller spheres to obtain the time evolution of the size ratio,

$$\frac{d\varepsilon}{dt} = k_0 r^{n-1} [(1 + \varepsilon)^n - (1 + \varepsilon)]; \quad \varepsilon \geq 0 .$$ (2.24)

The relative size difference ε increases with time for $n > 1$, in which case particle growth broadens the distribution. For $n = 1$, the size ratio between the spheres remains constant, whereas for $n < 1$ it monotonically decreases with time. Since this decrease holds for any pair of particles in the growing population, it follows that for $n < 1$ the relative size distribution is self-sharpening [1]. This condition is practically realistic. For example, when the growth rate is completely determined by a slow reaction of molecules at the sphere radius,

$$\frac{dr^3}{dt} = k_0 r^2 ,$$ (2.25)

implying that dr/dt is a constant, so $n = 0$. The opposite limiting case is growth governed by the rate at which molecules reach a colloid by diffusion. The diffusion flux for molecules with a diffusion coefficient D, relative to a sphere centred at the origin at $a = 0$, is given by equation (2.13). The saturation concentration is assumed to be maintained at the particle surface, neglecting the influence of particle size on c(sat), and keeping the bulk concentration of molecules constant [1],

$$c(a = r) = c(sat); \quad c(a \to \infty) = c(\infty) .$$ (2.26)

For these boundary conditions, the stationary flux towards the sphere equals,

$$J = 4\pi D r [c(\infty) - c(sat)] ,$$ (2.27)

showing that the rate at which the colloid intercepts the diffusing molecules is proportional to its radius and not to its surface area. If every molecule contributes a volume v_m to the growing colloid, then for a homogeneous sphere, the volume increases at a rate given by,

$$\frac{d}{dt} \frac{4}{3} \pi r^3 = J v_m ,$$ (2.28)

which on substitution in equation (2.27) leads to,

$$\frac{dr}{dt} = D v_m [c(\infty) - c(sat)] r^{-1} ,$$ (2.29)

with the typical scaling $r^2 \approx t$, as expected for a diffusion-controlled process. Thus the exponent in equation (2.22) for diffusion controlled growth is $n = -1$, and consequently the relative width of the size distribution decreases with time.

For charged species, an electrostatic interaction may be present between the growing colloids and the molecules they consume. This will either enhance or retard growth, depending on whether colloids and monomers attract or repel each other. In this case the diffusion coefficient D of the monomers in the diffusion flux J has to be replaced by an effective coefficient of the form,

$$D_{eff} = \frac{D}{r \int_r^{\infty} \exp -[u(a)/kT]a^{-2}\, da}, \tag{2.30}$$

where $u(a)$ is the interaction energy between molecule and colloid.

If the molecules are ions with charge ze and the colloid sphere has a surface potential ψ^0, then for low electrolyte where the interaction is unscreened (upper estimate of the ion-colloid interaction), $u(a)$ is given by,

$$\frac{u(a)}{kT} = u_0 \frac{r}{a}, \tag{2.31}$$

$$u_0 = \frac{ze\psi^0}{kT} = zy^0, \tag{2.32}$$

where u_0 is the colloid-ion interaction energy and $y^0 = (e\psi^0/kT)$.

Thus, Coulombic interaction, equation (2.30), gives,

$$D_{eff} = D \frac{zy^0}{\exp(zy^0) - 1}. \tag{2.33}$$

Thus, the growth rate is slowed down exponentially by Coulombic interaction. For example when $\psi^0 = 75$ mV, the effective diffusion coefficient for divalent ions is about 0.01 D. Addition of electrolyte screens the colloid-ion interaction and this moderates the effect of y^0 on the growth kinetics.

The interaction between charged monomers and the growing colloid within the approximation given by equation (2.33) does not change the growth equation (2.29) and, therefore, does not affect the conclusion that the diffusional growth sharpens the size distribution.

The effect of Ostwald ripening on the kinetics of particle growth can be analysed using the Gibbs–Kelvin [1, 6] equation that relates the solubility $c(r)$ of a particle with radius r to that of infinitely large particle $c(sat)$, i.e. a flat surface, by,

$$\ln\left[\frac{c(r)}{c(sat)}\right] = \frac{2\sigma v_m}{r\, kT} \tag{2.34}$$

The increased solubility according to equation (2.34) is referred to as the Gibbs–Kelvin effect. By considering the Gibbs energy maximum of Fig. 2.3, it is clear that it represents an unstable equilibrium, which can only be maintained only for particles of exactly the same size. For polydisperse particles (with the same interfacial tension), there is no single, common equilibrium solubility; particles either grow or dissolve. Clearly, the largest particles have the strongest tendency to grow owing to their low solubility. This coarsening of colloids is referred to as Ostwald ripening and it is an important ageing effect which occurs with most polydisperse systems with small particles.

In a polydisperse system, the bulk concentration c(bulk) is not constant, but slowly decreasing over time due to the gradual disappearance of small particles. At any moment in time there is one sphere with radius r_0 which is in metastable equilibrium with the bulk concentration,

$$c(\text{bulk}) = c(\text{sat}) \exp\left[\frac{2\sigma v_m}{r_0 kT}\right],\tag{2.35}$$

where c(sat) is the solubility of a flat surface. If the local solute concentration near a sphere with radius r_i is also fixed by the Gibbs–Kelvin equation, the steady state diffusion flux for sphere I is given by,

$$J_i = 4\pi D r_i c(\text{sat}) \left\{\exp\left[\frac{2\sigma v_m}{r_0 kT}\right] - \exp\left[\frac{2\sigma v_m}{r_i kT}\right]\right\}.\tag{2.36}$$

It is clear that particles with radii $r_i < r_0$ dissolve because $J < 0$, whereas for $r_i > r_0$ the particles grow. The average particle radius and the critical radius r_0 increase over time, so that the exponents in the diffusion flux can be linearized at a later stage of the ripening process. In this case, one can write for the growth or dissolution rate of sphere i the following approximate equation,

$$\frac{d}{dt}r_i^3 = 6D r_i c(\text{sat})\frac{\sigma v_m^2}{kT}\left[\frac{1}{r_0} - \frac{1}{r_i}\right].\tag{2.37}$$

One limiting case of Ostwald ripening allows for a simple analytical solution, namely a monodisperse sphere with radius r, from which dissolved matter is deposited on very large particles, or a flat substrate. If that substrate controls the bulk concentration, r_0 is infinitely large and consequently,

$$\frac{dr^3}{dt} = -6Dc(\text{sat})\frac{\sigma v_m^2}{kT}.\tag{2.38}$$

Thus, for this case the particle volume decreases at a constant rate.

The time evolution of a continuous size distribution was analysed by Lifshitz and Slesov [8] and Wagner [9] (referred to as the LSW theory) who predicted for large times the asymptotic result,

$$\frac{d\langle r\rangle^3}{dt} = \frac{8}{9}Dc(\text{sat})\frac{\sigma v_m^2}{kT},\tag{2.39}$$

which predicts that at a late stage of the ripening process, the average particle radius increases as $t^{1/3}$. The supersaturation falls as $t^{-1/3}$ and the number of spheres as t^{-1}. A remarkable finding of the LSW theory is that due to Ostwald ripening the size distribution approaches a certain universal, time-independent shape, irrespective of the initial distribution.

2.2.3 Seeded nucleation and growth

In the above analysis it is assumed that particle nucleation and growth occur in a solution of one solute. In practice this process of homogeneous nucleation is difficult to realize due to the presence of contaminants, dust, motes and irregularities on the vessel wall. The process of heterogeneous nucleation may have a dramatic effect on the kinetics. This process may be advantageous, resulting in particle size polydispersity. This process of seed nucleation was first exploited for preparation of quite monodisperse gold colloids by using a finely divided Faraday gold sol as the seed. The seed can also differ chemically from the precipitating material, leading to the formation of core-shell colloids [1]. Good examples are the growth of silica on gold cores, and other inorganic particles for the preparation of core-shell semiconductor particles [1]. Such well-defined composite colloids are increasingly important in materials science, in addition to their use in fundamental studies.

The efficiency of seeds or the container wall to catalyse nucleation is due to the reduction of the interfacial Gibbs energy of a precipitating particle. Steps and kinks on the seed substrate may act as active sites because they enable more of the surface of the nucleus to be in contact with the seed which lowers its surface excess Gibbs energy.

2.2.4 Surface modification

Surface modification is the deliberate attachment of a polymeric surfactant to the surface of the colloid to change its physical properties or chemical functionality. This modification is permanent if the attached polymer is not desorbed by thermal motion. Such surface modification occurs either via a chemical bond or significant adsorption energy (lack of desorption). The polymeric surfactant provides steric repulsion for the particles as will be discussed in Chapter 5. Surface modification is generally straightforward by choosing a molecule with suitable chemical linker. For example, for metal hydroxide particles, such as silica, one can use a linkage between the −OH group on the surface of the particles and a carboxylic group or alcohol. For example, under mild conditions, the surface silane groups on silica react with silane coupling agents (SCAs) and these materials are suitable for in situ modification of the colloid in a sol. The SCAs hydrolyse to reactive silanes, which graft themselves onto silica via the formation of a silixane linkage.

Once reactive oligomers or polymers attach to a colloidal core, the core-shell particle behaves as one kinetic unit with an average kinetic energy of $(3/2)kT$ (where k is the Boltzmann constant and T is the absolute temperature). This energy has to be weighed against the replacement of a large number of solvent molecules by the adsorbed species. Even a very small Gibbs energy penalty per replacement may suffice to produce aggregates that do not break apart by thermal motion. Such aggregation can also be induced by minute changes in the nature or composition of the solvent, a subtle effect that is difficult to predict or explain. Any small change in the composition involves a large number of low-molecular species, with a net enthalpy change that easily compensates the entropy loss due to aggregation of large colloids. One obvious counterexample is any solvent adsorption on modified or unmodified colloids. Water adsorption on silica is well known, but polar organic solvents such as dimethylformamide or triethylphosphate also adsorb in significant amounts on bare silica particles, often sufficient to prevent this aggregation. It should be noted that small particles also have a disadvantage since the coagulation rate is proportional to the square of the number density. For modified, stable colloids, the small particle size becomes a benefit in view of the many functional groups per gram. One attractive option is the simultaneous synthesis and modification of inorganic colloids by nucleation and growth in the presence of the modifying agent, which also influences and controls the particle size [1].

2.2.5 Other methods for preparation of suspensions by the bottom-up process

Several other methods can be applied for preparation of suspensions using the bottom-up processes of which the following are worth mentioning: (i) precipitation of particles by addition of a nonsolvent (containing a stabilizer for the particles formed) to a solution of the compound in question; (ii) preparation of an emulsion of the substance by using a solvent in which it is soluble following emulsification of the solvent in another immiscible solvent. This is then followed by removal of the solvent making the emulsion droplets by evaporation; (iii) preparation of the particles by mixing two microemulsions containing two chemicals that react together when the microemulsion droplets collide with each other; (iv) sol-gel process particularly used for preparation of silica particles; (v) production of polymer suspensions by emulsion or suspension polymerization; (vi) preparation of polymer suspensions by polymerization of microemulsions. A brief description of each process is given below.

(i) Solvent-antisolvent method [10]. In this method, the substance (e.g. a hydrophobic drug) is dissolved in a suitable solvent such as acetone. The resulting solution is carefully added to another miscible solvent in which the resulting compound is insoluble. This results in precipitation of the compound by nucleation and growth. The particle size distribution is controlled by using a polymeric surfactant that is strongly adsorbed on the particle surface and providing an effective repulsive barrier to prevent aggregation of the particles. The polymeric surfactant is chosen to have specific adsorption on the particle surface to prevent Ostwald ripening. This method can be adapted for preparation of low water solubility drug suspensions. In this case the drug is dissolved in acetone and the resulting solution is added to an aqueous solution of Poloxamer (an A-B-A block copolymer consisting of two A polyethylene oxide (PEO) chains and a B polypropylene oxide (PPO) chain, i.e. PEO-PPO-PEO). After precipitation of the particles the acetone is removed by evaporation. The main problem with this method is the possibility of formation of several unstable polymorphs that will undergo crystal growth. In addition, the resulting particles may be of needle shape structure. However, by proper choice of the polymeric surfactant one can control the particle morphology and shape. Another problem may be the lack of removal of the solvent after precipitation of the particles.

(ii) Use of an emulsion. In this case the compound is dissolved in a volatile organic solvent that is immiscible with water, such as methylene dichloride. The oil solution is emulsified in water using a high speed stirrer followed by high pressure homogenization [11]. A suitable emulsifier for the oil phase is used which has the same HLB number as the oil. The volatile oil in the resulting emulsion is removed by evaporation and the formed suspension particles are stabilized against aggregation by the use of an effective polymeric surfactant that could be dissolved in the aqueous phase. The main problem with this technique is the possible interaction with the emulsifier which may result in destabilization of the resulting suspension. However, by careful selection of the emulsifier/stabilizing system one can form a colloidally stable nanosuspension.

(iii) Preparation of suspensions by mixing two microemulsions [12]. Reverse microemulsions lend themselves as suitable "nonreactors" for the synthesis of particles. Inorganic salts can be dissolved in the water pools of a W/O microemulsion. Another W/O microemulsion with reducing agent dissolved in the water pools is then prepared. The two microemulsions are then mixed and the reaction between the inorganic salt and the reducing agent starts at the interface and proceeds towards the centre of the droplet. The rate limiting step appears to be the droplet diffusion. Control of the exchange can be achieved by tuning the film rigidity. This procedure has been applied for the preparation of noble metal particles that could be applied in electronics, catalysis and in potential medical applications.

(iv) Sol-gel process. This method is particularly applicable for preparation of silica particles [13]. This involves the development of networks through an arrangement of colloidal suspension (sol) and gelation to form a system in continuous liquid phase (gel). A sol is basically a dispersion of colloidal particles (1–100 nm) in a liquid and a gel is an interconnected rigid network with pores of submicron dimensions and polymeric chains. The sol-gel process, depending on the nature of the precursors, may be divided into two classes; namely inorganic precursors (chlorides, nitrates, sulphides, etc.) and alkoxide precursors. Extensively used precursors include tetramethyl silane and tetraethoxysilane. In this process, the reaction of metal alkoxides and water, in the presence of acid or base, forms a one phase solution that goes through a solution-to-gel transition to form a rigid, two phase system comprised of metal oxides and solvent-filled pores. The physical and electrochemical properties of the resultant materials largely depend on the type of catalyst used in the reaction. In the case of silica alkoxides, the acid catalysed reaction results in weakly crosslinked linear polymers. These polymers entangle and form additional branches leading to gelation. In base-catalysed reaction, due to rapid hydrolysis and condensation of the alkoxide

silanes, the system forms highly branched clusters. The difference in cluster formation is due to the solubility of the resulting metal oxide in the reaction medium. The
solubility of the silicon oxide is greater in alkaline medium, which favours the inter-
linking of the silica clusters, than in acidic medium. A general procedure for sol-gel
includes four stages, namely hydrolysis, condensation, growth and aggregation. The
complete hydrolysis to form $M(OH)_4$ is very difficult to achieve. Instead, condensation
may occur between two –OH or M–OH groups and an alkoxy group to form bridging oxygen and a water or alcohol molecule. The hydrolysis and polycondensation
reactions are initiated at numerous sites and the kinetics of the reaction can be very
complex. When a sufficient number of interconnected M–O–M bonds are formed in
a particular region, they interact cooperatively to form colloidal particles or a sol.
With time the colloidal particles link together to form three-dimensional networks.
The size, shape and morphological features of the silica nanoparticles can be controlled by reaction kinetics, use of templates such as cationic, nonionic surfactants,
polymers, electrolytes, etc.

(v) Preparation of polymer particles by emulsion or suspension polymerization
[14]. In emulsion polymerization, the monomer, e.g. styrene or methyl methacrylate
that is insoluble in the continuous phase, is emulsified using a surfactant that adsorbs
at the monomer/water interface [14]. The surfactant micelles in bulk solution solubilize some of the monomer. A water soluble initiator such as potassium persulphate
$K_2S_2O_8$ is added and this decomposes in the aqueous phase forming free radicals that
interacts with the monomers forming oligomeric chains. It has long been assumed
that nucleation occurs in the "monomer-swollen micelles". The reasoning behind this
mechanism is the sharp increase in the rate of reaction above the critical micelle concentration and that the number of particles formed and their size depend to a large
extent on the nature of the surfactant and its concentration (which determines the
number of micelles formed). However, later this mechanism was disputed and it was
suggested that the presence of micelles means that excess surfactant is available and
molecules will readily diffuse to any interface.

The most accepted theory of emulsion polymerization is referred to as the coagulative nucleation theory [15, 16]. A two-step coagulative nucleation model has been
proposed by Napper and co-workers [15, 16]. In this process the oligomers grow by
propagation and this is followed by a termination process in the continuous phase.
A random coil is produced which is insoluble in the medium and this produces a precursor oligomer at the θ-point. The precursor particles subsequently grow primarily
by coagulation to form true latex particles. Some growth may also occur by further
polymerization. The colloidal instability of the precursor particles may arise from their
small size, and the slow rate of polymerization can be due to reduced swelling of the
particles by the hydrophilic monomer [15, 16]. The role of surfactants in these processes is crucial since they determine the stabilizing efficiency and the effectiveness
of the surface active agent, ultimately determining the number of particles formed.
This was confirmed by using surface active agents of different nature. The effective-

ness of any surface active agent in stabilizing the particles was the dominant factor and the number of micelles formed was relatively unimportant.

A typical emulsion polymerization formulation contains water, 50 % monomer blended for the required glass transition temperature, T_g, surfactant (and often colloid), initiator, pH buffer and fungicide. Hard monomers with a high T_g used in emulsion polymerization may be vinyl acetate, methyl methacrylate and styrene. Soft monomers with a low T_g include butyl acrylate, 2-ethylhexyl acrylate, vinyl versatate and maleate esters. Most suitable monomers are those with low, but not too low, water solubility. Other monomers such as acrylic acid, methacrylic acid and adhesion promoting monomers may be included in the formulation. It is important that the latex particles coalesce as the diluents evaporate. The minimum film forming temperature (MFFT) of the latex is a characteristic of say a paint system and is closely related to the T_g of the polymer. However, the latter can be affected by materials present, such as surfactant, and the inhomogeneity of the polymer composition at the surface. High T_g polymers will not coalesce at room temperature and in this case a plasticizer ("coalescing agent") such as benzyl alcohol is incorporated in the formulation to reduce the T_g of the polymer thus reducing the MFFT of the paint. Several types of surfactants can be used in emulsion polymerization such as anionic surfactants (sulphates, sulphonates and phosphates), cationic surfactants (alkyl ammonium salts), amphoteric surfactants (such as alkyl betaine) and nonionic surfactants (such as alcohol ethoxylates). The role of surfactants is two-fold, firstly to provide a locus for the monomer to polymerize and secondly to stabilize the polymer particles as they form. In addition, surfactants aggregate to form micelles (above the critical micelle concentration) and these can solubilize the monomers. In most cases a mixture of anionic and nonionic surfactants is used for optimum preparation of polymer latexes. Cationic surfactants are seldom used, except for some specific applications where a positive charge is required on the surface of the polymer particles.

In addition to surfactants, most latex preparations require the addition of a polymer (sometimes referred to as "protective colloid") such as partially hydrolyzed polyvinyl acetate (commercially referred to as polyvinyl alcohol, PVA), hydroxyethyl cellulose or a block copolymer of polyethylene oxide (PEO) and polypropylene oxide (PPO). These polymers can be supplied with various molecular weights or proportions of PEO and PPO. When used in emulsion polymerization they can be grafted by the growing chain of the polymer being formed. They assist in controlling the particle size of the latex, enhancing the stability of the polymer dispersion and controlling the rheology of the final paint.

A typical emulsion polymerization process involves two stages known as the seed stage and the feed stage. In the seed stage, an aqueous charge of water, surfactant, and colloid is raised to the reaction temperature (85–90 °C) and 5–10 % of the monomer mixture is added along with a proportion of the initiator (a water soluble persulphate). In this seed stage, the formulation contains monomer droplets stabilized by surfactant, a small amount of monomer in solution as well as surfactant monomers and micelles. Radicals are formed in solution from the breakdown of the initiator and these radicals polymerize the small amount of monomer in solution. These oligomeric chains will grow to some critical size; their length depends on the solubility of the monomer in water. The oligomers build up to a limiting concentration and this is followed by a precipitous formation of aggregates (seeds), a process similar to micelle formation, except in this case the aggregation process is irreversible (unlike surfactant micelles which are in dynamic equilibrium with monomers).

In the feed stage, the remaining monomer and initiator are fed together and the monomer droplets become emulsified by the surfactant remaining in solution (or by extra addition of surfactant). Polymerization proceeds as the monomer diffuses from the droplets, through the water phase, into the already forming growing particles. At the same time radicals enter the monomer-swollen particles causing both termination and re-initiation of polymerization. As the particles grow, the remaining surfactant from the water phase is adsorbed onto the surface of particles to stabilize the polymer particles. The stabilization mechanism involves both electrostatic and steric repulsion. The final stage of polymerization may include a further shot of initiator to complete the conversion.

Most aqueous emulsion and dispersion polymerization reported in the literature is based on a few commercial products with a broad molecular weight distribution and varying block composition. The results obtained from these studies could not establish what effect the structural features of the block copolymer has on their stabilizing ability and effectiveness in polymerization. Fortunately, model block copolymers with well-defined structures could be synthesized and their role in emulsion polymerization has been carried out using model polymers and model latexes.

A series of well-defined A-B block copolymers of polystyrene-block-polyethylene oxide (PS-PEO) were synthesized [15] and used for emulsion polymerization of styrene. These molecules are "ideal" since the polystyrene block is compatible with the polystyrene formed and thus it forms the best anchor chain. The PEO chain (the stabilizing chain) is strongly hydrated with water molecules and it extends into the aqueous phase forming the steric layer necessary for stabilization. However, the PEO chain can become dehydrated at high temperature (due to the breaking of hydrogen bonds) thus reducing the effective steric stabilization. Thus, emulsion polymerization should be carried out at temperatures well below the theta (θ)-temperature of PEO.

The above method of emulsion polymerization was adapted for the preparation of polymer colloids. The main advantage of using miniemulsions (with diameters in the range 20–200 nm) in place of macroemulsions (with diameters > 500 nm) is the inherent greater surface area which may allow them to compete far more effectively for radicals than macroemulsions. The main problem with using miniemulsions is Ostwald ripening since the monomers with low molecular weight have finite solubility in water. This problem can be overcome by addition of a secondary disperse phase that is highly insoluble in water, such as hexadecane, hexadecanol, dodecanethiol and other monomer soluble polymers. In miniemulsion polymerization, nucleation occurs predominantly by radical entry into monomer in the interior of the miniemulsion droplet. Due to the improved physical stability of the miniemulsion droplets, it is possible to adjust the total surfactant concentration so as to limit the total number of micelles in solution, thereby limiting aqueous phase nucleation. Relative to micelles, the droplets can have higher radical numbers (due to their larger size) and can, therefore, significantly enhance the early stages of the polymerization process. The rate of polymerization per particle is faster, and the systems are converted faster. Following nucleation, the reaction proceeds by polymerization of the monomer in the miniemulsion droplets.

A novel graft copolymer of hydrophobically modified inulin (INUTEC® SP1) has been used to produce latex particles in emulsion polymerization of styrene, methyl methacrylate, butyl acrylate and several other monomers [17]. All lattices were prepared by emulsion polymerization using potassium persulphate as initiator. The z-average particle size was determined by photon correlation spectroscopy (PCS) and electron micrographs were also taken.

Emulsion polymerization of styrene or methylmethacrylate showed an optimum weight ratio of (INUTEC)/monomer of 0.0033 for PS and 0.001 for PMMA particles. The (initiator)/(monomer) ratio was kept constant at 0.00125. The monomer conversion was higher than 85 % in all cases. Latex dispersions of PS reaching 50 % and of PMMA reaching 40 % could be obtained using such a low concentration of INUTEC® SP1. Figure 2.4 shows the variation of particle diameter with monomer concentration.

Latex particles could be prepared up to 30 % styrene monomer, whereas for PMMA, the latex could be prepared up to 30 % monomer.

The suspension polymerization process can be adapted to prepare polymer particles. This is divided into three stages for both polymer soluble (A) and insoluble (B) in its monomer. In the first stage for the A system, when the viscosity of the disperse phase remains low, the bulk monomer phase is dispersed into small droplets due to the shear stress imposed by the stirring conditions. Simultaneously, through the reverse process of coalescence, the drops tend to reverse to the original monomer mass. The droplet size distribution results from break-up-coalescence dynamic equilibrium. The adsorption of polymeric stabilizers at the monomer droplet-water interface decreases the interfacial tension and the adsorbed layer prevents coalescence. During the second stage, the viscosity within the droplets increases with increasing conversion causing coalescence to overcome break-up. However, if the stabilizer is present in sufficient amount and gives strong repulsion between the droplets, this coalescence process is delayed resulting in a small increase in particle size. Towards the end of this stage, coalescence is stopped due to the elastic nature of the particle collisions. After this point, the particle size remains virtually constant. The degree of agitation and design of the stirrer/reactor system has a big influence on the dispersion of monomer droplets as well as on the overall process. An increase in agitation improves the mixing and the heat transfer and promotes the break-up of the droplets. However, an increase in agitation increases the frequency of collisions thus increasing coalescence. These

(a) PS latexes

(b) PMMA Latexes

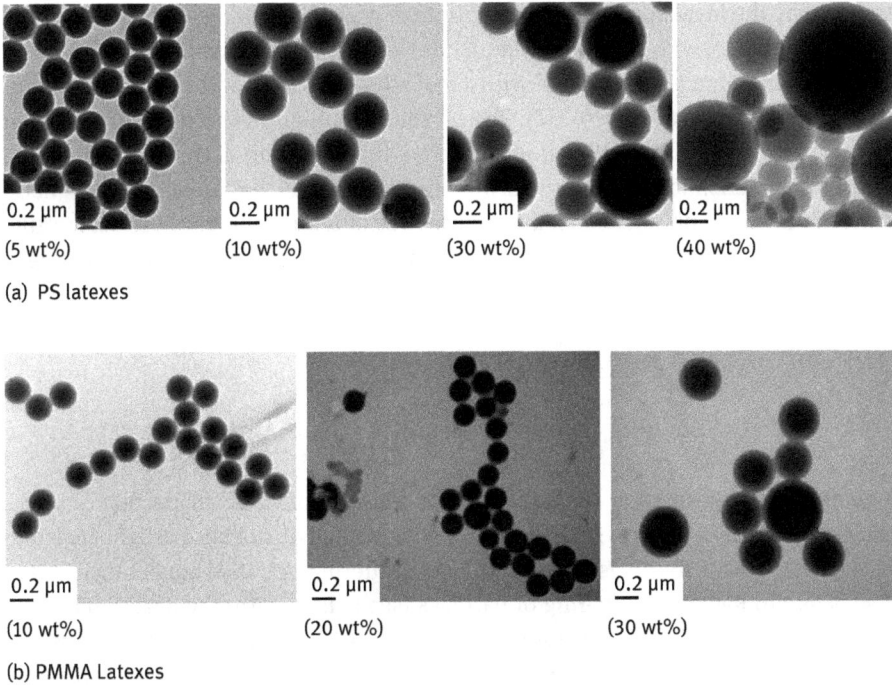

Fig. 2.4: Electron micrographs of the latexes.

conflicting mechanisms show a reduction in droplet size with increasing speed of agitation reaching a minimum at an optimum agitation speed followed by an increase in droplet size with a further increase in stirrer speed due to coalescence. The formation of nanoparticles is also determined by the concentration and nature of the stabilizer. In most cases a mixture of polymeric stabilizer such as poly(vinyl alcohol) or Pluronic (an A-B-A block copolymer of polyethylene oxide, A, and polypropylene oxide, PPO) with an anionic surfactant such as sodium dodecyl sulphate is used. In this case the stabilizing mechanism is a combination of the electrostatic and steric mechanisms, referred to as electrosteric.

(vi) Preparation of polymer latexes by polymerization of microemulsions. This method has attracted considerable attention and several comprehensive investigations have been carried out by Candau and co-workers [18, 19]. The polymerization of water-in-oil microemulsions containing acrylamide monomer in the aqueous droplets was investigated to prepare polyacrylamide with high molecular weight. The size of the nanolatex produced was larger than that of the water droplets of the microemulsion. This was explained by considering the possible fusion by coalescence of the droplets during polymerization. A systematic study of the polymerization of styrene-in-water and methylmethacrylate-in-water microemulsions was carried out by Larpent and Tadros [20] and Girard et al. [21]. The oil-in-water (O/W) microemulsion was prepared by the inversion method by addition of water containing a high HLB surfactant (nonyl phenol with 20 mol ethylene oxide) to an oil solution containing a low HLB nonionic surfactant (nonyl phenol with 4 mol ethylene oxide) and a small amount of an anionic surfactant (Aerosol OT). The oil soluble initiator (azobisisobutyronitrile; AIBN) was introduced into the oil phase before the formation of the microemulsion. Various polymerization methods were investigated, namely thermally induced, chemically induced and photochemically induced. The latter method gave the best results with latex size ($\approx$ 10 nm diameter) similar to that of the microemulsion droplet size.

2.3 Characterization of suspension particles [12]

2.3.1 Visual observations

A great deal of information can already be obtained from visual inspection of a suspension, aided by a torch or small laser. For colloids that do not absorb light at visible wavelengths, the turbidity is only due to light scattering. A bluish appearance in this case is due to Rayleigh scattering of particles (see below) with a typical diameter of 100 nm or smaller. This bluish Tyndall effect can be observed for dilute dispersions of latex particles and several metal hydroxide colloids such as boehmite and silica. Milky white appearance may be due to anything that shortens the mean free path of photons in the dispersion, namely large particle size, high refractive index and high colloid concentration. Multiple scattering is easy to demonstrate as it spreads an incoming narrow beam of laser light. A white appearance sometimes manifests aggregation; the bluish Tyndall effect for small aluminium hydroxide or silica colloids changes to white turbidity when the particles coagulate. Inspection of a (either stirred or shaken) sol with a light beam between cross polarizers reveals optical birefringence when the particles have an anisometric, plate or rod-like shape.

When the particles settle significantly within a few days, it is worthwhile to estimate the effective Stokes radius, which would produce the observed settling rate. If this radius is much larger than the expected colloid size, the colloids may be aggregated. The sediment on the bottom should also be observed when the vessel is tilted; stable colloids tend to flow like a liquid, but this flow can be slow if the particles are densely packed. Aggregated particles for sediments or gels with a yield stress can be measured using rheology.

Instability of colloidal dispersions with respect to aggregation or phase separation is easy to detect. Shaking a dilute, unstable sol usually produces visible specs of aggregated particles, which stick to the glass surface. The onset of coagulation or phase separation is illustrated by a strong increase in the light scattering on approach of a critical point due to the occurrence of large fluctuations in density and hence in refractive index. When such fluctuations can be observed in a gently shaken sol, one can be sure that the sol will gel or phase separate soon after.

2.3.2 Optical microscopy

This is by far the most valuable tool for a qualitative or quantitative examination of the suspension. Information on the size, shape, morphology and aggregation of particles can be conveniently obtained with minimum time required for sample preparation. Since individual particles can be directly observed and their shape examined, optical microscopy is considered the only absolute method for particle characterization. However, optical microscopy has some limitations: the minimum size that can be detected. The practical lower limit for accurate measurement of particle size is 1.0 μm, although some detection may be obtained down to 0.3 μm. Image contrast may not be good enough for observation, particularly when using a video camera which is mostly used for convenience. The contrast can be improved by decreasing the aperture of the iris diaphragm but this reduces the resolution. The contrast of the image depends on the refractive index of the particles relative to that of the medium. Hence the contrast can be improved by increasing the difference between the refractive index of the particles and the immersion medium. Unfortunately, changing the medium for the suspension is not practical since this may affect the state of the dispersion. Fortunately, water with a refractive index of 1.33 is a suitable medium for most organic particles with a refractive index usually > 1.4.

The ultramicroscope by virtue of dark field illumination extends the useful range of optical microscopy to small particles not visible in bright light illumination. Dark field illumination utilizes a hollow cone of light at a large angle of incidence. The image is formed by light scattered from the particles against a dark background. Particles about 10 times smaller than those visible by bright light illumination can be detected. However, the image obtained is abnormal and the particle size cannot be accurately measured. For that reason, the electron microscope (see below) has displaced the ultramicroscope, except for dynamic studies by flow ultramicroscopy.

Three main attachments to the optical microscope are possible:

2.3.2.1 Phase contrast

This utilizes the difference between the diffracted waves from the main image and the direct light from the light source. The specimen is illuminated with a light cone and this illumination is within the objective aperture. The light illuminates the specimen and generates zero order and higher orders of diffracted light. The zero order light beam passes through the objective and a phase plate which is located at the objective back focal plane. The difference between the optical path of the direct light beam and that of the beam diffracted by a particle causes a phase difference. The constructive and destructive interferences result in brightness changes which enhance the contrast. This produces sharp images allowing one to obtain particle size measurements more accurately. The phase contrast microscope has a plate in the focal plane of the objective back focus. Instead of a conventional iris diaphragm, the condenser is equipped with a ring matched in its dimension to the phase plate.

2.3.2.2 Differential Interference Contrast (DIC)

This gives a better contrast than the phase contrast method. It utilizes a phase difference to improve contrast, but the separation and recombination of a light beam into two beams is accomplished by prisms. DIC generates interference colours and the contrast effects indicate the refractive index difference between the particle and medium.

2.3.2.3 Polarized light microscopy

This illuminates the sample with linearly or circularly polarized light, either in a reflection or transmission mode. One polarizing element, located below the stage of the microscope, converts the illumination to polarized light. The second polarizer is located between the objective and the ocular and is used to detect polarized light. Linearly polarized light cannot pass the second polarizer in a crossed position, unless the plane of polarization has been rotated by the specimen. Various characteristics of the specimen can be determined, including anisotropy, polarization colours, birefringence, polymorphism, etc.

2.3.2.4 Sample preparation for optical microscopy

A drop of the suspension is placed on a glass slide and covered with a cover glass. If the suspension has to be diluted, the dispersion medium (that can be obtained by centrifugation and/or filtration of the suspension) should be used as the diluent in order to avoid aggregation. At low magnifications the distance between the objective and the sample is usually adequate for manipulating the sample, but at high magnification the objective may be too close to the sample. An adequate working distance can be obtained, while maintaining high magnification, by using a more powerful eyepiece with a low power objective. For suspensions encountering Brownian motion (when the particle size is relatively small), microscopic examination of moving particles can become difficult. In this case one can record the image on a photographic film or video tape or disc (using computer software).

2.3.2.5 Particle size measurements using optical microscopy

The optical microscope can be used to observe dispersed particles and flocs. Particle sizing can be carried out using manual, semiautomatic or automatic image analysis techniques. In the manual method (which is tedious) the microscope is fitted with a minimum of 10× and 43× achromatic or apochromatic objectives equipped with high numerical apertures (10×, 15× and 20×), a mechanical XY stage, a stage micrometre and a light source. The direct measurement of particle size is aided by a linear scale or globe-and-circle graticules in the ocular. The linear scale is mainly useful for spherical particles, with a relatively narrow particle size distribution. The globe-and-circle graticules are used to compare the projected particle area with a series of circles in the ocular graticule. The size of spherical particles can be expressed by the diameter, but for irregularly shaped particles various statistical diameters are used. One of the difficulties with the evaluation of dispersions by optical microscopy is the quantification of data. The number of particles in at least six different size ranges must be counted to obtain a distribution. This problem can be alleviated by the use of automatic image analysis which can also give an indication on the floc size and its morphology.

2.3.3 Electron microscopy

Electron microscopy is by far the most used quantitative technique for determination of particle size in the nanosize range. It utilizes an electron beam to illuminate the sample. The electrons behave as charged particles which can be focused by annular electrostatic or electromagnetic fields surrounding the electron beam. Due to the very short wavelength of electrons, the resolving power of an electron microscope exceeds that of an optical microscope by ≈ 200 times. The resolution depends on the accelerating voltage which determines the wavelength of the electron beam and magnifications as high as 200 000 can be reached with intense beams but this could damage the sample. Mostly the accelerating voltage is kept below 100–200 kV and the maximum magnification obtained is below 100 000. The main advantage of electron microscopy is the high resolution, sufficient for resolving details separated by only a fraction of a nanometre. The increased depth of field, usually by about 10 μm or about 10 times that of an optical microscope, is another important advantage of electron microscopy. Nevertheless, electron microscopy has also some disadvantages such as sample preparation, selection of the area viewed and interpretation of the data. The main drawback of electron microscopy is the potential risk of altering or damaging the sample that may introduce artefacts and possible aggregation of the particles during sample preparation. The suspension has to be dried or frozen and the removal of the dispersion medium may alter the distribution of the particles. If the particles do not conduct electricity, the sample has to be coated with a conducting layer, such as gold, carbon or platinum to avoid negative charging by the electron beam. Two main types of electron microscopes are used: transmission and scanning.

2.3.3.1 Transmission electron microscopy (TEM)
TEM displays an image of the specimen on a fluorescent screen and the image can be recorded on a photographic plate or film. The TEM can be used to examine particles

in the range 0.001–5 µm. The sample is deposited on a Formvar (polyvinyl formal) film resting on a grid to prevent charging of the sample. The sample is usually observed as a replica by coating with an electron transparent material (such as gold or graphite). The preparation of the sample for the TEM may alter the state of dispersion and cause aggregation. Freeze fracturing techniques have been developed to avoid some of the alterations of the sample during sample preparation. Freeze fracturing allows the dispersions to be examined without dilution and replicas can be made of dispersions containing water. It is necessary to have a high cooling rate to avoid the formation of ice crystals.

2.3.3.2 Scanning electron microscopy (SEM)

SEM can show particle topography by scanning a very narrowly focused beam across the particle surface. The electron beam is directed normally or obliquely at the surface. The backscattered or secondary electrons are detected in a raster pattern and displayed on a monitor screen. The image provided by secondary electrons exhibits good three-dimensional detail. The backscattered electrons, reflected from the incoming electron beam, indicate regions of high electron density. Most SEMs are equipped with both types of detectors. The resolution of the SEM depends on the energy of the electron beam which does not exceed 30 kV and hence the resolution is lower than that obtained by the TEM. A very important advantage of SEM is elemental analysis by energy dispersive X-ray analysis (EDX). If the electron beam impinging on the specimen has sufficient energy to excite atoms on the surface, the sample will emit X-rays. The energy required for X-ray emission is characteristic of a given element and since the emission is related to the number of atoms present, quantitative determination is possible.

Scanning transmission electron microscopy (STEM) coupled with EDX has been used for the determination of metal particle sizes. Specimens for STEM were prepared by ultrasonically dispersing the sample in methanol and one drop of the suspension was placed onto a Formvar film supported on a copper grid.

2.3.4 Confocal laser scanning microscopy (CLSM)

CLSM is a very useful technique for identification of suspensions. It uses a variable pinhole aperture or variable width slit to illuminate only the focal plane by the apex of a cone of laser light. Out-of-focus items are dark and do not distract from the contrast of the image. As a result of extreme depth discrimination (optical sectioning) the resolution is considerably improved (up to 40 % when compared with optical microscopy). The CLSM technique acquires images by laser scanning or uses computer software to subtract out-of-focus details from the in-focus image. Images are stored as the sample is advanced through the focal plane in elements as small as 50 nm. Three-dimensional images can be constructed to show the shape of the particles.

2.3.5 Scanning probe microscopy (SPM)

SPM can measure physical, chemical and electrical properties of the sample by scanning the particle surface with a tiny sensor of high resolution. Scanning probe microscopes do not measure a force directly; they measure the deflection of a cantilever which is equipped with a tiny stylus (the tip) functioning as the probe. The deflection of the cantilever is monitored by (i) a tunnelling current, (ii) laser deflection beam from the back side of the cantilever, (iii) optical interferometry, (iv) laser output controlled by the cantilever used as a mirror in the laser cavity, and (v) change in capacitance. SPM generates a three-dimensional image and allows calibrated measurements in three (x, y, z) coordinates. SPM not only produces a highly magnified image, but also provides valuable information on sample characteristics. Unlike EM, which requires vacuum for its operation, SPM can be operated under ambient conditions and, with some limitation, in liquid media.

2.3.6 Scanning tunnelling microscopy (STM)

STM measures an electric current that flows through a thin insulating layer (vacuum or air) separating two conductive surfaces. The electrons are visualized to "tunnel" through the dielectric and generate a current, I, that depends exponentially on the distance, s, between the tiny tip of the sensor and the electrically conductive surface of the sample. The STM tips are usually prepared by etching a tungsten wire in an NaOH solution until the wire forms a conical tip. Pt/Ir wire has also been used. In the contrast current imaging mode, the probe tip is raster-scanned across the surface and a feedback loop adjusts the height of the tip in order to maintain a constant tunnel current. When the energy of the tunnelling current is sufficient to excite luminescence, the tip surface region emits light and functions as an excitation source of subnanometre dimensions. In situ STM has revealed a two-dimensional molecular lamellar arrangement of long chain alkanes adsorbed on the basal plane of graphite. Thermally induced disordering of adsorbed alkanes was studied by variable temperature STM and atomic scale resolution of the disordered phase was claimed by studying the quenched high temperature phase

2.3.7 Atomic force microscopy (AFM)

AFM allows one to scan the topography of a sample using a very small tip made of silicon nitride. The tip is attached to a cantilever that is characterized by its spring constant, resonance frequency and a quality factor. The sample rests on a piezoceramic tube which can move the sample horizontally (x, y motion) and vertically (z motion). The displacement of the cantilever is measured by the position of a laser beam reflected from the mirrored surface on the top side of the cantilever. The reflected laser beam is detected by a photodetector. AFM can be operated in either a contact or a noncontact mode. In the contact mode the tip travels in close contact with the surface, whereas in the noncontact mode the tip hovers 5–10 nm above the surface.

2.3.8 Scattering techniques

These are by far the most useful methods for characterization of suspensions and in principle they can give quantitative information on the particle size distribution, floc size and shape. The only limitation of the methods is the need to use sufficiently dilute samples to avoid interference such as multiple scattering which makes interpreting the results difficult. However, recently backscattering methods have been designed to allow one to measure the sample without dilution. In principle, one can use any electromagnetic radiation such as light, X-ray or neutrons but in most industrial labs only light scattering is applied (using lasers).

2.3.8.1 Light scattering techniques

These can be divided into three main classes: (i) time-average light scattering, static or elastic scattering; (ii) dynamic (quasi-elastic) light scattering that is usually referred as photon correlation spectroscopy. This is a rapid technique that is very suitable for measuring submicron particles (nanosize range); (iii) backscattering techniques that are suitable for measuring concentrated samples. Application of any of these methods depends on the information required and availability of the instrument.

2.3.8.1.1 Time-average light scattering

In this method a dispersion that is sufficiently diluted to avoid multiple scattering is illuminated by a collimated light (usually laser) beam and the time-average intensity of scattered light is measured as a function of scattering angle θ. Static light scattering is termed elastic scattering. Three regimes can be identified:

Rayleigh regime. The particle radius R is smaller than $\lambda/20$ (where λ is the wavelength of incident light). The scattering intensity is given by the equation,

$$I(Q) = [\text{Instrument constant}] \, [\text{Material constant}] \, NV_p^2 . \qquad (2.40)$$

Q is the scattering vector that depends on the wavelength of light λ used and is given by,

$$Q = \left(\frac{4\pi n}{\lambda}\right) \sin\left(\frac{\theta}{2}\right) , \qquad (2.41)$$

where n is the refractive index of the medium.

The material constant depends on the difference between the refractive index of the particle and that of the medium. N is the number of particles and V_p is the volume of each particle. Assuming that the particles are spherical, one can obtain the average size using equation (2.40).

The Rayleigh equation reveals two important relationships: (i) The intensity of scattered light increases with the square of the particle volume and consequently with the sixth power of the radius R. Hence the scattering from larger particles may dominate the scattering from smaller particles. (ii) The intensity of scattering is inversely proportional to λ^4. Hence a decrease in the wavelength will substantially increase the scattering intensity.

Rayleigh–Gans–Debye regime (RGD) $\lambda/20 < R < \lambda$. The RGD regime is more complicated than the Rayleigh regime and the scattering pattern is no longer symmetrical about the line corresponding to the 90° angle but favours forward scattering ($\theta < 90°$) or backscattering ($180° > \theta > 90°$). Since the preference for forward scattering increases with increasing particle size, the ratio $I_{45°}/I_{135°}$ can indicate the particle size.

Mie regime $R > \lambda$. The scattering behaviour is more complex than the RGD regime and the intensity exhibits maxima and minima at various scattering angles depending on particle size and refractive index. The Mie theory for light scattering can be used to obtain the particle size distribution using numerical solutions. One can also obtain information on particle shape.

2.3.8.1.2 Turbidity measurements
Turbidity (total light scattering technique) can be used to measure particle size, flocculation and particle sedimentation. This technique is simple and easy to use; a single or double beam spectrophotometer or a nephelometer can be used.

For nonabsorbing particles the turbidity τ is given by,

$$\tau = \left(\frac{1}{L}\right) \ln\left(\frac{I_0}{I}\right) , \qquad (2.42)$$

where L is the path length, I_0 is the intensity of incident beam and I is the intensity of transmitted beam.

The particle size measurement assumes that the light scattered by a particle is singular and independent of other particles. Any multiple scattering complicates the analysis. According to the Mie theory the turbidity is related to the particle number N and their cross section πr^2 (where r is the particle radius) by,

$$\tau = Q\pi r^2 N , \tag{2.43}$$

where Q is the total Mie scattering coefficient. Q depends on the particle size parameter α (which depends on particle diameter and wavelength of incident light λ) and the ratio of the refractive index of the particles and medium m.

Q depends on α in an oscillatory mode that exhibits a series of maxima and minima whose position depends on m. For particles with $R < (1/20)\lambda$, $\alpha < 1$ and it can be calculated using the Rayleigh theory. For $R > \lambda$, Q approaches 2 and between these two extremes the Mie theory is used. If the particles are not monodisperse (as is the case with most practical systems), the particle size distribution must be taken into account. Using this analysis one can establish the particle size distribution using numerical solutions.

2.3.8.1.3 Light diffraction technique

This is a rapid and nonintrusive technique for determination of particle size distribution in the range 2–300 µm with good accuracy for most practical purposes. Light diffraction gives an average diameter over all particle orientations as randomly oriented particles pass the light beam. A collimated and vertically polarized laser beam illuminates a particle dispersion and generates a diffraction pattern with the undiffracted beam in the centre. The energy distribution of diffracted light is measured by a detector consisting of light sensitive circles separated by isolating circles of equal width. The angle formed by the diffracted light increases with decreasing particle size. The angle-dependent intensity distribution is converted by Fourier optics into a spatial intensity distribution $I(r)$. The spatial intensity distribution is converted into a set of photocurrents and the particle size distribution is calculated using a computer. Several commercial instruments are available, e.g. Malvern Mastersizer (Malvern, UK), Horriba (Japan) and Coulter LS Sizer (USA). A schematic illustration of the setup is shown in Fig. 2.5.

In accordance with the Fraunhofer theory (which was introduced by Fraunhofer over 100 years ago), the special intensity distribution is given by,

$$I(r) = \int_{X_{min}}^{X_{max}} N_{tot} q_0(x) I(r, x) \, dx , \tag{2.44}$$

where $I(r, x)$ is the radial intensity distribution at radius r for particles of size x, N_{tot} is the total number of particles and $q_0(x)$ describes the particle size distribution.

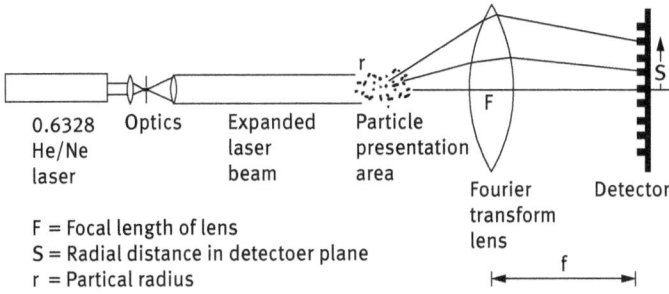

0.6328 He/Ne laser | Optics | Expanded laser beam | Particle presentation area

F = Focal length of lens
S = Radial distance in detectoer plane
r = Partical radius

Fourier transform lens

Detector

f

Fig. 2.5: Schematic illustration of light diffraction particle sizing system.

The radial intensity distribution $I(r, x)$ is given by,

$$I(r, x) = I_0 \left(\frac{\pi x^2}{2f} \right)^2 \left(\frac{J_i(k)}{k} \right)^2 ,$$ (2.45)

with $k = (\pi x r)/(\lambda f)$, where r is the distance to the centre of the disc, λ is the wavelength, f is the focal length, and J_i is the first order Bessel function.

The Fraunhofer diffraction theory applies to particles whose diameter is considerably larger than the wavelength of illumination. As shown in Fig. 2.5, an He/Ne laser is used with $\lambda = 632.8$ nm for particle sizes mainly in the 2–120 µm range. In general, the diameter of the sphere-shaped particle should be at least four times the wavelength of the illumination light. The accuracy of particle size distribution determined by light diffraction is not very good if a large fraction of particles with diameter < 10 µm is present in the suspension. For small particles (diameter < 10 µm) the Mie theory is more accurate if the necessary optical parameters, such as refractive index of particles and medium and the light absorptivity of the dispersed particles, are known. Most commercial instruments combine light diffraction with forward light scattering to obtain a full particle size distribution covering a wide range of sizes.

As an illustration, Fig. 2.6 shows the result of particle sizing using a six component mixture of standard polystyrene lattices (using a Mastersizer).

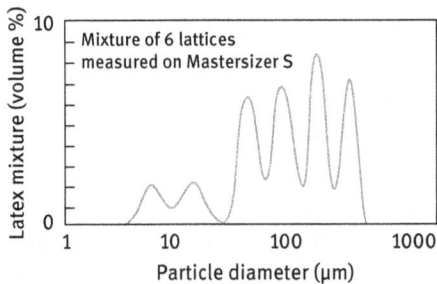

Fig. 2.6: Single measurement of a mixture of six standard lattices using the Mastersizer.

Most practical suspensions are polydisperse and generate a very complex diffraction pattern. The diffraction pattern of each particle size overlaps with diffraction patterns of other sizes. The particles of different sizes diffract light at different angles and the energy distribution becomes a very complex pattern. However, manufacturers of light diffraction instruments (such as Malvern, Coulters and Horriba) have developed numerical algorithms relating diffraction patterns to particle size distribution.

Several factors can affect the accuracy of Fraunhofer diffraction: (i) particles smaller than the lower limit of Fraunhofer theory; (ii) nonexistent "ghost" particles in a particle size distribution obtained by Fraunhofer diffraction that was applied to systems containing particles with edges, or a large fraction of small particles (below 10 µm); (iii) computer algorithms that are unknown to the user and vary with the manufacturer software version; (iv) the composition-dependent optical properties of the particles and dispersion medium; (v) if the density of all particles is not the same, the result may be inaccurate.

2.3.8.1.4 Dynamic light scattering – Photon Correlation Spectroscopy (PCS)

Dynamic light scattering (DLS) is a method that measures the time-dependent fluctuation of scattered intensity. It is also referred to as quasi-elastic light scattering (QELS) or photon correlation spectroscopy (PCS). The latter is the most commonly used term for describing the process since most dynamic scattering techniques employ autocorrelation.

PCS is a technique that utilizes Brownian motion to measure the particle size. As a result of the Brownian motion of dispersed particles, the intensity of scattered light undergoes fluctuations that are related to the velocity of the particles. Since larger particles move less rapidly than the smaller ones, the intensity fluctuation (intensity versus time) pattern depends on particle size as is illustrated in Fig. 2.7. The velocity of the scatterer is measured in order to obtain the diffusion coefficient.

In a system where Brownian motion is not interrupted by sedimentation or particle-particle interaction, the movement of particles is random. Hence, the intensity fluctuations observed after a large time interval do not resemble those fluctuations observed initially, but represent a random distribution of particles. Consequently, the fluctuations observed at large time delay are not correlated with the initial fluctuation pattern. However, when the time differential between the observations is very small (a nanosecond or a microsecond) both positions of particles are similar and the scattered intensities are correlated. When the time interval is increased, the correlation decreases. The decay of correlation is particle size dependent. The smaller the particles are, the faster the decay.

The fluctuations in scattered light are detected by a photomultiplier and are recorded. The data containing information on the particle motion are processed by a digital correlator. The latter compares the intensity of scattered light at time t, I(t), to the intensity at a very small time interval τ later, I(t + τ), and it constructs the

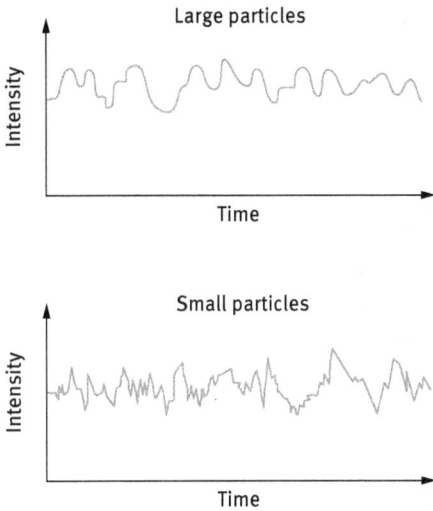

Fig. 2.7: Schematic representation of the intensity fluctuation for large and small particles.

second-order autocorrelation function $G_2(\tau)$ of the scattered intensity,

$$G_2(\tau) = \langle I(t)I(t + \tau)\rangle. \tag{2.46}$$

The experimentally measured intensity autocorrelation function $G_2(\tau)$ depends only on the time interval τ, and is independent of t, the time when the measurement started.

PCS can be measured in a homodyne mode where only scattered light is directed to the detector. It can also be measured in heterodyne mode where a reference beam split from the incident beam is superimposed on scattered light. The diverted light beam functions as a reference for the scattered light from each particle.

In the homodyne mode, $G_2(\tau)$ can be related to the normalized field autocorrelation function $g_1(\tau)$ by,

$$G_2(\tau) = A + Bg_1^2(\tau), \tag{2.47}$$

where A is the background term designated as the baseline value and B is an instrument-dependent factor. The ratio B/A is regarded as a quality factor for the measurement or the signal-to-noise ratio and expressed sometimes as the % merit.

The field autocorrelation function $g_1(\tau)$ for a monodisperse suspension decays exponentially with τ,

$$g_1(\tau) = \exp(-\Gamma\tau), \tag{2.48}$$

where Γ is the decay constant (s^{-1}).

Substituting equation (2.44) into equation (2.43) yields the measured autocorrelation function,

$$G_2(\tau) = A + B\exp(-2\Gamma\tau). \tag{2.49}$$

The decay constant Γ is linearly related to the translational diffusion coefficient D_T of the particle,

$$\Gamma = D_T q^2 . \tag{2.50}$$

The modulus q of the scattering vector is given by,

$$q = \frac{4\pi n}{\lambda_0} \sin\left(\frac{\theta}{2}\right) , \tag{2.51}$$

where n is the refractive index of the dispersion medium, θ is the scattering angle and λ_0 is the wavelength of the incident light in vacuum.

PCS determines the diffusion coefficient and the particle radius R is obtained using the Stokes–Einstein equation,

$$D = \frac{kT}{6\pi\eta R} , \tag{2.52}$$

where k is the Boltzmann constant, T is the absolute temperature and η is the viscosity of the medium.

The Stokes–Einstein equation is limited to noninteracting, spherical and rigid spheres. The effect of particle interaction at relatively low particle concentration c can be taken into account by expanding the diffusion coefficient into a power series of concentration,

$$D = D_0(1 + k_D c) , \tag{2.53}$$

where D_0 is the diffusion coefficient at infinite dilution and k_D is the virial coefficient that is related to particle interaction. D_0 can be obtained by measuring D at several particle number concentrations and extrapolating to zero concentration.

For polydisperse suspensions, the first-order autocorrelation function is an intensity-weighted sum of autocorrelation functions of particles contributing to the scattering,

$$g_1(\tau) = \int_0^\infty C(\Gamma) \exp(-\Gamma\tau)\, d\Gamma . \tag{2.54}$$

$C(\Gamma)$ represents the distribution of decay rates.

For a narrow particle size distribution, the cumulant analysis is usually satisfactory The cumulant method is based on the assumption that for monodisperse suspensions $g_1(\tau)$ is monoexponential. Hence $\log g_1(\tau)$ versus τ yields a straight line with a slope equal to Γ,

$$\ln g_1(\tau) = 0.5 \ln(B) - \Gamma\tau , \tag{2.55}$$

where B is the signal-to-noise ratio.

The cumulant method expands the Laplace transform about an average decay rate,

$$\langle \Gamma \rangle = \int_0^\infty \Gamma C(\Gamma)\, d\Gamma . \tag{2.56}$$

The exponential in equation (2.52) is expanded about an average and integrated term,

$$\ln g_1(\tau) = \langle \Gamma \rangle \tau + (\mu_2 \tau^2)/2! - (\mu_3 \tau^3)/3! + \cdots . \tag{2.57}$$

An average diffusion coefficient is calculated from $\langle \Gamma \rangle$ and the polydispersity (termed the polydispersity index) is indicated by the relative second moment, $\mu_2/\langle \Gamma \rangle^2$. A constrained regulation method (CONTIN) yields several numerical solutions to the particle size distribution and this is normally included in the software of the PCS machine.

PCS is a rapid, absolute and nondestructive method for particle size measurements. It has some limitations. The main disadvantage is the poor resolution of particle size distribution. Also it suffers from the limited size range (absence of any sedimentation) that can be accurately measured. Several instruments are commercially available, e.g. by Malvern, Brookhaven, Coulters, etc. The most recent instrument that is convenient to use is supplied by Malvern (UK) and this allows one to measure the particle size distribution without the need of too much dilution (which may cause some particle dissolution).

2.3.8.1.5 Backscattering technique

This method is based on the use of fibre optics, sometimes referred to as fibre optic dynamic light scattering (FODLS), and it allows one to measure at high particle number concentrations. The FODLS employs either one or two optical fibres. Alternatively, fibre bundles may be used. The exit port of the optical fibre (optode) is immersed in the sample and the scattered light in the same fibre is detected at a scattering angle of 180° (i.e. backscattering).

The technique is suitable for on-line measurements during manufacture of a nanosuspension or nanoemulsion. Several commercial instruments are available, e.g. Lesentech (USA).

2.4 Measurement of charge and zeta potential

Many suspension particles acquire a charge either by dissociation of surface groups (such as metal oxides or latexes) or by adsorption of ionic surfactants as will be discussed in Chapter 4. A surface potential can be ascribed to the particle surface which, as will be discussed in Chapter 4, determines the electrostatic repulsion between the particles. In most cases it is not easy to determine the surface potential which is replaced by the measurable zeta potential. The latter can be obtained by measuring the particle mobility on application of an electric field (electrophoresis). This results in charge separation at the interface between two phases and on application of an electric field one of the phases is caused to move tangentially past the second phase. One measures the particle velocity v (m s^{-1}) from which the particle or droplet mobility u

44 —— 2 Preparation of suspension concentrates by the bottom-up process

$(m^2 \, V^{-1} \, s^{-1})$ is calculated by dividing v by the field strength (E/l, where E is the applied potential in volts and l is the distance between the two electrodes used).

The main problem in any analysis of electrophoresis is defining the plane at which the liquid begins to move past the surface of the particle or droplet [23]. This is defined as the "shear plane", which is at some distance from the surface. One usually defines an "imaginary" surface close to the particle surface within which the fluid is stationary. The point just outside this imaginary surface is described as the surface of shear and the potential at this point is described as the zeta potential (ζ). A schematic representation of the surface of shear, the surface and zeta potential is shown in Fig. 2.8.

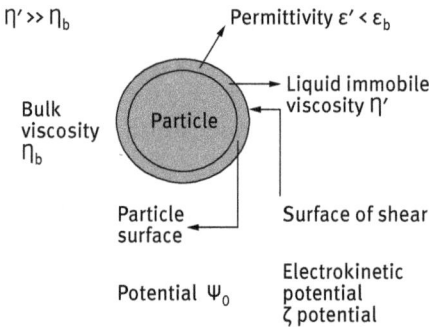

Fig. 2.8: Surface (plane) of shear.

The exact position of the plane of shear is not known; it is usually in the region of few A. In some cases one may equate the shear plane with the Stern plane (the centre of specifically adsorbed ions) although this may be an underestimate of its location. Several layers of liquid may be immobilized at the particle surface (which means that the shear plane is farther apart from the Stern plane). The particle or droplet plus its immobile liquid layer form the kinetic unit that moves under the influence of the electric field. The viscosity of the liquid in the immobile sheath around the particles (η') is much larger than the bulk viscosity η. The permittivity of the liquid in this liquid sheath ε' is also lower than the bulk permittivity (due to dielectric saturation in this layer). In the absence of specific adsorption, the assumption is usually made that $\zeta \approx \psi_0$. The latter potential is the value that is commonly used to calculate the repulsive energy between two particles.

The zeta potential ζ is calculated from the mobility u using the Smoluckowski equation [24] that is applicable when the particle radius R is much higher than the double layer thickness $(1/\kappa)$ where κ is the Debye–Huckel parameter (see Chapter 4) that is the case with most suspensions with $R > 0.5$ µm and $1/\kappa < 10$ nm, i.e. with $1:1$ electrolyte concentration $> 10^{-3}$ mol dm^{-3})

$$u = \frac{\varepsilon_r \varepsilon_0 \zeta}{\eta} . \tag{2.58}$$

The most suitable method for measuring zeta potential is the laser velocimetry method that is suitable for particles that undergo Brownian motion. As discussed above, the light scattered by particles will show intensity fluctuations as a result of Brownian diffusion (Doppler shift). If an electric field is placed at right angles to the incident light and in the plane defined by the incident and observation beam, the line broadening is unaffected but the centre frequency of the scattered light is shifted to an extent determined by the electrophoretic mobility. The shift is very small compared to the incident frequency (≈ 100 Hz for and incident frequency of $\approx 6 \times 10^{14}$ Hz) but with a laser source it can be detected by heterodyning (i.e. mixing) the scattered light with the incident beam and detecting the output of the difference frequency. A homodyne method may be applied in which case a modulator to generate an apparent Doppler shift at the modulated frequency is used. To increase the sensitivity of the laser Doppler method, the electric fields are much higher than those used in conventional electrophoresis. Joule heating is minimized by pulsing the electric field in opposite directions. The Brownian motion of the particles also contributes to the Doppler shift and an approximate correction can be made by subtracting the peak width obtained in the absence of an electric field from the electrophoretic spectrum. An He-Ne laser is used as the light source and the output of the laser is split into two coherent beams that are crossfocused in the cell to illuminate the sample. The light scattered by the particle together with the reference beam is detected by a photomultiplier. The output is amplified and analysed to transform the signals to a frequency distribution spectrum. At the intersection of the beam, interferences of known spacing are formed.

The magnitude of the Doppler shift Δv is used to calculate the electrophoretic mobility u using the following expression,

$$\Delta v = \left(\frac{2n}{\lambda_0} \right) \sin \left(\frac{\theta}{2} \right) uE , \tag{2.59}$$

where n is the refractive index of the medium, λ_0 is the incident wavelength in vacuum, θ is the scattering angle and E is the field strength.

Several commercial instruments are available for measuring electrophoretic light scattering: (i) The Coulter DELSA 440SX (Coulter Corporation, USA) is a multi-angle laser Doppler system employing heterodyning and autocorrelation signal processing. Measurements are made at four scattering angle (8, 17, 25 and 34°) and the temperature of the cell is controlled by a Peltier device. The instrument reports the electrophoretic mobility, zeta potential, conductivity and particle size distribution. (ii) Malvern (Malvern Instruments, UK) has two instruments: The ZetaSizer 3000 and ZetaSizer 5000. The ZetaSizer 3000 is a laser Doppler system using crossed beam optical configuration and homodyne detection with photon correlation signal processing. The zeta potential is measured using laser Doppler velocimetry and the particle size is measured using photon correlation spectroscopy (PCS). The ZetaSizer 5000 uses PCS to measure both (a) movement of the particles in an electric field for zeta potential determination and (b) random diffusion of particles at different measuring angles for size measurement on the same sample. In both instruments, a Peltier device is used for temperature control.

An alternative method that can be applied on more concentrated suspensions is the electroacoustic method which is described below.

The mobility of a particle in an alternating field is termed dynamic mobility, to distinguish it from the electrophoretic mobility in a static electric field. The principle of the technique is based on the creation of an electric potential by a sound wave transmitted through an electrolyte solution, as described by Debye [25]. The potential, termed the ionic vibration potential (IVP), arises from the difference in the frictional forces and the inertia of hydrated ions subjected to ultrasound waves. The effect of the ultrasonic compression is different for ions of different masses and the displacement amplitudes are different for anions and cations. Hence the sound waves create periodically changing electric charge densities. This original theory of Debye was extended to include electrophoretic, relaxation and pressure gradient forces [26, 27].

A much stronger effect can be observed in colloidal dispersions. The sound waves transmitted by the suspension of charged particles generate an electric field because the relative motion of the two phases is different. The displacement of a charged particle from its environment by the ultrasound waves generates an alternating potential, termed colloidal vibration potential (CVP). The IVP and CVP are both called ultrasound vibration potential (UVP).

The converse effect, namely the generation of sound waves by an alternating electric field [28] in a colloidal dispersion, can be measured and is termed the electrokinetic sonic amplitude (ESA). The theory for the ESA effect has been developed by O'Brian and co-workers [29–34]. Dynamic mobility can be determined by measuring either UVP or ESA, although in general the ESA is the preferred method. Several commercial instruments are available for measurement of the dynamic mobility: (i) the ESA-8000 system from Matec Applied Sciences that can measure both CVP and ESA signals; (ii) the Pen Kem System 7000 Acoustophoretic titrator that measures the CVP, conductivity, pH, temperature, pressure amplitude and sound velocity.

In the ESA system (from Matec) and the AcoustoSizer (from Colloidal Dynamics) the dispersion is subjected to a high frequency alternating field and the ESA signal is measured. The ESA-8000 operates at constant frequency of $\approx 1\,\text{MHz}$ and the dynamic mobility and zeta potential (but not particle size) are measured. The AcoustoSizer operates at various frequencies of the applied electric field and can measure the particle mobility, zeta potential and particle size.

The frequency synthesizer feeds a continuous sinusoidal voltage into a grated amplifier that creates a pulse of sinusoidal voltage across the electrodes in the dispersion. The pulse generates sound waves which appear to emanate from the electrodes. The oscillation, the back-and-forth movement of the particle caused by an electric field, is the product of the particle charge times the applied field strength. When the direction of the field is alternating, particles in the suspension between the electrodes are driven away towards the electrodes. The magnitude and phase angle of the ESA signal created is measured with a piezoelectric transducer mounted on a solid nonconductive (glass) rod attached to the electrode as illustrated in Fig. 2.9. The purpose of this nonconductive acoustic delay line is to separate the transducer from the high-frequency electric field in the cell. Three pulses of the voltage signal are recorded as schematically shown in Fig. 2.10. The first pulse of the signal, shown on the left, is generated when the voltage pulse is applied to the sample and is unrelated to the ESA effect. This first pulse of the signal is received before the sound has sufficient time to pass down the glass rod and is an electronic cross-talk deleted from data processing. The second and third pulses are ESA signals. The second pulse is detected by the nearest electrode. This pulse is used for data processing to determine the particle size and zeta potential. The third pulse originates from the other electrode and is deleted.

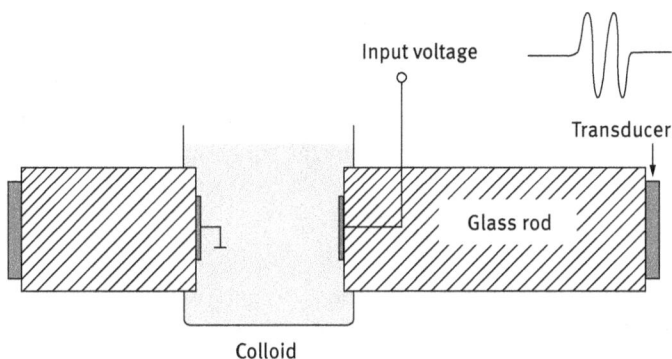

Fig. 2.9: Schematic representation of the AcoustoSizer cell.

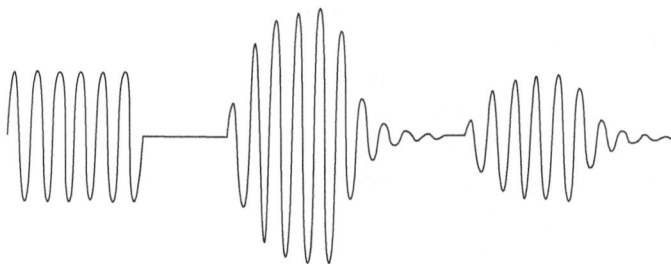

Fig. 2.10: Signals from the right-hand transducer.

In addition to the electrodes, the sample cell of the ESA instruments also houses sensors for pH, conductivity and temperature measurements. It is also equipped with a stirrer and the system is linked to a digital titrator for dynamic mobility and zeta potential measurements as a function of pH.

To convert the ESA signal to dynamic mobility one needs to know the density of the disperse phase and dispersion medium, the volume fraction of the particles and the velocity of sound in the solvent. As shown before, to convert mobility to zeta potential one needs to know the viscosity of the dispersion medium and its relative permittivity. Because of the inertia effects in dynamic mobility measurements, the weight average particle size has to be known.

For dilute suspensions with a volume fraction $\phi = 0.02$, the dynamic mobility ud can be calculated from the electrokinetic sonic amplitude $A_{ESA}(\omega)$ using the following expression [26, 27],

$$A_{ESA}(\omega) = Q(\omega)\phi\left(\frac{\Delta\rho}{\rho}\right)(u_d), \tag{2.60}$$

where ω is the angular frequency of the applied field, $\Delta\rho$ is the density difference between the particle (with density ρ) and the medium. $Q(\omega)$ is an instrument-related coefficient independent of the system being measured.

For a dilute dispersion of spherical particles with $\phi < 0.1$, a thin double layer ($\kappa R > 50$) and narrow particle size distribution (with standard deviation < 20 % of the mean size), u_d can be related to the zeta potential ζ by the equation [26],

$$u_d = \frac{2\varepsilon\zeta}{3\eta}G\left(\frac{\omega R^2}{\nu}\right)[1 + f(\lambda, \omega)], \tag{2.61}$$

where ε is the permittivity of the liquid (that is equal to $\varepsilon_r\varepsilon_0$, defined before), R is the particle radius, η is the viscosity of the medium, λ is the double layer conductance and ν is the kinematic viscosity (= η/ρ). G is a factor that represents particle inertia, which reduces the magnitude of ud and increases the phase lag in a monotonic fashion as the frequency increases. This inertia factor can be used to calculate the particle size from electroacoustic data. The factor $[1 + f(\lambda, \omega)]$ is proportional to the tangential component of the electric field and dependent on particle permittivity and a surface conductance parameter λ. For most suspensions with large κR, the effect of surface conductance is insignificant and the particle permittivity/liquid permittivity $\varepsilon_p/\varepsilon$ is small. In most cases where the ionic strength is at least 10^{-3} mol dm^{-3} and a zeta potential < 75 mV, the factor $[1 + f(\lambda, \omega)]$ assumes the value 0.5. In this case the dynamic mobility is given by the simple expression,

$$u_d = \frac{\varepsilon\zeta}{\eta}G(\alpha). \tag{2.62}$$

Equation (2.56) is identical to the Smoluchowski equation, except for the inertia factor $G(\alpha)$.

The equation for converting the ESA amplitude, A_{ESA}, to dynamic mobility is given by,

$$u_d = \frac{A_{ESA}}{\phi v_s \Delta\rho}G(\alpha)^{-1}. \tag{2.63}$$

The zeta potential ζ is given by,

$$\zeta = \frac{u_d \eta}{\varepsilon} G(\alpha)^{-1} = \frac{A_{ESA}}{\phi v_s \Delta \rho} G(\alpha)^{-1}.$$ (2.64)

For a polydisperse system $\langle u_d \rangle$ is given by,

$$\langle u_d(\omega) \rangle = \int_0^\infty u(\omega, R) p(R) \, dR,$$ (2.65)

where $u(\omega, R)$ is the average dynamic mobility of particles with radius R at a frequency ω, and $p(R) \, dR$ is the mass fraction of particles with radii in the range $R \pm dR/2$.

The ESA measurements can also be applied for determining the particle size in a suspension from particle mobilities. The electric force acting upon a particle is opposed by the hydrodynamic friction and inertia of the particles. At low frequencies of alternating electric field, the inertial force is insignificant and the particle moves in the alternating electric field with the same velocity as it would have moved in a constant field. The particle mobility at low frequencies can be measured to calculate the zeta potential. At high frequencies the inertia of the particle increases causing the velocity of the particle to decrease and the movement of the particle to lag behind the field. This is illustrated in Fig. 2.11 which shows the variation of applied field and particle velocity with time. Since inertia depends on particle mass, both of these effects depend on the particle mass and consequently on its size. Hence, both zeta potential and particle size can be determined from the ESA signal, if the frequency of the alternating field is sufficiently high. This is the method that is provided by the AcoustoSizer from Colloidal Dynamics.

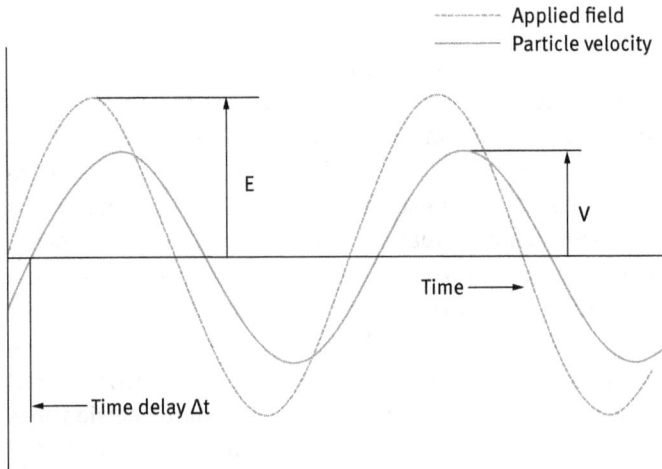

Fig. 2.11: Variation of applied field and particle velocity with time at high frequency.

Several variables affect the ESA measurements and these are listed below.

(i) Particle concentration range: Very dilute suspensions generate a weak signal and are not suitable for ESA measurements. The magnitude of the ESA signal is proportional to the average particle mobility, the volume fraction of the particles ϕ and the density difference between the particles and the medium $\Delta\rho$. To obtain a signal that is at least one order of magnitude higher than the background electrical noise (≈ 0.002 mPa M/V) the concentration and/or the density difference have to be sufficiently large. If the density difference between the particles and medium is small, e.g. polystyrene latex with $\Delta\rho \approx 0.05$, then a sufficiently high concentration ($\phi > 0.02$) is needed to obtain a reasonably strong ESA signal. The accuracy of the ESA measurement is also not good at high ϕ values. This is due to the nonlinearity of the ESA amplitude-ϕ relationship at high ϕ values. Such deviation becomes appreciable at $\phi > 0.1$. However, reasonable values of zeta potential can be obtained from ESA measurements up to $\phi = 0.2$. Above this concentration, the measurements are not sufficiently accurate and the results obtained can only be used for qualitative assessment.

(ii) Electrolyte effects: Ions in the dispersion generate electroacoustic (IVP) potential and the ESP signal is therefore a composite of the signals created by the particles and ions. However, the ionic contribution is relatively small, unless the particle concentration is low, their zeta potential is low and the ionic concentration is high. The ESA system is therefore not suitable for dynamic mobility and zeta potential measurements in systems with electrolyte concentration higher than 0.3 mol dm^{-3} KCl.

(iii) Temperature: Since the viscosity of the dispersion decreases by $\approx 2\%$ per °C and its conductivity increases by about the same amount, it is important that temperature be accurately controlled using a Peltier device. Temperature control should also be maintained during sample preparation, for example when the suspension is sonicated. To avoid overheating the sample should be cooled in an ice bath at regular intervals during sonication.

(iv) Calibration and accuracy: The electroacoustic probe should be calibrated using a standard reference dispersion such as polystyrene latex or colloidal silica (Ludox). The common sources of error are unsuitable particle concentration (too low or too high), irregular particle shape, polydispersity, electrolyte signals, temperature variations, sedimentation, coagulation and entrained air bubbles. The latter in particular can cause erroneous ESA signal fluctuations resulting from weakening of the sound by the air bubbles. In many cases the zeta potential results obtained using the ESA method do not agree with those obtained using other methods such as microelectrophoresis or laser velocimetry. However, the difference seldom exceeds 20% and this makes the ESA method more convenient for measurement of many industrial methods. The main advantages are the speed of measurement and the dispersion does not need to be diluted, in which case the state of the suspension could be changed.

References

[1]   Philipse, A., "ParticulateColloids: Aspects of Preparation and Characterisation", in "Fundamentals of Interface and Colloid Science", Vol. IV, J. Lyklema (ed.), Elsevier, Amsterdam (2005).
[2]   Tadros, Th. F., "Nanodispersions", De Gruyter, Germany (2016).
[3]   Deryaguin, B. V. and Landau, L., Acta Physicochem. USSR, **14**, 633 (1941).
[4]   Verwey, E. J. W. and Overbeek, J. Th. G., "Theory of Stability of Lyophobic Colloids", Elsevier, Amsterdam (1948).
[5]   Napper, D. H., "Polymeric Stabilisation of Colloidal Dispersions", Academic Press, London (1983).
[6]   Gibbs, J. W., Collected Work, Vol. I, Longman, New York, p. 219.
[7]   Smoluchowski, M. V., Z. Phys. Chem, **92**, 129 (1927).
[8]   Lifshitz, I. M. and Slesov, V. V., Sov. Phys. JETP, **35**, 331 (1959).
[9]   Wagner, C., Z. Electrochem., **35**, 581 (1961).
[10] I. Capek, in "Encyclopedia of Colloid and Interface Science", Th. F. Tadros (ed.), Springer, Germany (2013), p. 748.
[11] Tadros, Th. F., "Emulsions", De Gruyter, Germany (2016).
[12] Eastoe, J., Hatzopolous, M. H. and Tabor, R., in "Encyclopedia of Colloid and Interface Science", Th. F. Tadros (ed.), Springer, Germany (2013), p. 688.
[13] Stober, W., Fink, A. and Bohn, E., J. Colloid Interface Sci., **26**, 62 (1968).
[14] Blakely, D. C., "Emulsion Polymerization", Elsevier, Applied Science, London (1975).
[15] Litchi, G., Gilbert R. G. and Napper, D. H., J. Polym. Sci., **21**, 269 (1983).
[16] Feeney, P. J., Napper, D. H. and Gilbert, R. G., Macromolecules, **17**, 2520 (1984); **20**, 2922 (1987).
[17] Nestor, J., Esquena, J., Solans, C., Luckham, P. F., Levecke, B. and Tadros, Th. F., J. Colloid Interface Sci., **311**, 430 (2007).
[18] Leong, Y. S. and Candau, F., J. Phys. Chem., **86**, 2269 (1982).
[19] Candau, F., Leong, Y. S., Candau, G. and Candau, S., in Progr. SIF Course XV, V. Degiorgi and M. Corti (eds.), (1985), p. 830
[20] Larpent, C. and Tadros, Th. F., Colloid Polym. Sci., **269**, 1171 (1991).
[21] Girard, N., Tadros, Th. F. and Bailey, A. I., Colloid Polym. Sci., **276**, 999 (1998).
[22] Tadros, Th. F., "Dispersion of Powders in Liquids and Stabilisation of Suspensions", Wiley-VCH, Germany (2012).
[23] Hunter, R. J., "Zeta Potential in Colloid Science: Principles and Application", Academic Press, London (1981).
[24] Smoluchowski, M. V., Physik. Z., **17**, 557, 585 (1916).
[25] Debye, P., J. Chem. Phys., **1**, 13 (1933).
[26] Bugosh, J., Yeager, E. and Hovarka, F., J. Chem. Phys., **15**, 542 (1947).
[27] Yeager, E., Bugosh, J., Hovarka, F. and McCarthy, J., J. Chem. Phys., **17**, 411 (1949).
[28] Dukhin, A. S. and Goetz, P. J., Colloids and Surfaces, **144**, 49 (1998).
[29] Oja, T., Petersen, G. L. and Cannon, D. C., US Patent 4,497,208 (1985).
[30] O'Brian, R. W., J. Fluid Mech., **190**, 71 (1988).
[31] O'Brian, R. W., J. Fluid Mech., **212**, 81 (1990).
[32] O'Brian, R. W., Garaside, P. and Hunter, R. J., Langmuir, **10**, 931 (1994).
[33] O'Brian, R. W., Cannon, D. W. and Rowlands, W. N., J. Colloid Interface Sci., **173**, 406 (1995).
[34] Rowlands, W. N. and O'Brian, R. W., J. Colloid Interface Sci., **175**, 190 (1995).

3 Preparation of suspensions using the top-down process

As mentioned in Chapter 1, in the top-down process one starts with the bulk material (which may consist of aggregates and agglomerates) that is dispersed into single particles (using a wetting/dispersing agent) using high speed stirrers followed by subdivision of the large particles into smaller units that fall within the required size [1–3]. This process requires the application of intense mechanical energy that can be applied using bead milling. Finally, the resulting suspension must remain colloidally stable under all conditions (such as temperature changes, vibration, etc.) with absence of any flocculation and/or crystal growth.

A schematic representation of the dispersion process is shown in Fig. 3.1.

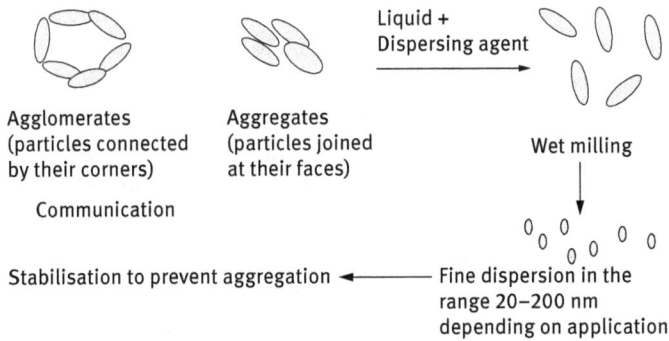

Fig. 3.1: Schematic representation of the dispersion process.

The above process requires wetting of the aggregates and agglomerates (both external and internal surfaces) by molecules of the dispersion medium. This is particularly the case with hydrophobic solids dispersed in aqueous media. As will be discussed below, wetting of hydrophobic solids in aqueous media requires the presence of a wetting agent (surfactant) that lowers the surface tension of water and adsorbs very quickly at the solid/liquid interface, thus reducing the solid/liquid interfacial tension. Once the powder is completely wetted, the aggregates and agglomerates are dispersed into single particles using high speed stirrers and addition of a dispersing agent (surfactant and/or polymer). The resulting dispersion of single particles (referred to as the "mill base") is then subjected to a comminution (milling or particle size reduction), mostly achieved using bead mills. The final suspension must remain colloidally stable using electrostatic and/or steric stabilization as will be discussed in Chapters 4 and 5. These processes of dispersion and milling are described below.

DOI 10.1515/9783110486872-004

3.1 Wetting of the bulk powder

As mentioned above, wetting of powders is a prerequisite for dispersion of powders in liquids. Most chemicals are supplied as powders consisting of aggregates where the particles are joined together with their "faces" (compact structures), or agglomerates where the particles are connected at their corners (loose aggregates) as illustrated in Fig. 3.1. It is essential to wet both the external and internal surface (in the pores within the aggregate or agglomerate structures) and this requires the use of an effective wetting agent (surfactant) [1–3]. Wetting of a solid by a liquid (such as water) requires the replacement of the solid/vapour interfacial tension, γ_{SV}, by the solid/liquid interfacial tension, γ_{SL}.

The equilibrium aspects of wetting can be studied at a fundamental level using interfacial thermodynamics. A useful parameter to describe wetting is the contact angle θ of a liquid drop on a solid substrate [4, 5] which is the angle between planes tangent to the surfaces of solid and liquid at the wetting perimeter. If the liquid makes no contact with the solid, i.e. $\theta = 180°$, the solid is referred to as non-wettable by the liquid in question. This may be the case for a perfectly hydrophobic surface with a polar liquid such as water. However, when $180° > \theta > 90°$, one may refer to a case of poor wetting. When $0° < \theta < 90°$, partial (incomplete) wetting is the case, whereas when $\theta = 0°$ complete wetting occurs and the liquid spreads on the solid substrate forming a uniform liquid film.

The utility of contact angle measurements depends on equilibrium thermodynamic arguments (static measurements) using the well-known Young's equation [4]. The value depends on: (i) the history of the system; (ii) whether the liquid is tending to advance across or recede from the solid surface (Advancing angle θ_A, Receding angle θ_R; usually $\theta_A > \theta_R$). Under equilibrium, the liquid drop takes the shape that minimizes the free energy of the system. Three interfacial tensions can be identified: γ_{SV}, solid/vapour area A_{SV}; γ_{SL}, solid/liquid area A_{SL}; γ_{LV}, liquid/vapour area A_{LV}. A schematic representation of the balance of tensions at the solid/liquid/vapour interface is shown in Fig. 3.2. Here, solid and liquid are simultaneously in contact with each other and the surrounding phase (air or vapour of the liquid). The wetting perimeter is referred to as the three phase line or wetting line. In this region there is an equilibrium between vapour, liquid and solid.

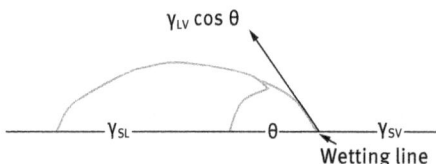

Fig. 3.2: Schematic representation of the contact angle and wetting line.

$\gamma_{SV} A_{SV} + \gamma_{SL} A_{SL} + \gamma_{LV} A_{LV}$ should be a minimum at equilibrium and this leads to the well-known Young's equation [4],

$$\gamma_{SV} = \gamma_{SL} + \gamma_{LV} \cos \theta \qquad (3.1)$$

$$\cos \theta = \frac{\gamma_{SV} - \gamma_{SL}}{\gamma_{LV}} . \qquad (3.2)$$

The contact angle θ depends on the balance between the solid/vapour (γ_{SV}) and solid/liquid (γ_{SL}) interfacial tensions. The angle which a drop assumes on a solid surface is the result of the balance between the adhesion force between solid and liquid and the cohesive force in the liquid,

Wetting of a powder is achieved by the use of surface active agents (wetting agents) of the ionic or nonionic type which are capable of diffusing quickly (i.e. lower the dynamic surface tension) to the solid/liquid interface and displace the air entrapped by rapid penetration through the channels between the particles and inside any "capillaries". For wetting of hydrophobic powders in water, anionic surfactants, e.g. alkyl sulphates or sulphonates or nonionic surfactants of the alcohol ethoxylates are usually used [1–3].

$$\gamma_{LV} \cos \theta = \gamma_{SV} - \gamma_{SL} . \qquad (3.3)$$

A useful concept for choosing wetting agents of the ethoxylated surfactants is the hydrophilic-lipophilic balance (HLB) concept,

$$HLB = \frac{\% \text{ of hydrophilic groups}}{5} . \qquad (3.4)$$

Most wetting agents of this class have an HLB number in the range 7–9.

The process of wetting a solid of unit surface area by a liquid involves three types of wetting [1–3]: adhesion wetting, W_a; immersion wetting W_i; spreading wetting W_s. In every step one can apply Young's equation,

$$W_a = \gamma_{SL} - (\gamma_{SV} + \gamma_{LV}) = -\gamma_{LV}(\cos \theta + 1) , \qquad (3.5)$$

$$W_i = 4\gamma_{SL} - 4\gamma_{SV} - 4\gamma_{LV} \cos \theta , \qquad (3.6)$$

$$W_s = (\gamma_{SL} + \gamma_{LV}) - \gamma_{SV} = -\gamma_{LV}(\cos \theta - 1) . \qquad (3.7)$$

The work of dispersion of a solid with unit surface area W_d is the sum of W_a, W_i and W_s,

Wetting and dispersion depends on: γ_{LV}, liquid surface tension; θ, contact angle between liquid and solid. W_a, W_i and W_s are spontaneous when $\theta < 90°$. W_d is spontaneous when $\theta = 0$. Since surfactants are added in sufficient amounts ($\gamma_{dynamic}$ is lowered sufficiently), spontaneous dispersion is the rule rather than the exception.

$$W_d = W_a + W_i + W_s = 6\gamma_{SV} - \gamma_{SL} = -6\gamma_{LV} \cos \theta . \qquad (3.8)$$

The work of dispersion of a powder with surface area A, W_d, is given by [1–3],

$$W_d = A(\gamma_{SL} - \gamma_{SV}) . \qquad (3.9)$$

Using Young's equation,

$$\gamma_{SV} = \gamma_{SL} + \gamma_{LV} \cos \theta , \qquad (3.10)$$

where γ_{LV} is the liquid/vapour interfacial tension and θ is the contact angle of the liquid drop at the wetting line.

$$W_d = -A\gamma_{LV} \cos \theta \qquad (3.11)$$

Equation (3.11) shows that W_d depends on γ_{LV} and θ, both of which are lowered by addition of surfactants (wetting agents). If $\theta < 90°$, W_d is negative and dispersion is spontaneous.

Wetting of the internal surface requires penetration of the liquid into channels between and inside the agglomerates. The process is similar to forcing a liquid through fine capillaries. To force a liquid through a capillary with radius r, a pressure p is required that is given by,

$$p = -\frac{2\gamma_{LV} \cos \theta}{r} = \left[\frac{-2(\gamma_{SV} - \gamma_{SL})}{r\gamma_{LV}} \right] . \qquad (3.12)$$

γ_{SL} has to be made as small as possible; rapid surfactant adsorption to the solid surface, low θ. When $\theta = 0$, $p \propto \gamma_{LV}$. Thus for penetration into pores one requires a high γ_{LV}. Thus, wetting of the external surface requires low contact angle θ and low surface tension γ_{LV}. Wetting of the internal surface (i.e. penetration through pores) requires low θ but high γ_{LV}. These two conditions are incompatible and a compromise has to be made: $\gamma_{SV} - \gamma_{SL}$ must be kept at a maximum. γ_{LV} should be kept as low as possible but not too low.

The above conclusions illustrate the problem of choosing the best dispersing agent for a particular powder. This requires measurement of the above parameters as well as testing the efficiency of the dispersion process.

The contact angle of liquids on solid powders can be measured by application of the Rideal–Washburn equation. For horizontal capillaries (gravity neglected), the depth of penetration l in time t is given by the Rideal–Washburn equation [6, 7],

$$l = \left[\frac{rt\gamma_{LV} \cos \theta}{2\eta} \right]^{1/2} . \qquad (3.13)$$

To enhance the rate of penetration, γ_{LV} has to be made as high as possible, θ as low as possible and η as low as possible. For dispersion of powders into liquids one should use surfactants that lower θ while not reducing γ_{LV} too much. The viscosity of the liquid should also be kept at a minimum. Thickening agents (such as polymers) should not be added during the dispersion process. It is also necessary to avoid foam formation during the dispersion process.

For a packed bed of particles, r may be replaced by K, which contains the effective radius of the bed and a tortuosity factor, which takes into account the complex path formed by the channels between the particles, i.e.

$$l = \left(\frac{Kt\gamma_{LV} \cos \theta}{2\eta} \right)^{1/2} . \qquad (3.14)$$

Thus a plot of l^2 versus t gives a straight line and from the slope of the line one can obtain θ. The Rideal–Washburn equation can be applied to obtain the contact angle of liquids (and surfactant solutions) in powder beds. K should first be obtained using a liquid that produces zero contact angle. A packed bed of powder is prepared, say in a tube fitted with a sintered glass at the end (to retain the powder particles). It is essential to pack the powder uniformly in the tube (a plunger may be used in this case). The tube containing the bed is immersed in a liquid that gives spontaneous wetting (e.g. a lower alkane), i.e. the liquid gives a zero contact angle and $\cos θ = 1$. By measuring the rate of penetration of the liquid (this can be carried out gravimetrically using for example a microbalance or a Kruss instrument) one can obtain K. The tube is then removed from the lower alkane liquid and left to stand for evaporation of the liquid. It is then immersed in the liquid in question and the rate of penetration is measured again as a function of time. Using equation (3.14), one can calculate $\cos θ$ and hence θ.

For efficient wetting of hydrophobic solids in water, a surfactant is needed that lowers the surface tension of water very rapidly (within few ms) and quickly adsorbs at the solid/liquid interface [1–3]. To achieve rapid adsorption, the wetting agent should be either a branched chain with central hydrophilic group or a short hydrophobic chain with hydrophilic end group. The most commonly used wetting agents are the following:

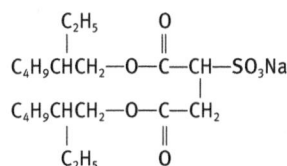

$$
\begin{array}{l}
C_2H_5 O \\
| \| \\
C_4H_9CHCH_2-O-C-CH-SO_3Na \\
C_4H_9CHCH_2-O-C-CH_2 \\
| \| \\
C_2H_5 O
\end{array}
$$

The above molecule has a low critical micelle concentration (cmc) of $0.7\,g\,dm^{-3}$ and at and above the cmc the water surface tension is reduced to $\approx 25\,mN\,m^{-1}$ in less than 15 s.

Several nonionic surfactants, such as the alcohol ethoxylates, can also be used as wetting agents. These molecules consist of a short hydrophobic chain (mostly C_{10}) which is also branched. A medium chain polyethylene oxide (PEO) mostly consisting of 6 EO units or lower is used. These molecules also reduce the dynamic surface tension within a short time (< 20 s) and they have reasonably low cmc. In all cases one should use the minimum amount of wetting agent to avoid interference with the dispersant that needs to be added to maintain the colloid stability during dispersion and on storage.

3.2 Breaking of aggregates and agglomerates into individual units

This usually requires the application of mechanical energy. High speed mixers (which produce turbulent flow) of the rotor-stator type [8] are efficient in breaking up the aggregates and agglomerates, e.g. Silverson mixers, Ultra-Turrax. These are the most commonly used mixers for dispersion of powders in liquids. Two main types are available. The most commonly used toothed device (schematically illustrated in Fig. 3.3) is the Ultra-Turrax (IKA works, Germany).

Fig. 3.3: Schematic representation of a toothed mixer (Ultra-Turrax).

Toothed devices are available both as in-line as well as batch mixers, and because of their open structure they have a relatively good pumping capacity. Therefore, in batch applications they frequently do not need an additional impeller to induce bulk flow even in relatively large mixing vessels.

Batch radial discharge mixers, such as Silverson mixers (Fig. 3.4), have a relatively simple design with a rotor equipped with four blades pumping the fluid through a stationary stator perforated with differently shaped/sized holes or slots.

They are frequently supplied with a set of easily interchangeable stators enabling the same machine to be used for a range of operations e.g. blending, particle size reduction and de-agglomeration. Changing from one screen to another is quick and simple. Different stators/screens used in batch Silverson mixers are shown in Fig. 3.5. The general purpose disintegrating stator (Fig. 3.5 (a)) is recommended for preparation of thick suspensions ("gels") whilst the slotted disintegrating stator (Fig. 3.5 (b)) is designed for suspensions containing elastic materials such as polymers. Square holed screens (Fig. 3.5 (c)) are recommended for the preparation of suspensions whereas the standard screen (Fig. 3.5 (d)) is used for solid/liquid dispersion.

Fig. 3.4: Schematic representation of batch radial discharge mixer (Silverson mixer).

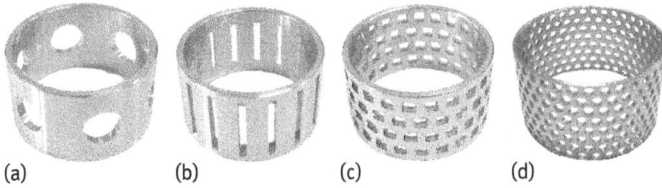

(a) (b) (c) (d)

Fig. 3.5: Stators used in batch Silverson radial discharge mixers.

In all methods there is liquid flow, unbounded and strongly confined flow. In unbounded flow any particle is surrounded by a large amount of flowing liquid (the confining walls of the apparatus are far away from most of the particles). The forces can be frictional (mostly viscous) or inertial. Viscous forces cause shear stresses to act on the interface between the particles and the continuous phase (primarily in the direction of the interface). The shear stresses can be generated by laminar flow (LV) or turbulent flow (TV); this depends on the dimensionless Reynolds numbers Re,

$$Re = \frac{vl\rho}{\eta}, \tag{3.15}$$

where v is the linear liquid velocity, ρ is the liquid density and η is its viscosity. l is a characteristic length that is given by the diameter of flow through a cylindrical tube and by twice the slit width in a narrow slit.

For laminar flow, $Re \lesssim 1000$, whereas for turbulent flow $Re \gtrsim 2000$. Thus whether the regime is linear or turbulent depends on the scale of the apparatus, the flow rate and the liquid viscosity [1–3].

Batch toothed and radial discharge rotor-stator mixers are manufactured in different sizes ranging from the laboratory to the industrial scale. In lab applications, mixing heads (assembly of rotor and stator) can be as small as 0.01 m (Turrax, Silverson) and the volume of processed fluid can vary from several millilitres to few litres. In models used in industrial applications, mixing heads might have up to 0.5 m diameter enabling processing of several cubic meters of fluids in one batch.

In practical applications the selection of the rotor-stator mixer for a specific dispersion process depends on the required morphology of the product, frequently quantified in terms of average particle size or in terms of particle size distributions, and by the scale of the process. The selection of an appropriate mixer and processing conditions for a required formulation is frequently carried out by trial and error. Initially, one can carry out lab scale dispersion of given formulations testing different type/geometries of mixers. Once the type of mixer and its operating parameters are determined at the lab scale the process needs to be scaled up. The majority of lab tests of dispersion are carried out in small batch vessels as it is easier and cheaper than running continuous processes. Therefore, prior to scaling up of the rotor-stator mixer, it has to be decided whether industrial dispersion should be run as a batch or as a continuous process. Batch mixers are recommended for processes where formulation of a product requires long processing times typically associated with slow chemical reactions. They require simple control systems, but spatial homogeneity may be an issue in large vessels which could lead to a longer processing time. In processes where quality of the product is controlled by mechanical/hydrodynamic interactions between continuous and dispersed phases or by fast chemical reactions, but large amounts of energy are necessary to ensure adequate mixing, in-line rotor-stator mixers are recommended. In-line mixers are also recommended to efficiently process large volumes of fluid.

In the case of batch processing, rotor-stator devices immersed as top entry mixers are mechanically the simplest arrangement, but in some processes bottom entry mixers ensure better bulk mixing; however, in this case sealing is more complex. In general, the efficiency of batch rotor-stator mixers decreases as the vessel size increases and as the viscosity of the processed fluid increases because of limited bulk mixing by rotor-stator mixers. Whilst the open structure of Ultra-Turrax mixers frequently enables sufficient bulk mixing even in relatively large vessels, if the suspension has a low apparent viscosity, processing of very viscous suspensions requires an additional impeller (typically anchor type) to induce bulk flow and to circulate the dispersion through the rotor-stator mixer. On the other hand, batch Silverson rotor-stator mixers have a very limited pumping capacity and even at the lab scale they are mounted off the centre of the vessel to improve bulk mixing. At the large scale there is always need for at least one additional impeller and in the case of very large units more than one impeller is mounted on the same shaft.

Problems associated with the application of batch rotor-stator mixers for processing large volumes of fluid discussed above can be avoided by replacing batch mixers with in-line (continuous) mixers. There are many designs offered by different suppliers (Silverson, IKA, etc.) and the main differences are related to the geometry of the rotors and stators with stators and rotors designed for different applications. The main difference between batch and in-line rotor-stator mixers is that the latter have a strong pumping capacity, therefore they are mounted directly in the pipeline. One of the main advantages of in-line over batch mixers is that for the same power duty, a much smaller mixer is required, therefore they are better suited for processing large volumes of fluid. When the scale of the processing vessel increases, a point is reached where it is more efficient to use an in-line rotor-stator mixer rather than a batch mixer of a large diameter. Because power consumption increases sharply with rotor diameter (to the fifth power) an excessively large motor is necessary at large scales. This transition point depends on the fluid rheology, but for a fluid with a viscosity similar to water, it is recommended to change from a batch to an in-line rotor-stator process at a volume of approximately 1 to 1.5 t. The majority of manufacturers supply both single and multistage mixers for the emulsification of highly viscous liquids.

As mentioned above, in all methods there is liquid flow, unbounded and strongly confined flow. In unbounded flow any particle is surrounded by a large amount of flowing liquid (the confining walls of the apparatus are far away from most of the droplets); the forces can be frictional (mostly viscous) or inertial. Viscous forces cause shear stresses to act on the interface between the particles and the continuous phase (primarily in the direction of the interface). The shear stresses can be generated by laminar flow (LV) or turbulent flow (TV); this depends on the Reynolds number Re as given by equation (3.15). For laminar flow, Re $\lesssim$ 1000, whereas for turbulent flow Re $\gtrsim$ 2000. Thus, whether the regime is linear or turbulent depends on the scale of the apparatus, the flow rate and the liquid viscosity. If the turbulent eddies are much larger than the particles, they exert shear stresses on the particles. If the turbulent eddies are much smaller than the particles, inertial forces will cause disruption (TI). In bounded flow other relations hold; if the smallest dimension of the part of the apparatus in which the particles are disrupted (say a slit) is comparable to particle size, other relations hold (the flow is always laminar).

Within each regime, an essential variable is the intensity of the acting forces; the viscous stress during laminar flow $\sigma_{viscous}$ is given by,

$$\sigma_{viscous} = \eta G , \tag{3.16}$$

where G is the velocity gradient.

The intensity in turbulent flow is expressed by the power density ε (the amount of energy dissipated per unit volume per unit time); for turbulent flow,

$$\varepsilon = \eta G^2 . \tag{3.17}$$

The most important regimes are: Laminar/Viscous (LV) – Turbulent/Viscous (TV) – Turbulent/Inertial (TI). For water as the continuous phase, the regime is always TI. For higher viscosity of the continuous phase ($\eta_C = 0.1$ Pa s), the regime is TV. For still higher viscosity or a small apparatus (small l), the regime is LV. For very small apparatus (as is the case with most laboratory homogenizers), the regime is nearly always LV.

The mixing conditions have to be optimized: Heat generation at high stirring speeds must be avoided. This is particularly the case when the viscosity of the resulting dispersion increases during dispersion (note that the energy dissipation as heat is given by the product of the square of the shear rate and the viscosity of the suspension). One should avoid foam formation during dispersion; proper choice of the dispersing agent is essential and antifoams (silicones) may be applied during the dispersion process.

Rotor-stator mixers can be characterized as energy-intensive mixing devices. The main feature of these mixers is their ability to focus high energy/shear in a small volume of fluid. They consist of a high speed rotor enclosed in a stator, with the gap between them ranging from 100 to 3000 µm. Typically, the rotor speed is between 10 and $50 \, \text{m s}^{-1}$, which, in combination with a small gap, generates very high shear rates. By operating at high speed, the rotor-stator mixers can significantly reduce the processing time. In terms of energy consumption per unit mass of product, the rotor-stator mixers require high power input over a relatively short time. However, as the energy is uniformly delivered and dissipated in a relatively small volume, each element of the fluid is exposed to a similar intensity of processing. Frequently, the quality of the final product is strongly affected by its structure/morphology and it is essential that the key ingredients are uniformly distributed throughout the whole mixer volume.

The most common application of rotor-stator mixers is in dispersion of powders in liquids and they are used in manufacture of particle-based products with sizes between 1 and 20 µm, e.g. in pharmaceuticals, paints, agrochemicals and cosmetics.

There are a wide range of designs of rotor-stator mixers, of which the Ultra-Turrax (IKA works, Germany) and Silverson (UK) are the most commonly used. They are broadly classified according to their mode of operation such as batch or in-line (continuous) mixers. In-line radial-discharge mixers are characterized by high throughput and good pumping capacity at low energy consumption. The disperse phase can be injected directly into the high shear/turbulent zone, where mixing is much faster than by injection into the pipe or into the holding tank. They are used for manufacturing very fine solid particles of relatively narrow dispersed size distribution. They are typically supplied with a range of interchangeable screens, making them reliable and versatile in different applications. Toothed devices are available as in-line as well as batch mixers. Due to their open structure they have a relatively good pumping capacity and they frequently do not need an additional impeller to induce bulk flow even in relatively large vessels.

In rotor-stator mixers, both shear rate in laminar flow and energy dissipate flow depend on the position inside the mixer. In laminar flow in stirred vessels, the average shear rate is proportional to the rotor speed N with the proportionality constant K dependent on the type of the impeller [8],

$$\dot{\gamma} = KN. \tag{3.18}$$

In stirred vessels the proportionality constant cannot be calculated and has to be determined experimentally. In rotor-stator mixers, the average shear rate in the gap between the rotor and stator can be calculated if the rotor speed and geometry of the mixer are known,

$$\dot{\gamma} = \frac{\pi D N}{\delta} = K_1 N \,, \tag{3.19}$$

where D is the outer rotor diameter and δ is the rotor-stator gap width.

The average energy dissipation rate ε in turbulent flow in rotor-stator mixers can be calculated from [8],

$$\varepsilon = \frac{P}{\rho_c V} \,, \tag{3.20}$$

where P is the power draw, V is the swept rotor volume and ρ_c is the continuous phase density.

The power draw in batch rotor-stator mixers is calculated in the same way as in stirred vessels,

$$P = P_0 \rho_c N^3 D^5 \tag{3.21}$$

where P_0 is the power number constant for in-line rotor-stator mixers zero flow.

The power draw in in-line rotor-stator mixers in turbulent flow is given by,

$$P = P_{0z} \rho_c N^3 D^5 + k_1 M N^2 D^2 + P_L \,, \tag{3.22}$$

where M is the mass flow rate and P_L is the power losses term. The first term in equation (3.22) is analogous to power consumption in a batch rotor-stator mixer and the second term takes into account the effect of pumping action on total power consumption. The third term accounts for mechanical losses and is typically a few percent, and therefore can be ignored.

While in turbulent flow P_{0z} in equation (3.22) is approximately independent of the Reynolds number Re, in laminar flow there is a strong dependency of power number on Re and in this case the power draw can be calculated from,

$$P = k_0 N^2 D^3 \eta_c + k_1 M N^2 D^2 + P_L \,, \tag{3.23}$$

where η_c is the viscosity of the continuous phase and k_0 is a constant that depends on the Reynolds number Re,

$$k_0 = P_{0z} Re \,. \tag{3.24}$$

From equations (3.18)–(3.22), the average energy dissipation rate in the rotor-stator mixer can be calculated.

3.3 Wet milling or comminution

The primary dispersion (sometimes referred to as the mill base) may then be subjected to a bead milling process to produce nanoparticles. Subdivision of the primary particles into much smaller units (<1 μm) requires application of intense energy. In some cases high pressure homogenizers (such as the Microfluidizer, USA) may be sufficient to produce small particles. This is particularly the case with many drugs. In some cases, the high pressure homogenizer is combined with application of ultrasound to produce the small particles [9]. It has been shown that high pressure homogenization is a simple technique, well established on large scale for the production of fine suspensions and already available in the pharmaceutical industry. High pressure homogenization is also an efficient technique that has been utilized to prepare stable suspensions of several drugs such as carbazepin, bupravaquone, aphidicolin, cyclosporine, paclitaxel, prednisolone, etc. During homogenization, cavitation forces as well as collision and shear forces determine breakdown of the drug particles down to the nanometre range. Process conditions lead to an average particle size that remains constant as a result of continuous fragmentation and reaggregation processes. These high energetic forces can also induce a change of crystal structure and/or partial or total amorphization of the sample, which further enhances the solubility. For long-term storage stability of the nanosuspension formulation, the crystal structure modification must be maintained over the storage time.

Microfluidization is a milling technique which results in minimal product contamination. Besides minimal contamination, this technique can be easily scaled up. In this method a sample dispersion containing large particles is made to pass through specially designed interaction chambers at high pressure. The specialized geometry of the chambers along with the high pressure causes the liquid stream to reach extremely high velocities and these streams then impinge against each other and against the walls of the chamber resulting in particle size reduction. The shear forces developed at high velocities due to attrition of particles against one another and against the chamber walls, as well as the cavitation fields generated inside the chamber are the main mechanisms of particle size reduction with this technique. In the interaction chambers the liquid feed is divided into two parts which are then made to impinge against each other and against the walls of the chambers. Particle size reduction occurs due to attrition between the particles and against the chamber walls at high velocities. Cavitation fields generated inside the chambers also contribute to particle size reduction [9].

The process of microfluidization for the preparation of suspensions varies in a complex way with the various critical processes and formulation parameters. Milling time, microfluidization pressure, stabilizer type, processing temperature and stabilizer concentration were identified as critical parameters affecting the formation of stable particles. Both ionic as well as steric stabilization were effective in stabilizing the suspensions. Microfluidization and precipitation under sonication can also be used for suspension preparation.

The extreme transient conditions generated in the vicinity and within the collapsing cavitational bubbles have been used for size reduction of the material to the nanoscale. Particles synthesis techniques include sonochemical processing and cavitation processing. In sonochemistry, an acoustic cavitation process can generate a transient localized hot zone with extremely high temperature gradient and pressure.

Such sudden changes in temperature and pressure assist the destruction of the sono-chemical precursor and the formation of nanoparticles [9].

A dimensionless number known as cavitation number (C_v) is used to relate the flow conditions to the cavitation intensity [9],

$$C_v = \frac{P_2 - P_v}{0.5\rho V_0^2},$$

(3.25)

where P_2 is the recovered downstream pressure; P_v is the vapour pressure of the liquid, ρ is the density of dispersed media and V_0 is the liquid velocity at the orifice. The cavitation number at which the inception of cavitation occurs is known as the cavitation inception number C_{vi}. Ideally speaking, the cavitation inception should occur at 1.0. It was also reported that generally the inception of cavitation occurs from 1.0 to 2.5. This has been attributed to the presence of the dissolved gases in the flowing liquid. C_v is a function of the flow geometry and usually increases with an increase in the size of the opening in a constriction such as an orifice in a flow.

Cavitation can be used, for example, for the formation of iron oxide particles. Iron precursor, either as a neat liquid or in a decalin solution, was sonicated and this produced 10–20 nm sized amorphous iron particles. Similar experiments have been reported for the synthesis of the particles of many other inorganic materials using acoustic cavitation. To understand the mechanism of the formation of the particles during the cavitation phenomenon, the hotspot theory has been successfully applied. It explains the adiabatic collapse of a bubble, producing the hotspots. This theory claims that very high temperatures (5000–25 000 K) are obtained upon the collapse of the bubble. Since this collapse occurs in few microseconds, very high cooling rates have been obtained. These high cooling rates hinder the organization and crystallization of the products. While the explanation for the creation of amorphous products is well understood, the reason for the formation of nanostructured products under cavitation is not yet clear. The products are sometimes nanoamorphous particles, and in other cases, nanocrystalline. This depends on the temperature in the fluid ring region where the reaction takes place. The temperature in this liquid ring is lower than that inside the collapsing bubble, but higher than the temperature of the bulk liquid. In summary, in sonochemical reactions leading to inorganic products, nanomaterials have been obtained. They vary in size, shape, structure, and in their solid phase (amorphous or crystalline), but they were always of nanometre size. Cavitation being a nuclei dominated (statistical in nature) phenomenon, such variations are expected. In hydrodynamic cavitation, nanoparticles are generated through the creation and release of gas bubbles inside the sol-gel solution. By rapidly pressurizing in a supercritical drying chamber and exposing it to the cavitational disturbance and high temperature heating, the sol-gel solution is rapidly mixed. The erupting hydrodynamically generated cavitating bubbles are responsible for the nucleation, the growth of the nanoparticles, and also for their quenching to the bulk operating temperature. Particle size can be controlled by adjusting the pressure and the solution retention time in the cavitation chamber. Cavitation methods can be used to reduce the size of the rubber latex particles (styrene butadiene rubber, SBR), present in the form of aqueous suspension with micrometre particle initial size, to the nanoscale [9].

An alternative method of size reduction to produce nanoparticles, commonly used in many industrial applications, is wet milling, also referred to as comminution (the generic term for size reduction). It is a complex process and there is little fundamental information on its mechanism. For the breakdown of single crystals or particles into smaller units, mechanical energy is required. This energy in a bead mill is supplied by impaction of the glass or ceramic beads with the particles. As a result, permanent deformation of the particles and crack initiation occur. This will eventually lead to the fracture of particles into smaller units. Since the milling conditions are random, some particles receive impacts far in excess of those required for fracture whereas others receive impacts that are insufficient for the fracture process. This makes the milling operation grossly inefficient and only a small fraction of the applied energy is used in comminution. The rest of the energy is dissipated as heat, vibration, sound, interparticulate friction, etc.

The role of surfactants and dispersants in grinding efficiency is far from being understood. In most cases the choice of surfactants and dispersant is made by trial and error until a system is found that gives the maximum grinding efficiency. Rehbinder and his collaborators [10] investigated the role of surfactants in the grinding process. As a result of surfactant adsorption at the solid/liquid interface, the surface energy at the boundary is reduced and this facilitates the process of deformation or destruction. The adsorption of surfactants at the solid/liquid interface in cracks facilitates their propagation. This mechanism is referred to as the Rehbinder effect.

Several factors affect the efficiency of dispersion and milling [11]: (i) the volume concentration of dispersed particles (i.e. the volume fraction); (ii) the nature of the wetting/dispersing agent; (iii) the concentration of wetter/dispersant (which determines the adsorption characteristics).

For optimization of the dispersion/milling process the above parameters need to be systematically investigated. From the wetting performance of a surfactant, which can be evaluated using contact angle measurements, one can establish the nature and concentration of the wetting agent. The nature and concentration of the dispersing agent required is determined by adsorption isotherm and rheological measurements.

Once the concentration of wetting/dispersing agent is established, dispersions are prepared at various volume fractions keeping the ratio of wetting/dispersing agent to the solid content constant. Each system is then subjected to dispersion/milling process keeping all parameters constant: (i) speed of the stirrer (normally one starts at lower speed and gradually increases the speed in increments at fixed time); (ii) volume and size of beads relative to the volume of the dispersion (an optimum value is required); (iii) speed of the mill.

The change of average particle size with time of grinding is established using for example the Mastersizer (Malvern, UK). Figure 3.6 shows a schematic representation

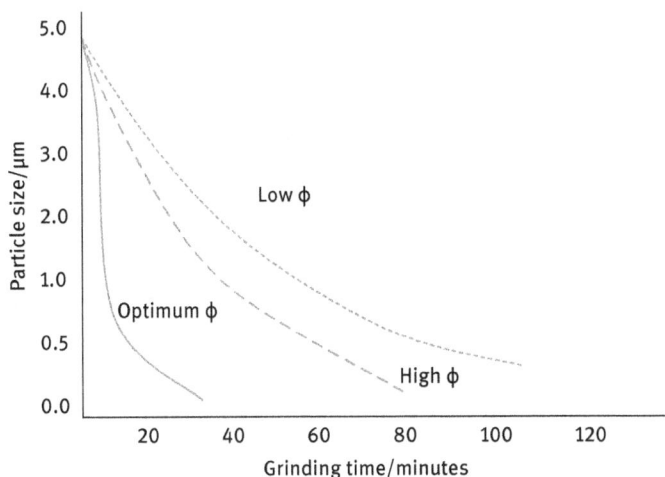

Fig. 3.6: Variation of particle size with grinding time in a typical bead mill.

of the reduction of particle size with grinding time in minutes using a typical bead mill (see below) at various volume fractions.

The presentation in Fig. 3.6 is only schematic and is not based on experimental data. It shows the expected trend. When the volume fraction ϕ is below the optimum (in this case the relative viscosity of the dispersion is low) one requires a long time to achieve size reduction. In addition, the final particle size may be large and outside the nanorange. When ϕ is above the optimum value, the dispersion time is prolonged (due to the relatively high relative viscosity of the system) and the grinding time is also longer. In addition, the final particle size is larger than that obtained at the optimum ϕ. At the optimum volume fraction, both dispersion and grinding time are shorter and also the final particle size is smaller [11].

For preparation of suspensions, bead mills are most commonly used. The beads are mostly made of glass or ceramics (which are preferred due to minimum contamination). The operating principle is to pump the premixed, preferably predispersed (using a high speed mixer), mill base through a cylinder containing a specified volume of say ceramic beads (normally 0.5–1 mm diameter to achieve nanosize particles). The dispersion is agitated by a single or multidisc rotor. The disc may be flat or perforated. The mill base passing through the shear zone is then separated from the beads by a suitable screen located at the opposite end of the feedport [11].

Generally speaking, bead mills may be classified to two types: (i) vertical mills with open or closed top, or (ii) horizontal mills with closed chambers. The horizontal mills are more efficient and the most commonly used one are: Netzsch (Germany) and Dyno Mill (Switzerland). These bead mills are available in various sizes from 0.5 to 500 l. The factors affecting the general dispersion efficiency are known reasonably well (from the manufacturer). The selection of the correct diameter of the beads is important for maximum utilization. In general, the smaller the size of the beads and the higher their density, the more efficient the milling process [11].

To understand the operating principle of the bead mill, one must consider the centrifugal force transmitted to the grinding beads at the tip of the rotating disc which increases considerably with its weight. This applies greater shear to the mill base. This explains why denser beads are more efficient in grinding. The speed transmitted to the individual chambers of the beads at the tip of the disc assumes that speed and the force can be calculated [11].

The centrifugal force F is simply given by,

$$F = \frac{v^2}{rg},$$ (3.26)

where v is the velocity, r is the radius of the disc and g is the acceleration due to gravity.

3.4 Stabilization of the suspension during dispersion and milling and the resulting nanosuspension

In order to maintain the particles as individual units during dispersion and milling, it is essential to use a dispersing agent that must provide an effective repulsive barrier preventing aggregation of the particles by van der Waals forces. This dispersing agent must be strongly adsorbed on the particle surface and should not be displaced by the wetting agent. As will be discussed in detail in Chapter 4, the repulsive barrier can be electrostatic in nature, whereby electrical double layers are formed at the solid/liquid interface [12, 13]. These double layers must be extended (by maintaining low electrolyte concentration) and strong repulsion occurs on double layer overlap. Alternatively, the repulsion can be produced by the use of nonionic surfactant or polymer layers which remain strongly hydrated (or solvated) by the molecules of the continuous medium [14] as will be discussed in detail in Chapter 5. On approach of the particles to a surface-to-surface separation distance that is lower than twice the adsorbed layer thickness strong repulsion occurs as a result of two main effects: (i) unfavourable mixing of the layers when these are in good solvent conditions; (ii) loss of configurational entropy on significant overlap of the adsorbed layers. This process is referred to as steric repulsion [14]. A third repulsive mechanism is that in which both electrostatic and steric repulsion are combined, for example when using polyelectrolyte dispersants.

The particles of the resulting suspension may undergo aggregation (flocculation) on standing as a result of the universal van der Waals attraction. This will be discussed in detail in Chapter 4 and only a summary is given in this chapter This attractive energy becomes very large at short distances of separation between the particles. This attractive energy, G_A, is given by the following expression,

$$G_A = -\frac{A_{11(2)}R}{12h},$$ (3.27)

where $A_{11(2)}$ is the effective Hamaker constant of two identical particles with Hamaker constant A_{11} in a medium with Hamaker constant A_{22}. The Hamaker constant of any material is given by the following expression,

$$A = \pi^2 q^2 \beta,$$ (3.28)

q is the number of atoms or molecules per unit volume, and β is the London dispersion constant. Equation (3.25) shows that A_{11} has the dimension of energy.

As mentioned in Chapter 4, to overcome the permanent van der Waals attraction energy, it is essential to have a repulsive energy between the particles. The first mechanism is electrostatic repulsive energy produced by the presence of electrical double layers around the particles, produced by charge separation at the solid/liquid interface. The dispersant should be strongly adsorbed to the particles, produce high charge (high surface or zeta potential) and form an extended double layer (that can be achieved at low electrolyte concentration and low valency) [12, 13].

When charged colloidal particles in a dispersion approach each other such that the double layers begin to overlap (particle separation becomes less than twice the double layer extension), repulsion occurs. The individual double layers can no longer develop unrestrictedly, since the limited space does not allow complete potential decay [13]. The potential $\psi_{H/2}$ half way between the plates is no longer zero (as would be the case for isolated particles at $x \to \infty$).

Combining G_{elec} and G_A results in the well-known theory of stability of colloids (DLVO theory) [12, 13],

$$G_T = G_{elec} + G_A.$$ (3.29)

A plot of G_T versus h is shown in Fig. 3.7, which represents the case at low electrolyte concentrations, i.e. strong electrostatic repulsion between the particles. G_{elec} decays exponentially with h, i.e. $G_{elec} \to 0$ as h becomes large. G_A is $\propto 1/h$, i.e. G_A does not decay to 0 at large h.

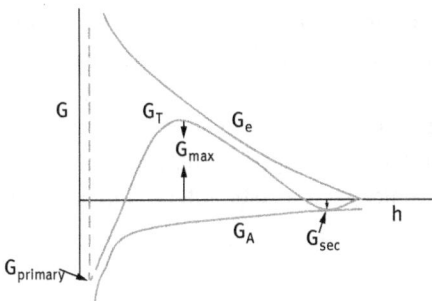

Fig. 3.7: Schematic representation of the variation of G_T with h according to the DLVO theory.

At long distances of separation, $G_A > G_{elec}$ resulting in a shallow minimum (secondary minimum), which for nanosuspensions is very low ($< kT$). At very short distances, $G_A \gg G_{elec}$ resulting in a deep primary minimum. At intermediate distances, $G_{elec} > G_A$ resulting in an energy maximum, G_{max}, whose height depends on ψ_0 (or ψ_d or zeta potential) and the electrolyte concentration and valency. At low electrolyte concentrations ($< 10^{-2}$ mol dm^{-3} for a $1:1$ electrolyte), G_{max} is high (> 25 kT) and this prevents particle aggregation into the primary minimum. The higher the electrolyte concentration (and the higher the valency of the ions), the lower the energy maximum.

The second stabilization mechanism is referred to as steric repulsive energy produced by the presence of adsorbed (or grafted) layers of surfactant or polymer molecules [14] as will be discussed in detail in Chapter 5. In this case the nonionic surfactant or polymer (referred to as polymeric surfactant) should be strongly adsorbed to the particle surface and the stabilizing chain should be strongly solvated (hydrated in the case of aqueous suspensions) by the molecules of the medium [14]. The most effective polymeric surfactants are those of the A-B, A-B-A block or BA$_n$ graft copolymer. The "anchor" chain B is chosen to be highly insoluble in the medium and has strong affinity to the surface. The A stabilizing chain is chosen to be highly soluble in the medium and strongly solvated by the molecules of the medium. For nanosuspensions of hydrophobic solids in aqueous media, the B chain can be polystyrene, poly(methylmethacrylate) or poly(propylene oxide). The A chain could be poly(ethylene oxide) which is strongly hydrated by the medium.

When two particles, each with a radius R and containing an adsorbed polymer layer with a hydrodynamic thickness δ_h, approach each other to a surface-surface separation distance h that is smaller than $2\delta_h$, the polymer layers interact with each other resulting in two main situations [14]: (i) the polymer chains may overlap with each other, or (ii) the polymer layer may undergo some compression. In both cases there will be an increase in the local segment density of the polymer chains in the interaction region. The real situation is perhaps in between the above two cases, i.e. the polymer chains may undergo some interpenetration and some compression.

Provided the dangling chains (the A chains in A-B, A-B-A block or BA$_n$ graft copolymers) are in a good solvent, this local increase in segment density in the interaction zone will result in strong repulsion as a result of two main effects [14]: (i) An increase in the osmotic pressure in the overlap region as a result of the unfavourable mixing of the polymer chains, when these are in good solvent conditions. This is referred to as osmotic repulsion or mixing interaction and it is described by a free energy of interaction G_{mix}. (ii) Reduction of the configurational entropy of the chains in the interaction zone; this entropy reduction results from the decrease in the volume available for the chains when these are either overlapped or compressed. This is referred to as volume restriction interaction, entropic or elastic interaction and it is described by a free energy of interaction G_{el}.

The combination of G_{mix} and G_{el} is usually referred to as the steric interaction free energy, G_s, i.e.,

$$G_s = G_{mix} + G_{el} . \qquad (3.30)$$

The sign of G_{mix} depends on the solvency of the medium for the chains. If in a good solvent, i.e. the Flory–Huggins interaction parameter χ is less than 0.5, then G_{mix} is positive and the mixing interaction leads to repulsion (see below). In contrast, if $\chi > 0.5$ (i.e. the chains are in a poor solvent condition), G_{mix} is negative and the mixing interaction becomes attractive. G_{el} is always positive and hence in some cases one can produce stable nanosuspensions in a relatively poor solvent (enhanced steric stabilization).

Combining G_{mix} and G_{el} with G_A gives the total energy of interaction G_T (assuming there is no contribution from any residual electrostatic interaction), i.e.,

$$G_T = G_{mix} + G_{el} + G_A . \qquad (3.31)$$

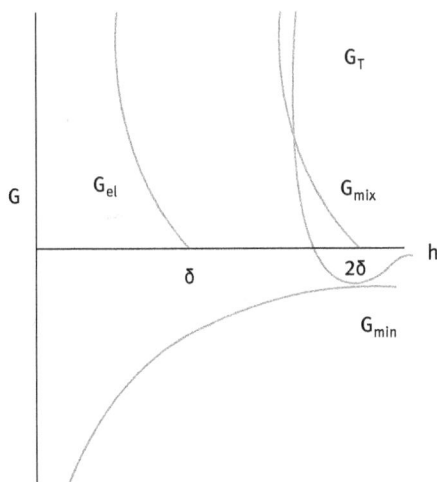

Fig. 3.8: Energy-distance curves for sterically stabilized systems.

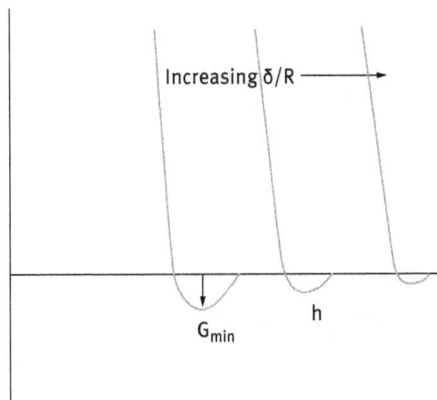

Fig. 3.9: Variation of G_{min} with δ/R.

A schematic representation of the variation of G_{mix}, G_{el}, G_A and G_T with surface-surface separation distance h is shown in Fig. 3.8. G_{mix} increases very sharply with decreasing h; when h < 2δ. G_{el} increases very sharply with decreasing h; when h < δ. G_T versus h shows a minimum, G_{min}, at separation distances comparable to 2δ. When h < 2δ, G_T shows a rapid increase with decreasing h. The depth of the minimum depends on the Hamaker constant A, the particle radius R and adsorbed layer thickness δ. G_{min} decreases with decreasing A and R. At a given A and R, G_{min} decreases with increasing δ (i.e. with increasing molecular weight, M_w, of the stabilizer). This is illustrated in Fig. 3.9, which shows the energy-distance curves as a function of δ/R. The larger the value of δ/R, the smaller the value of G_{min}. In this case the system may approach thermodynamic stability, as is the case with nanosuspensions.

3.5 Prevention of Ostwald ripening (crystal growth)

As will be discussed in Chapter 7, the driving force for Ostwald ripening is the difference in solubility between the small and large particles (the smaller particles have higher solubility than the larger ones). The difference in chemical potential between different sized particles was given by Lord Kelvin [15]

$$S(r) = S(\infty) \exp\left(\frac{2\gamma V_m}{rRT}\right), \tag{3.32}$$

where $S(r)$ is the solubility of a particle with radius r and $S(\infty)$ is the solubility of a particle with infinite radius (the bulk solubility), γ is the S/L interfacial tension, R is the gas constant and T is the absolute temperature. Equation (3.33) shows a significant increase of solubility of the particle with reduction of particle radius, particularly when the latter becomes significantly smaller than 1 μm.

For two particles with radii r_1 and r_2 ($r_1 < r_2$),

$$\frac{RT}{V_m} \ln\left[\frac{S(r_1)}{S(r_2)}\right] = 2\sigma\left[\frac{1}{r_1} - \frac{1}{r_2}\right]. \tag{3.33}$$

Equation (3.33) shows that the larger the difference between r_1 and r_2, the higher the rate of Ostwald ripening.

Ostwald ripening can be quantitatively assessed from plots of the cube of the radius versus time t [16, 17],

$$r^3 = \frac{8}{9}\left[\frac{S(\infty)\sigma V_m D}{\rho RT}\right]t, \tag{3.34}$$

D is the diffusion coefficient of the disperse phase in the continuous phase.

Several factors affect the rate of Ostwald ripening and these are determined by surface phenomena, although the presence of surfactant micelles in the continuous phase can also play a major role. Trace amounts of impurities that are highly insoluble in the medium and have strong affinity to the surface can significantly reduce Ostwald ripening by blocking the active sites on the surface on which the molecules of the active ingredient can deposit. Many polymeric surfactants, particularly those of the block and graft copolymer types, can also reduce the Ostwald ripening rate by strong adsorption on the surface of the particles, thus making it inaccessible for molecular deposition. Surfactant micelles that can solubilize the molecules of the active ingredient may enhance the rate of crystal grow by increasing the flux of transport by diffusion.

References

[1] Tadros, Th. F., "Dispersion of Powders in Liquids and Stabilisation of Suspensions", Wiley-VCH, Germany (2012).
[2] Tadros, Th. F., "Formulation of Disperse Systems", Wiley-VCH, Germany (2014).
[3] Tadros, Th. F., "Nanodispersions", De Gruyter, Germany (2016).
[4] Young, T., Phil. Trans. Royal Soc. (London), **95**, 65 (1805).
[5] Blake, T. B., "Wetting", in "Surfactants", Th. F. Tadros (ed.), Academic Press, London, (1984).
[6] Rideal, E. K., Phil. Mag., **44**, 1152 (1922).
[7] Washburn, E. D., Phys. Rev., **17**, 273 (1921).
[8] Pacek, A. W., Hall, S., Cooke, M., and Kowalski, A. J., "Emulsification in Rotor-Stator Mixers", in "Emulsion Formation and Stability, Th. F. Tadros (ed.), Wiley-VCH, Germany (2013).
[9] Capek, I., in "Encyclopedia of Colloid and Interface Science", Th. F. Tadros (ed.), Springer, Germany (2013), p. 748.
[10] Rhebinder, P. A., Colloid J. USSR, **20**, 493 (1958).
[11] Tadros, Th. F., "Colloids in Paints", Wiley-VCH, Germany (2010).
[12] Deryaguin, B. V. and Landau, L., Acta Physicochem. USSR, **14**, 633 (1941).
[13] Verwey, E. J. W. and Overbeek, J. Th. G., "Theory of Stability of Lyophobic Colloids", Elsevier, Amsterdam (1948).
[14] Napper, D. H., "Polymeric Stabilisation of Colloidal Dispersions", Academic Press, London (1983).
[15] Thompson, W. (Lord Kelvin), Phil. Mag., **42**, 448 (1871).
[16] Lifshitz, I. M. and Slesov, V. V., Sov. Phys. JETP, **35**, 331 (1959).
[17] Wagner, C., Z. Electrochem., **35**, 581 (1961).

4 Electrostatic stabilization of suspensions

4.1 Introduction

As mentioned in Chapter 3, after preparation of a suspension by wetting, dispersion and milling, it is necessary to stabilize it against aggregation. This can be achieved by the solid particles acquiring an electrostatic charge, e.g. by dissociation of ionogenic groups or adsorption of ionic surfactant [1–3]. This surface charge will be compensated in bulk solution by unequal distribution of counterions (with opposite sign to the surface charge) and co-ions (with the same charge sign as the surface). As will be discussed below, the surface and counter charges form an electrical double layer and when the suspension particles approach a separation distance where the doubly layers begin to overlap, electrostatic repulsion is produced which counteracts the van der Waals attraction resulting in colloid stability.

This chapter starts with a section on the structure of the solid/liquid interface, the origin of charge on surfaces and the structure of the electrical double layer. This is followed by a section on the origin of repulsion caused by double layer overlap with particular reference to the effect of surface (or zeta) potential, electrolyte concentration and valency of ions in the bulk liquid. The next section deals with the origin of van der Waals attraction and its variation with particle separation, particle radius and the medium in which the particles are dispersed. The combination of double layer repulsion with van der Waals attraction forms the basis of the theory of colloid stability. The variation of electrostatic repulsion, van der Waals attraction and total interaction with separation distance produces energy-distance curves with an energy maximum (barrier) that prevents particle aggregation. The dependency of the energy barrier on surface potential, electrolyte concentration and valency can be used to distinguish between stable and unstable suspensions.

4.2 Structure of the solid/liquid interface

4.2.1 Origin of charge on surfaces

A great variety of processes occur to produce a surface charge [1–3].

4.2.1.1 Surface ions
These are ions that have such a high affinity to the surface of the particles that they may be taken as part of the surface, e.g. Ag^+ and I^- for AgI. If silver iodide, that is sparingly soluble in water, is brought into water, dissolution occurs until the concentration of the ionic components in solution corresponds to the solubility product of

DOI 10.1515/9783110486872-005

the compound K_s. The latter $K_s = [Ag^+][I^-] = 10^{-16}$. If the precipitation is produced under conditions which have an excess of I^-, say 1×10^{-4} mol dm^{-3}, one forms a dispersion with negatively charged surface. Using $p[I^-] - \log[I^-]$, the solution has a $p[I^-] = 4$ and then $p[Ag^+] = 12$. In this case the solution of Ag^+ is suppressed with the surface of the particles consisting of I^- species. In contrast, if the precipitation is produced under conditions of excess Ag^+ ions, say 1×10^{-4} mol dm^{-3}, or $p[Ag^+] = 4$, one forms a dispersion that is positively charged with the surface of the particles consisting of Ag^+ ion species. At $p[I^-] = 10.5$ or $p[Ag^+] = 5.5$, the surface populations of the two types are equal and the point of zero charge (pzc) is reached as the net charge is zero.

For AgI in a solution of KNO_3, the surface charge σ_0 is given by the following expression,

$$\sigma_0 = F(\Gamma_{Ag^+} - \Gamma_{I^-}) = F(\Gamma_{AgNO_3} - \Gamma_{KI}), \tag{4.1}$$

where F is the Faraday constant ($96\,500$ C mol^{-1}) and Γ is the surface excess of ions (mol m^{-2}).

The surface potential ψ_0 can be expressed using the Nernst equation (as for an electrode surface),

$$\psi_0 = \frac{RT}{F} \ln \left(\frac{[Ag^+]}{[Ag^+]_{pzc}} \right), \tag{4.2}$$

where R is the gas constant and T is the absolute temperature. This gives a surface potential of about ± 60 mV for a factor of ± 10 in the silver ion concentration from that corresponding to the pzc.

4.2.1.2 Ionization of surface groups

Carboxylic groups that are chemically bound to the surface of synthetic latexes provide an example of this. The charge is a function of pH as the degree of dissociation is a function of pH. Although the pK_a of an isolated $-COOH$ group is pH ≈ 4. This is not the situation with a surface with many $-COOH$ groups in close proximity. The dissociation of one group makes it more difficult for the immediate neighbours to dissociate. This means that the surface has variable pK_a and pH values as high as 9 may be required to ensure dissociation of all surface groups.

Another example where the charge is produced by dissociation is that of oxides which have surface hydroxyl groups. At high pH, these can ionize to give $-O^-$ and at low pH the lone pair of electrons can hold a proton to give $-OH_2^+$. This process can be represented as follows,

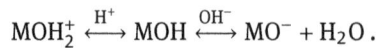

$$MOH_2^+ \xleftrightarrow{H^+} MOH \xleftrightarrow{OH^-} MO^- + H_2O.$$

The surface charge follows from,

$$\sigma_0 = F([MOH_2^+] - [MO^-]) = F(\Gamma_{H^+} - \Gamma_{OH^-}), \tag{4.3}$$

where Γ refers to the surface concentration in moles per unit area.

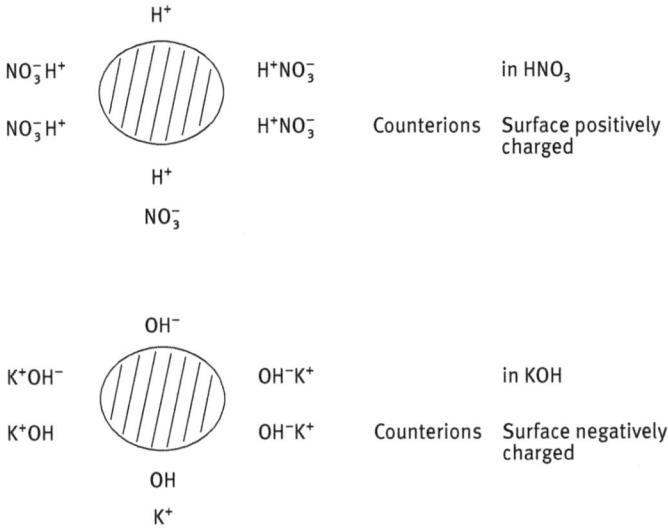

Fig. 4.1: Schematic representation of an oxide surface.

A schematic representation of the process of charge formation in an oxide is shown in Fig. 4.1 where HNO_3 and KOH are used to provide the H^+ and OH^- ions respectively. In this case one may write equation (4.3) as,

$$\sigma_0 = F(\Gamma_{HNO_3} - \Gamma_{KOH}).\qquad(4.4)$$

The charge depends on the pH of the solution: below a certain pH the surface is positive and above a certain pH the surface is negative. At a specific pH ($\Gamma_H = \Gamma_{OH}$) the surface is uncharged; this is referred to as the point of zero charge (pzc). The pzc depends on the type of oxide: for an acidic oxide such as silica the pzc is $\approx$ pH 2–3. For a basic oxide such as alumina the pzc is $\approx$ pH 9. For an amphoteric oxide such as titania the pzc $\approx$ pH 6. Some typical values of pzc for various oxides are given in Tab. 4.1.

4.2.1.3 Isomorphic substitution

This process occurs in clay minerals, with the basic structure of a clay particle consisting of an aluminosilicate layer lattice. As the clay is formed, it crystallizes with a layer of silicon atoms tetrahedrally coordinated to oxygen atoms, i.e. it forms an SiO_2 layer. The next layer of the lattice is aluminium with octahedrally coordinated oxygen atoms (i.e. an Al_2O_3), some of which are shared with the tetrahedral silica layer. This layer structure is repeated throughout the crystal. With kaolinite a 1:1 layer lattice structure is produced, whereas for montmorillonite a 2:1 structure, with the alumina layer sandwiched between two silica layers, is formed. A schematic representation of one unit cell of a 2:1 layer mineral is shown in Fig. 4.2.

Tab. 4.1: pzc values for some oxides.

Oxide	pzc
SiO_2 (precipitated)	2–3
SiO_2 (quartz)	3.7
SnO_2 (cassiterite)	5–6
TiO_2 (anatase)	6.2
TiO_2 (rutile)	5.7–5.8
RuO_2	5.7
$\alpha\text{-}Fe_2O_3$ (haematite)	8.5–9.5
$\alpha\text{-}FeO{\cdot}OH$ (goethite)	8.4–9.4
ZnO	8.5–9.5
$\gamma\text{-}Al(OH)_3$ (gibbsite)	8–9

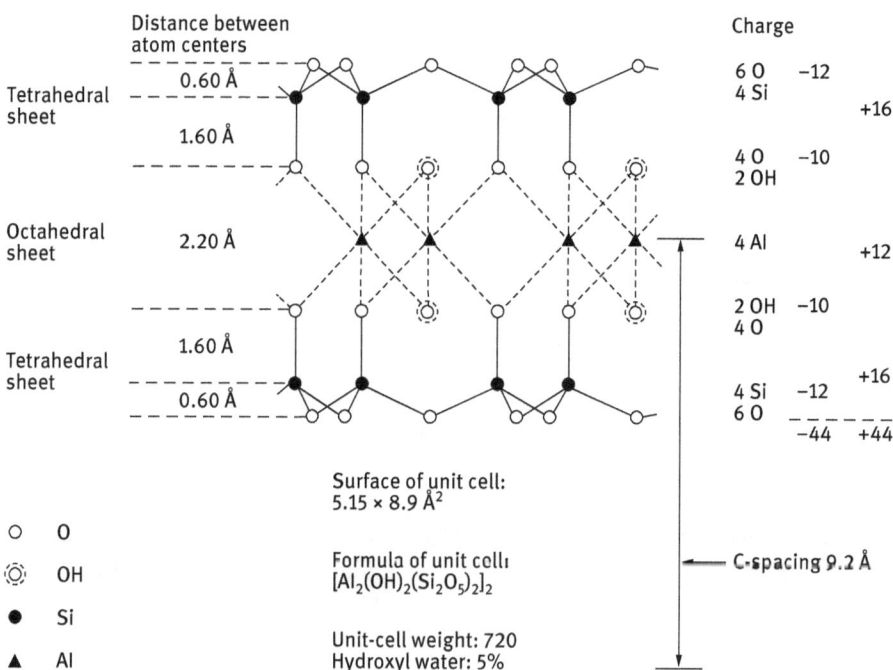

Fig. 4.2: Schematic representation of a 2 : 1 layer of montmorillonite.

During the crystallization process an occasional tetravalent Si atom is substituted by a trivalent Al atom and more frequently a trivalent Al atom is substituted by a bivalent Mg atom. This results in a deficit of positive charges and formation of negative charges, a process referred to as isomorphic substitution. This net negative charge on the surface of the clay particle is balanced by soluble cations (counterions), e.g. Na^+ ions. In the dry state, the counterions are located on the surface of the crystal, whereas in the hydrated state they are in solution near the surface. The 2:1 layer lattice clays have a unit cell $\approx 1\,nm$ thick and, on hydration, water penetrates between the layers and the negative charges of the surfaces repel, causing the clays to swell. This swelling can result in separation between (original) unit cells of a greater dimension than that of the (original) unit cell. A schematic representation of a clay particle is shown in Fig. 4.3. The surface charge + counterions form the electrical double layer.

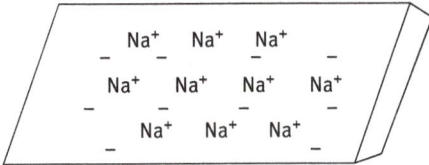

Fig. 4.3: Schematic representation of a clay particle.

4.2.2 Specific adsorption of ions

In some cases, specifically adsorbed ions (that have nonelectrostatic affinity to the surface) "enrich" the surface but may not be considered as part of the surface, e.g. bivalent cations on oxides, cationic and anionic surfactants on most surfaces. In particular, ionic surfactants are often added as a component to disperse systems, as wetting and dispersing agents for powders to produce solid/liquid dispersions (suspensions). In most cases the surfactant also acts as stabilizer for the final suspension. By adsorption at the solid/liquid or liquid/liquid interface the surfactant ions produce a charge (negative for anionic surfactants and positive for cationic surfactants) on the surface of the particle. This charge is compensated by unequal distribution of counterions (with opposite charge to the surface) and co-ions (with the same charge as the surface) forming an electrical double layer. In some cases the charge on the particle is produced by adsorption of anionic or cationic polyelectrolytes. An example is polyacrylates that are used to disperse many pigments such as titania.

4.3 Structure of the electrical double layer

4.3.1 Diffuse double layer (Gouy and Chapman)

Gouy and Chapman [4, 5] assumed that the charge is smeared out over a plane surface immersed in an electrolyte solution. This surface has a uniform potential ψ_0 and the compensating ions are regarded as point charges immersed in a continuous dielectric medium. The surface charge σ_0 is compensated by unequal distribution of counterions (opposite in charge to the surface) and co-ions (same sign as the surface) which extend to some distance from the surface [4, 5]. The counterion and co-ion concentration n_i near the surface can be related to the value in the bulk n_{io} using the Boltzmann distribution principle,

$$n_i = n_{i0} \exp\left[-\frac{Z_i e \psi_x}{kT}\right],$$ (4.5)

where Z_i is the valency of the ion, e is the electronic charge, ψ_x is the potential at a distance x from the surface, k is the Boltzmann constant and T is the absolute temperature. Since the charge on the counterion is always opposite to that of the surface, the exponent in equation (4.5) will always be negative for the counterion concentration and positive for the co-ion concentration. Equation (4.5) shows that the concentration of counterions increases close to the surface (positively adsorbed) whereas the co-ion concentration is reduced near the surface (negative adsorption). Figure 4.4 shows the local ion concentration profiles according to equation (4.5).

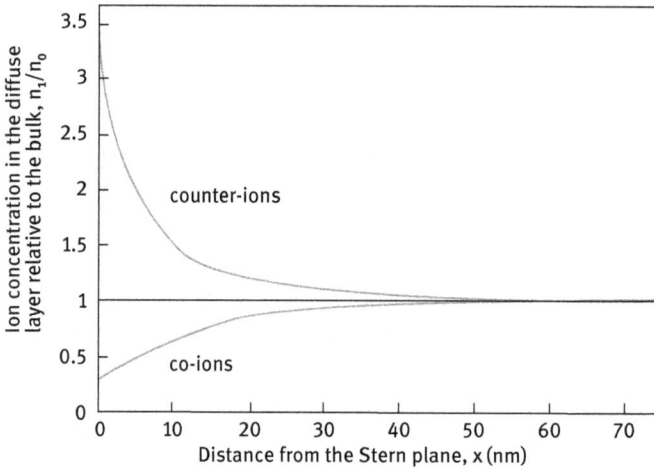

Fig. 4.4: Local ion concentration profiles: $\psi_0 = -30$ mV, in 10^{-3} mol dm^{-3} NaCl.

The number of charges per unit volume, i.e. the space charge density ρ is given by,

$$\rho = \sum_i n_i Z_i e = -2n_0 Z e \sinh\left[\frac{Ze\psi_x}{kT}\right].$$ (4.6)

Note that $\sinh x = (\exp(x) - \exp(-x))/2$.

A schematic picture of the diffuse double layer according to Gouy [4] and Chapman [5] is shown in Fig. 4.5. The potential decays exponentially with distance x.

For small potentials ($\psi(x) < 25$ mV), $Ze\psi(x)/kT < 1$ and $\sinh x \approx x$ so that equation (4.6) becomes,

$$\frac{d^2\psi}{dx^2} \approx \frac{2n_0(Ze)^2}{\varepsilon_0 \varepsilon_r kT}\psi(x).$$ (4.7)

Equation (4.7) is the well-known Debye–Huckel approximation whose solution is,

$$\psi = \psi_0 \exp{-(\kappa x)}.$$ (4.8)

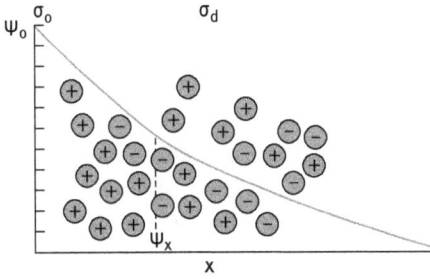

Fig. 4.5: Schematic representation of the diffuse double layer according to Gouy [4] and Chapman [5].

Note that when $x = 1/\kappa$, $\psi_x = \psi_0/e$; $1/\kappa$ is referred to as the "thickness of the double layer".

The double layer extension depends on electrolyte concentration and valency of the counterions,

$$\left(\frac{1}{\kappa}\right) = \left(\frac{\varepsilon_r \varepsilon_0 kT}{2 n_0 Z_i^2 e^2}\right)^{1/2} . \tag{4.9}$$

The double layer extension increases with decreasing electrolyte concentration. It also depends on the valency of the ions. An expression for $(1/\kappa)$ in terms of the electrolyte ionic strength I is,

$$\left(\frac{1}{\kappa}\right) = \left(\frac{\varepsilon_r \varepsilon_0 RT}{2000\, F^2}\right)^{1/2} I^{-1/2} \text{ in } m^{-1} = 0.304\, I^{-1/2} \text{ in nm} \tag{4.10}$$

where,

$$I = \sum c_i Z_i^2 , \tag{4.11}$$

c_i is the electrolyte concentration in mol dm^{-3}.

Increasing the ionic strength causes a decrease in $(1/\kappa)$ that is referred to as compression of the double layer. The distance $(1/\kappa)$ is referred to as the thickness of the double layer. For example, for KCl in water at 25 °C, $(1/\kappa) = 96.17$ nm at $I = 10^{-5} \text{ mol dm}^{-3}$ decreasing to 3.04 nm at $I = 10^{-2} \text{ mol dm}^{-3}$. Approximate values of $(1/\kappa)$ for KCl are given in Tab. 4.2.

Tab. 4.2: Approximate values of $(1/\kappa)$ for 1:1 electrolyte (KCl).

C / mol dm^{-3}	10^{-5}	10^{-4}	10^{-3}	10^{-2}	10^{-1}
$(1/\kappa)$ / nm	100	33	10	3.3	1

Equations (4.10) and (4.11) show that $(1/\kappa)$ depends on the valency of the counter- and co-ions.

4.3.2 Stern–Grahame model of the double layer

Stern [6] recognized that the assumption in the Gouy–Chapman theory [4, 5] that the electrolyte ions are regarded as point charges is unsatisfactory. Also the assumption that the solvent can be treated as a structureless dielectric of constant permittivity is also unsatisfactory. Consequently, he introduced the concept of the nondiffuse part of the double layer for specifically adsorbed ions, the rest being diffuse in nature. This is schematically illustrated in Fig. 4.6.

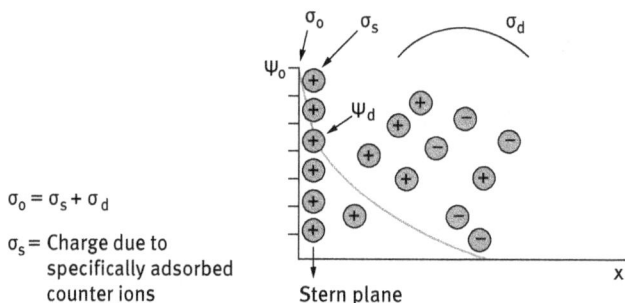

$\sigma_0 = \sigma_s + \sigma_d$

σ_s = Charge due to specifically adsorbed counter ions

Fig. 4.6: Schematic representation of the double layer according to Stern and Grahame.

The potential drops linearly in the Stern region and then exponentially. Grahame [7] distinguished two types of ions in the Stern plane, physically adsorbed counterions (outer Helmholtz plane) and chemically adsorbed ions (that lose part of their hydration shell) (inner Helmholtz plane). The outer Helmholtz plane is considered as the plane of closest approach of hydrated counterions, i.e. the Stern plane. The inner Helmholtz plane is that of specifically adsorbed counterions which may have lost part of or its complete hydration shell. The number of these specifically adsorbed ions may exceed the number of surface charges causing a reversal of the sign of the potential as is illustrated in Fig. 4.7 for a positively charged surface with specifically adsorbed anions.

For the specifically adsorbed ions the range of interaction is short, i.e. these ions must reside at the distance of closest approach, possibly within the hydration shell. For the indifferent ions the situation is different and these ions are subjected to an attractive (for the counterions) or repulsive (for the co-ions) potential (energy = $\pm ZF\psi(x)/RT$). The space charge density due to these ions is high near the surface and decreases gradually with distance to its bulk value. Such a layer is the diffuse double layer described by Gouy–Chapman [4, 5]. Generally speaking, a double layer contains a part that is specifically adsorbed and a diffuse part. Because of the finite size of the counterions there is always a charge-free layer near the surface.

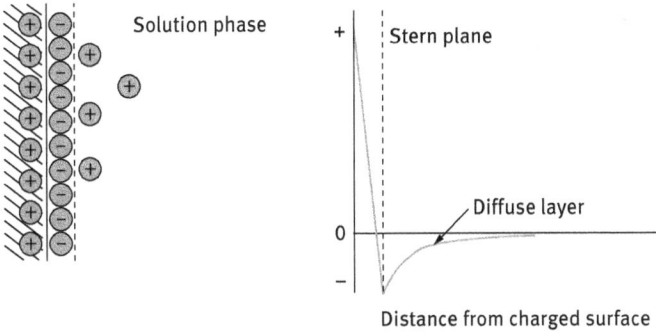

Fig. 4.7: Schematic representation of charge reversal by specifically adsorbed counterions.

4.4 Electrical double layer repulsion

The origin of electrostatic interactions is due to the double layer overlap that occurs when the surface-to-surface distance becomes smaller than twice the double layer thickness. One of the most important features of double layers is their strong dependence on the concentration of indifferent electrolytes. As discussed above, an increase in electrolyte concentration causes a reduction in the diffuse double layer potential, ψ_{diffuse}, and compression of the double layer, i.e. reduction of the Debye length κ^{-1}. Both of these effects affect the colloid stability of lyophobic colloids as will be discussed below.

A general expression can be written to describe the electrostatic repulsion, G_{el}, in terms of the double layer property f(DL) determined by the relative permittivity, particle size, the double layer potential, ψ_{diffuse}, and an exponential function determined by the Debye length and separation distance h [1],

$$G_{el} = f(DL)\psi_{\text{diffuse}}^2 \exp(-\kappa h). \tag{4.12}$$

$(1/\kappa)$ is the Debye length that is referred to as the "thickness of the double layer" that is given by equation (4.9).

For two particles of different, but not too high, diffuse double layer potential, at large κh,

$$G_{el} = f(DL)\psi_1^d\psi_2^d \exp(-\kappa h). \tag{4.13}$$

In the original theory developed by Deryaguin and Landau [8], and Verwey and Overbeek [9] (referred to as DLVO theory), the assumption was made that on the approach of surfaces the surface potentials remain constant and equal to that at infinite separation of the surfaces. This requires adjustment of the surface charge by surface charge regulation. In this case the isolated double layers form spontaneously by adsorption and/or desorption of potential determining ions. This is referred to as the "constant potential" case. However, with many surfaces the charge is fixed, e.g. in clay particles, and on surface approach the surface potential has to be adjusted to keep the surface charge constant. This situation is referred to as the "constant charge" case.

For two flat plates where the surface separation becomes less than twice the double layer extension [10], the individual double layers can no longer develop unrestrictedly, since the limited space does not allow complete potential decay (Fig. 4.8). The potential $\psi_{H/2}$ half way between the plates is no longer zero (as would be the case for isolated particles at $x \to \infty$). The potential distribution at an interparticle distance H is schematically depicted by the full line in Fig. 4.8. The Stern potential ψ_d is considered to be independent of the particle distance. The dashed curves show the potential as a function of distance x to the Helmholtz plane, had the particles been at infinite distance.

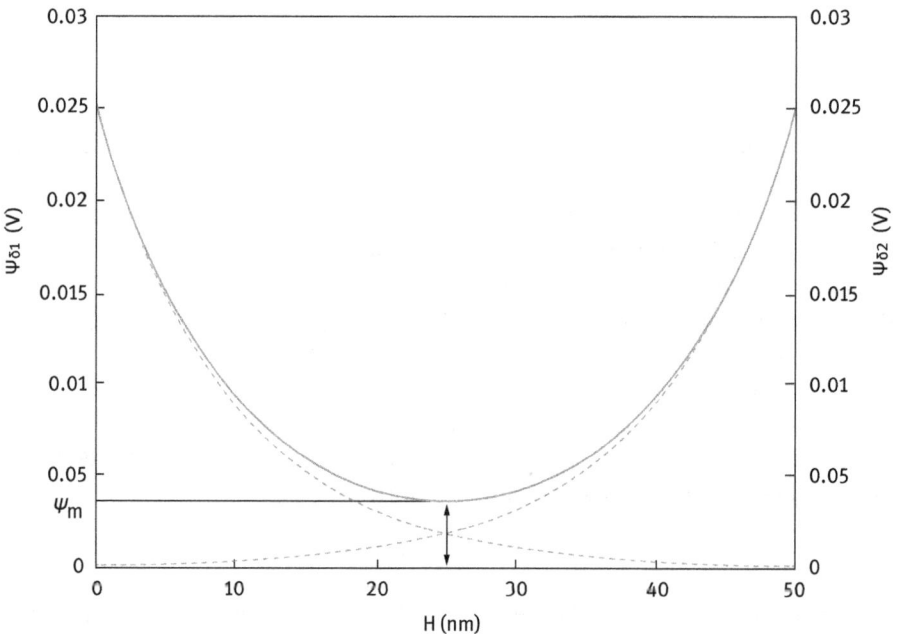

Fig. 4.8: Potential profile between two planer surfaces at 25 mV, in 10^{-3} mol dm^{-3} NaCl.

A simple model to obtain G_{el} is to calculate the ion concentration between the surfaces in order to obtain the osmotic pressure difference between the surfaces and the bulk electrolyte, i.e. the excess osmotic pressure at the mid-plane position between the surfaces separated by a distance h. At h/2, $d\psi_x/dx = 0$ and $\psi_x = \psi_m$.

The difference in ionic concentration at the mid-plane and the bulk electrolyte gives the osmotic pressure $\Pi(H)$,

$$\Pi(H) = kT(n_+ + n_- - 2n_0). \tag{4.14}$$

Using the Boltzmann distribution,

$$n_i = n_{i0} \exp\left[\frac{Z_i e\psi(x)}{kT}\right],$$ (4.15)

where Z_i is the ion valency, e is the electronic charge, k is the Boltzmann constant and T is the absolute temperature, equation (4.14) becomes,

$$\Pi(H) = 2n_0 kT\left[\cosh\left(\frac{Ze\psi_m}{kT}\right) - 1\right].$$ (4.16)

For small potentials equation (4.16) reduces to,

$$\Pi(H) \approx \frac{\kappa^2 \varepsilon_r \varepsilon_0}{2}\psi_m,$$ (4.17)

where ε_r is the relative permittivity of the medium, ε_0 is the permittivity of free space. The mid-plane potential ψ_m is equal to twice the potential at the mid-plane,

$$\psi_m = 2\psi_{(H/2)}.$$ (4.18)

Equation (4.16) becomes,

$$\Pi(H) \approx \frac{\kappa^2 \varepsilon_r \varepsilon_0}{2}\left[2\psi_d \exp\left(-\frac{\kappa H}{2}\right)\right]^2 = 2\kappa^2 \varepsilon_r \varepsilon_0 \psi_d^2 \exp(-\kappa H).$$ (4.19)

The electrostatic repulsion is then given by,

$$G_{el} = -\int_\infty^H \Pi(H)\,dH.$$ (4.20)

Using the boundary conditions: $\Pi(H) \to 0$ as $H \to \infty$,

$$G_{el} = 2\kappa \varepsilon_r \varepsilon_0 \psi_d^2 \exp(-\kappa H).$$ (4.21)

Replacing the Stern potential ψ_d with the zeta potential ζ,

$$G_{el} \approx 2\kappa \varepsilon_r \varepsilon_0 \zeta^2 \exp(-\kappa H).$$ (4.22)

For higher potentials, the Debye–Huckel approximation cannot be justified, but for weak overlap, i.e. $\kappa H > 1$, the local potentials, estimated from the isolated surfaces, can still be simply added and this results in equation (4.23) for the double layer repulsion,

$$G_{el} = \frac{64 n_0 kT}{\kappa}\tanh^2\left(\frac{Ze\psi_d}{4kT}\right)\exp(-\kappa H).$$ (4.23)

Deryaguin, Landau [8], Verwey and Overbeek [9] calculated the electrostatic interaction by computing the Gibbs energy G of the system at any distance h by an isothermal reversible charging process. The change in the Gibbs energy per unit area of the interface when the distance between the plates is reduced from ∞ to h is given by,

$$G_{a,el} = 2[\Delta G_a^\sigma(h) - \Delta G_a^\sigma(\infty)] = 2[G_a^\sigma(h) - G_a^\sigma(\infty)] . \tag{4.24}$$

The Δ is dropped because both terms refer to the same reference state. G_a^σ is related to the surface excess Γ_I (amount of adsorption in moles per unit area) and the chemical potential μ_I by the Gibbs equation,

$$G_a^\sigma = -\sum \mu_i \Gamma_i . \tag{4.25}$$

The most common procedure to obtain the interaction between spherical particles is to use the Deryaguin approximation [11]. The interaction energy is considered to be built up of contributions of parallel rings and to approximate the total interaction as the integral over that set of infinitesimal parallel rings. A schematic illustration of the Deryaguin approximation is shown in Fig. 4.9 [1].

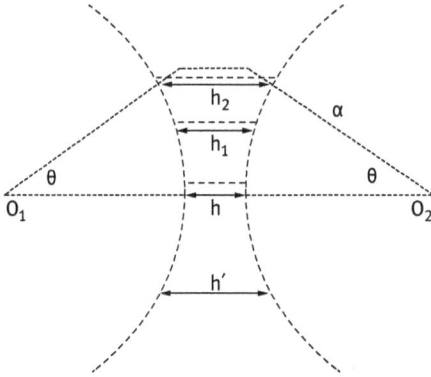

Fig. 4.9: Illustration of the Deryaguin approximation [11].

The above approximation is rigorous if there is no lateral interaction between the rings. For electrostatic interaction, this means that the field lines should run parallel to the O_1O_2 axis. This is exact for $h' = h$, but the condition becomes increasingly poorer in the direction $h' \to h_1 \to h_2$ as indicated by the dashed lines in Fig. 4.9. The Deryaguin approximation [11] is expected to work well for low κh and large κa (where a is the particle radius).

For identical spheres at $\kappa a < 5$, i.e. the diffuse layer is of similar magnitude as the particle radius,

$$G_{el}^\psi = 2\pi\varepsilon_r\varepsilon_0 a\psi_d^2 \exp(-\kappa h) . \tag{4.26}$$

For identical spheres at $\kappa a > 10$, i.e. the double layer is thin compared to particle radius, the result at low potential, for the constant potential case, is,

$$G_{el}^\psi = 2\pi\varepsilon_r\varepsilon_0 a\psi_d^2 \ln[1 + \exp(-\kappa h)] . \tag{4.27}$$

h is the closest distance between the particles, i.e. $h = r - 2a$, where r is the centre-centre distance between the particles.

The constant charge expression is,

$$G_{el}^{\sigma} = -2\pi\varepsilon_r\varepsilon_0 a\psi_d^2 \ln[1 + \exp(-\kappa h)].$$

(4.28)

The constant charge case, equation (4.28), should be used with caution, especially at close approach as a large overestimate can be obtained for the repulsive potential.

Increasing the electrolyte concentration results in compression of the double layer and this causes a reduction in double layer repulsion [1–3]. This is illustrated in Fig. 4.10 which shows the repulsive energy versus $H/2$ for two flat plates at a Stern potential of 154 mV and three values of κ for 1:1 electrolyte. In general, increasing the electrolyte concentration results in a decrease in electrostatic repulsion.

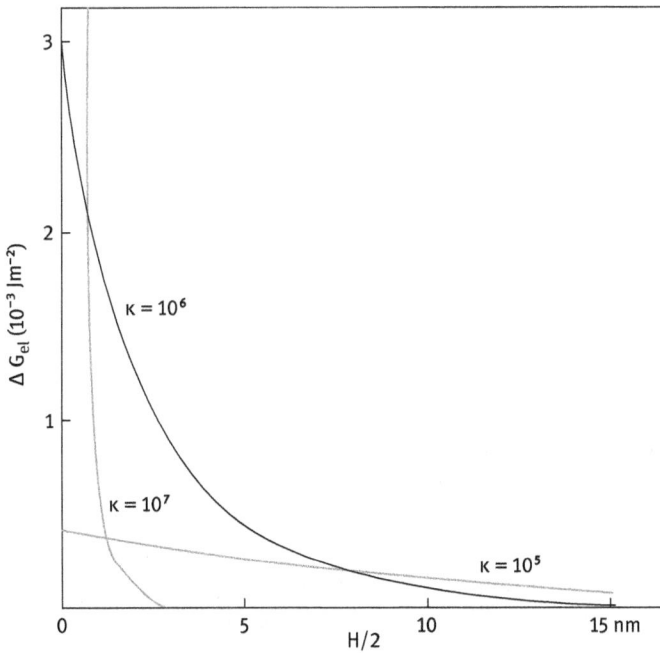

Fig. 4.10: Increase in the electrical free energy of two flat plates as a function of decreasing mutual distance for three values of the reciprocal Debye length: ψ_d = 154 mV; 1:1 electrolyte.

The effect of valency of counterions is shown in Fig. 4.11 which compares the result for $Z = 1$ with that for $Z = 3$. As expected, increasing Z results in a decrease in electrostatic repulsion.

The effect of increasing the Stern potential is illustrated in Fig. 4.12 which shows the values for three dimensionless $y_d (= F\psi_d/RT)$. An increase in the Stern potential at a given electrolyte concentration results in an increase in electrostatic repulsion.

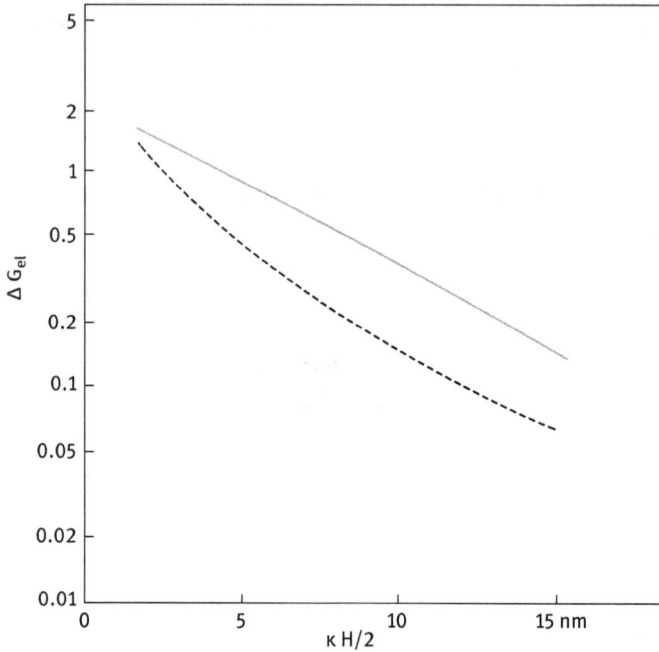

Fig. 4.11: Effect of counterion valency on the increase in free energy as a function of $\kappa H/2$ for two flat plates. Full line $y_d = 2$, $Z = 1$; dashed line $y_d = 6$, $Z = 3$. ($y_d = F\psi_d/RT$). The units on the y-axis are only relative.

In all the above treatments the interaction is considered for isolated particles where the time-average separation of the particles is much larger than the range of the diffuse layer. However, with most practical dispersions such a condition is seldom satisfied since in this case the separation of the particles becomes similar in magnitude to the range of the diffuse layer [10]. The background electrolyte now contains other particles with their counterions. Each charged particle is a "macro-ion", but the number is very much smaller than the number of corresponding counterions. Thus, the particle contribution can be ignored without introducing a large error. However, when the volume occupied by the particles become significant, i.e. at high volume fraction, the ionic concentration in the liquid becomes larger since the particle volume is excluded to the ions. Russel et al. [12] gave a convenient expression for κ by considering the volume fraction of the particles ϕ,

$$\kappa = \left(\frac{e^2}{\varepsilon_r \varepsilon_0 kT} \frac{2Z^2 n_0 - \frac{3\sigma_0 Z\phi}{ae}}{1 - \phi} \right). \tag{4.29}$$

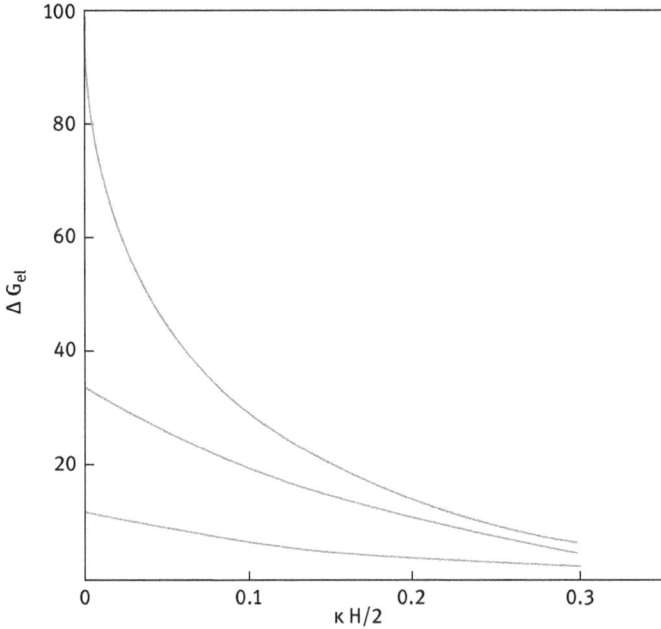

Fig. 4.12: Increase in electrical free energy of two flat plates as a function κH/2. From top to bottom the lines correspond to dimensionless potential y_d of 8, 6 and 4, respectively. The units on the y-axis are only relative.

The $(1 - \phi)$ term in the denominator corrects the ion concentration for the volume occupied by the particles, σ_0 is the surface charge density which can be calculated from the Stern or zeta potential. Figure 4.13 shows the variation of Debye length with particle volume fraction in 10^{-4} mol dm^{-3} NaCl for particles of radius a = 200 nm and charge density σ_0 of $-10 \,\mu C \,cm^{-2}$.

It is clear from Fig. 4.13 that the Debye length decreases gradually with increasing particle volume fraction and at $\phi > 0.01$ it shows a more rapid decrease, reaching about 4 nm when $\phi \approx 0.3$. This reduction in Debye length is the consequence of the increase in ion concentration produced by the counterions of the high number concentration of the particles. This reduction in Debye length is reflected in the reduction of electrostatic repulsion as is illustrated in Fig. 4.14. It can be clearly seen that by increasing the particle volume fraction the electrostatic repulsion decreases due to the reduction in the Debye length (compression of the electrical double layer). This explains why in many practical dispersions prepared at high volume fraction ($\phi > 0.3$) electrostatic repulsion is insufficient for the long-term colloid stability of the dispersion.

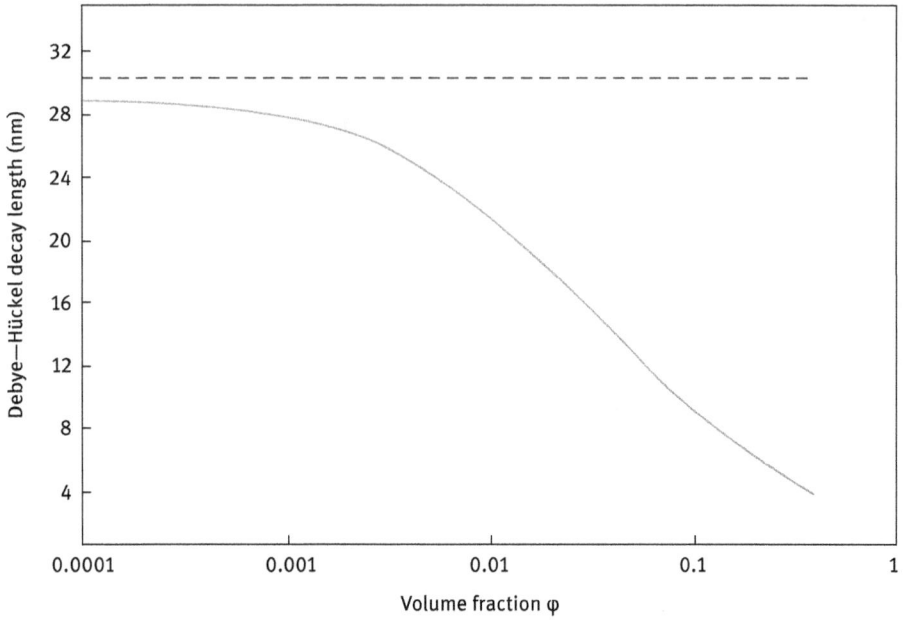

Fig. 4.13: Debye length as a function of particle volume fraction in 10^{-4} mol dm^{-3} NaCl, a = 200 nm; $\sigma_0 = -10$ µC cm^{-2}. The dashed line represents the limiting value for zero volume fraction.

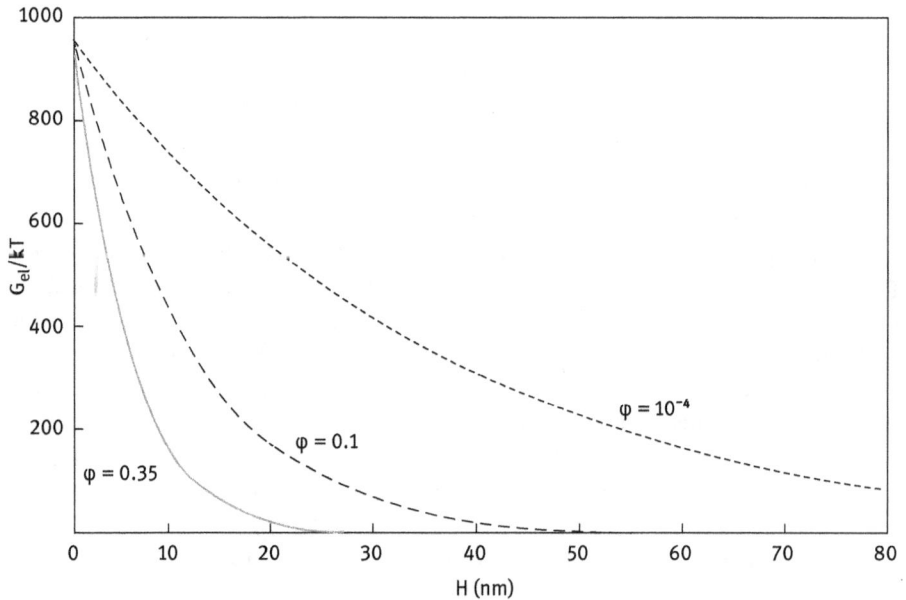

Fig. 4.14: Electrostatic repulsion for three particle volume fractions in 10^{-3} mol dm^{-3} NaCl, a = 200 nm; $\psi_d = -80$ mV; $\sigma_d = -10$ µC cm^{-2}.

4.5 Van der Waals Attraction

The interaction between two particles can in general be considered in terms of the potential energy or the work required to separate them from a centre-to-centre distance r to some large distance apart [10]. The interaction energy includes both enthalpic and entropic contributions of all components in the system, namely particles, solvent, ions, surfactants, polymers, etc. This means that the interaction energy between any two particles should take into consideration the contribution from all other components, i.e. the potential of mean force. A detailed account of both intermolecular forces and interaction between the particles was given in the text by Israelachvili [13] to which the reader should refer for more detail.

Before describing the van der Waals attraction between particles or droplets in a dispersion, one must first consider the intermolecular attraction between atoms or molecules. Three main contributions to intermolecular interaction can be considered: (i) Dipole-dipole interaction (Keesom–van der Waals interaction). This interaction arises from the presence of permanent dipoles [14], e.g. with HCl gas that has a strong dipole due to polarization of the covalent bond. Keesom interaction is short range in nature, being proportional to $1/r^6$. Due to marked dipole alignment, the rotational motion of the molecules is restricted and this results in a "long time" interaction that occurs at low frequency. (ii) Dipole-induced dipole interaction (Debye–van der Waals interaction). This occurs between a polar and nonpolar molecule [15]. The dipole on the polar molecule polarizes the electron clouds of the nonpolar molecule. The interaction free energy can be described similarly to that of the dipole-dipole interaction. (iii) London–van der Waals interaction (dispersion interaction). The London dispersion force is the most important, since it occurs for polar and nonpolar molecules. It arises from fluctuations in the electron density distribution [16]. It is due to the movement of the electron cloud around the atomic nucleus resulting in a dipole that fluctuates. When two atoms come in close proximity, the temporary dipoles become aligned, i.e. the fluctuations become coupled and this is a preferred (or lower) energy state. The range of interaction is similar to that of Keesom and Debye, but the timescale is now that of the electronic transition, near the visible-ultraviolet part of the electromagnetic spectrum. The London–van der Waals interaction energy is given by,

$$u(r) = -\frac{\beta_{11L}}{r^6} .$$

$$(4.30)$$

The London dispersion constant β_{11L} for two identical atoms is proportional to the ionization energy of the outer electrons, $h\nu_1$, and the polarizability α, where h is Planck's constant and ν is the frequency of radiation in Hz (= $\omega/2\pi$, where ω is the frequency in rad s^{-1}),

$$\beta_{11L} \propto h\nu_1\alpha^2 .$$

$$(4.31)$$

For two different atoms,

$$\beta_{12L} \propto h \frac{\nu_{11}\nu_{12}}{\nu_{11} + \nu_{12}} \alpha_1 \alpha_2 . \tag{4.32}$$

The London–van der Waals intermolecular interaction is mostly much larger than the Keesom–van der Waals or Debye–van der Waals contributions. Water is exceptional since in this case the dispersion contribution is only one-quarter of the total interaction.

As described above, all three interactions are based on the attraction between dipoles and have the same distance separation. Therefore, it is possible to describe the attraction in a general way. The London constant in equations (4.31) and (4.32) depends on the ionization potential of the outer electrons which is in the region of the visible-ultraviolet part of the electromagnetic spectrum. Other electronic transitions also take place so that contributions at other frequencies can also occur. For a general approach one must consider the full range of frequencies, ranging from those of a few Hertz up to the ultraviolet region at $\approx 10^{16}$ Hz.

The above description considered the case of interaction of atoms or molecules in the absence of any intervening medium. The relative permittivity of the medium is an important factor. The relative permittivity $\varepsilon(\nu)$ as a function of frequency describes the dielectric behaviour at low frequencies and the refractive index $n(\nu)$ is a viable measure of the dielectric behaviour at the higher end of the spectrum.

There are generally two approaches for describing the van der Waals attraction between macroscopic bodies such as suspension particles. The first approach considers the London–van der Waals attraction to be the sum of the forces acting between isolated molecules. This approach, referred to as the microscopic approach, was suggested by de Boer and Hamaker [17, 18]. The second approach, developed by Lifshitz [19], is based on the correlation between electric fluctuations of two macroscopic bodies. This is referred to as the macroscopic approach.

The approach starts with the finding that the Gibbs energy of interaction can be given by the product of a material constant, $A_{ij(k)}$, and a function of geometry and distance h, $f(a, h)$ [10]. $A_{ij(k)}$, referred to as the Hamaker constant, refers to the interaction between two particles or macrobodies i and j across a medium k,

$$G_{VDW}(h) = -A_{ij(k)}f(a, h) . \tag{4.33}$$

For two identical particles in an aqueous medium, $A_{ij(k)}$ becomes $A_{11(w)}$. The dimensions of the distance function, $f(a, h)$, depend on the geometry of the system. For semifinite plates, $f(a, h) = f(h)$ is proportional to h^{-2} and $G_{VDW}(h)$ is in $J\,m^{-2}$. For two spheres, $f(a, h)$ is dimensionless and $G_{VDW}(h)$ is in J.

Consider two slabs of interacting materials as schematically represented in Fig. 4.15.

The starting point is to consider the interaction between a single molecule and a slab of material and then extend that to two slabs interacting [10]. In the microscopic approach due to Hamaker [18], the energies are assumed to be additive. In this case

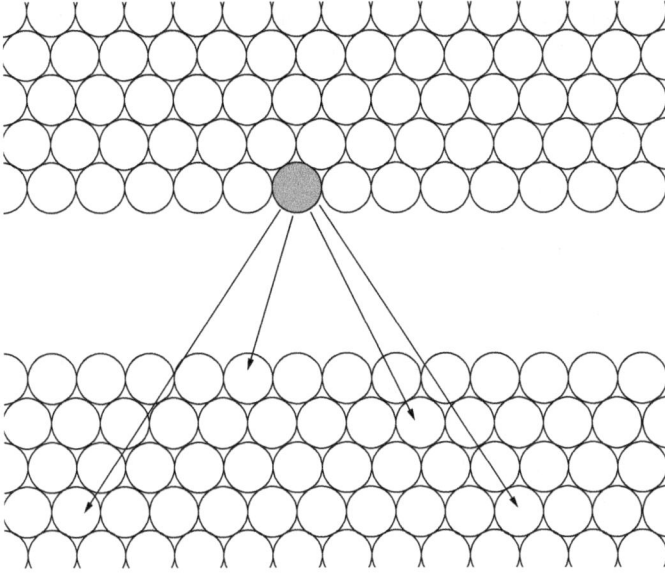

Fig. 4.15: Schematic representation of atomic dipolar interactions between two slabs of material.

one can simply use equation (4.30) and add all interactions between the reference molecule in the upper slab (shown by a dark circle in Fig. 4.15) and all other molecules in the material of the lower slab. The interaction energy is given by the sum for each molecule in the lower slab. One then adds this up for all molecules in the upper slab. The interaction is then given by the sum for all molecules in the upper slab multiplied by the sum for each molecule in the lower slab.

To calculate the intermolecular distance, r, one can simply use the number density of molecules in the slab, $\rho_N (= N/V)$, and integrate over the volumes. This "semi-continuum" approach is the basis of the additivity assumption and it would only become a problem at very close approach. The interaction energy then follows from equation (4.33),

$$G_{VDW}(h) = -\pi^2 \beta_{11L} \rho_N^2 f(a, h).$$ (4.34)

Equation (4.34) gives a general expression for the van der Waals attraction in terms of the material constant (that is given by the electronic polarizability, the ionization potential and the square of the product of the density and molar mass) and the shape factor and separation distance between the two bodies.

For two infinitely thick slabs, the energy of unit area of one slab interacting with the whole of the other slab is given by,

$$G_{VDW}(h) = -\frac{\pi \beta_{11L} \rho_N^2}{12h^2},$$ (4.35)

where $G_{VDW}(h)$ represents the dispersion energy of two slabs of the same material a distance h apart and it is the energy per unit area of the surface. The numerator represents the material property whereas the denominator arises from the geometry. It is common in colloid science to express the material property as a single constant, referred to as the Hamaker constant that is simply given by,

$$A_{11} = \pi^2 \beta_{11L} \rho_N^2 .$$ (4.36)

Equation (4.35) can then be written as,

$$G_{VDW}(h) = -\frac{A_{11}}{12\pi h^2} .$$ (4.37)

If one has a slab of material 1 interacting with material 2,

$$G_{VDW}(h) = -\frac{A_{12}}{12\pi h^2} .$$ (4.38)

And,

$$A_{12} = \pi^2 \beta_{12L} \rho_{N1} \rho_{N2} .$$ (4.39)

If the "semi-infinite" slabs are replaced by plates with thickness t, equation (4.38) has to be modified,

$$G_{VDW}(h) = -\frac{A_{11}}{12\pi} \left(\frac{1}{h^2} + \frac{1}{(h+2t)^2} - \frac{2}{(h+t)^2} \right)$$ (4.40)

Hamaker constants can be computed from equation (4.40) if the molecular properties of the materials under consideration are known. Alternatively, they can be derived from the macroscopic theory as will be discussed below. Direct measurement of van der Waals attraction can also be used to obtain the Hamaker constant. Its magnitude is in the order of $10^{-20} - 10^{-19}$ J ($\approx$ 5–50 kT) at room temperature. As we will see later, the Hamaker constants for interaction across a medium are much lower.

For curved interfaces (as is the case for spherical particles), the expression for the van der Waals attraction is more complex. However, a useful method to derive such an expression is to use the Deryaguin approximation [11], where the curved surface is replaced by a stepped one as illustrated in Fig. 4.16. The total interaction between the macroscopic bodies is considered to be built up of contributions of parallel rings where each pair contributes an amount $G_A(x) \, dA$, where $G_A(x)$ is given by equation (4.37). From the energy per ring, the total interaction energy is obtained by integration over y (as indicated in Fig. 4.16) after replacing dA by $2\pi y \, dy$. As the approximation is limited to short distances, the contributions of layers with large y are negligible, so that for convenience the integration may be carried out from $y = 0$ to $y = \infty$. After relating y to x, the Deryaguin formula for two spheres of radii R_1 and R_2 becomes,

$$G_A = -\frac{2\pi R_1 R_2}{R_1 + R_2} \int_h^\infty G_A(x) \, dx .$$ (4.41)

Substituting equation (4.37) into (4.41) and carrying out the integration gives,

$$G_{VDW}(h) = -\frac{A_{12} R_1 R_2}{6h(R_1 + R_2)} .$$ (4.42)

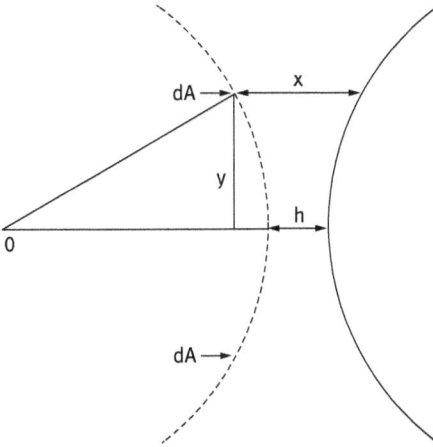

Fig. 4.16: Schematic representation of Deryaguin's approximation [11].

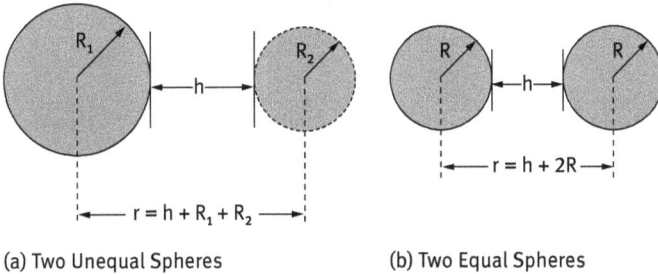

(a) Two Unequal Spheres   (b) Two Equal Spheres

Fig. 4.17: Schematic representation of the interaction between spheres of different radii (a) and with the same radius (b).

For two spheres of equal radius,

$$G_{VDW}(h) = -\frac{A_{11}R}{12h}. \tag{4.43}$$

Equations (4.42) and (4.43) apply only for the case $h < R_1, R_2$ and $h < R$ respectively. For the general case of a wide range of h, expressions can be derived both for particles with different radii and of equal radii as schematically shown in Fig. 4.17.

For two spheres with different radii R_1 and R_2,

$$G_{VDW}(h) = -\frac{A_{12}}{12}\left[\frac{y}{x^2+xy+x}+\frac{y}{x^2+xy+y}=2\ln\left\{\frac{x^2+xy+y}{x^2+xy+x+y}\right\}\right], \tag{4.44}$$

where $x = h/2R_1$ and $y = R_2/R_1$.

For two spheres with equal radii,

$$G_{VDW}(h) = -\frac{A_{11}}{6}\left[\frac{2R^2}{r^2-4R^2}+\frac{2R^2}{r^2}+\ln\frac{r^2-4R^2}{r^2}\right]. \tag{4.45}$$

Substituting $s = r/R$,

$$G_{VDW}(h) = -\frac{A_{11}}{6}\left(\frac{2}{s^2-4} + \frac{2}{s^2} + \ln\frac{s^2-4}{s^2}\right).\tag{4.46}$$

For very short distances ($h \ll R$), equation (4.46) may be approximated by equation (4.43).

When the particles are dispersed in a liquid medium, the van der Waals attraction has to be modified to take the medium effect into account. When two particles are brought from infinite distance to h in a medium, an equivalent amount of medium has to be transported the other way round. Hamaker forces in a medium are excess forces.

Consider two identical spheres 1 at a large distance apart in a medium 2, as illustrated in Fig. 4.18 (a). In this case the attractive energy is zero. Figure 4.18 (b) gives the same situation with arrows indicating the exchange of 1 against 2. Figure 4.18 (c) shows the complete exchange which now shows the attraction between the two particles 1 and 1 and equivalent volumes of the medium 2 and 2.

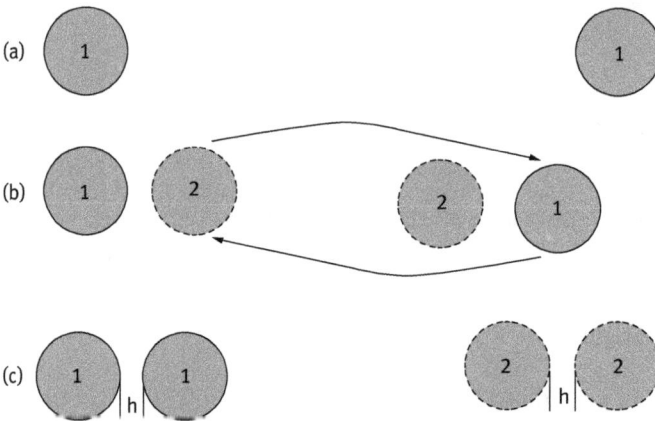

Fig. 4.18: Schematic representation of interaction of two particles in a medium.

The effective Hamaker constant for two identical particles 1 and 1 in a medium 2 is given by,

$$A_{11(2)} = A_{11} + A_{22} - 2A_{12} = (A_{11}^{1/2} - A_{22}^{1/2})^2.\tag{4.47}$$

Equation (4.47) shows that two particles of the same material attract each other unless their Hamaker constants exactly match each other. Equation (4.43) now becomes,

$$G_{VDW}(h) = -\frac{A_{11(2)}R}{12h},\tag{4.48}$$

where $A_{11(2)}$ is the effective Hamaker constant of two identical particles with Hamaker constant A_{11} in a medium with Hamaker constant A_{22}.

In most cases, the Hamaker constant of the particles is higher than that of the medium. Examples of the Hamaker constant for some materials in vacuum are given in Tab. 4.3. The Hamaker constant for some liquids is given in Tab. 4.4. Table 4.5 gives values of the effective Hamaker constant for some particles in some liquids. Generally speaking, the effect of the liquid medium is to reduce the Hamaker constant of the particles below its value in vacuum (air).

Tab. 4.3: Hamaker constant in vacuum A_{11} for some materials.

Material	$A_{11} \times 10^{20}$ J
Fused quartz (SiO_2)	6.5
Al_2O_3	15.6
Silver	50.0
Copper	40.0
Poly(methylmethacrylate)	7.1
Poly(vinylchloride)	7.8

Tab. 4.4: Hamaker constant of some liquids.

Material	$A_{22} \times 10^{20}$ J
Water	3.7
Ethanol	4.2
Decane	4.8
Hexadecane	5.2
Cyclohexane	5.2

Tab. 4.5: Effective Hamaker constant $A_{11(2)}$ of some particles in water.

Material	$A_{11(2)} \times 10^{20}$ J
Fused quartz/water	0.83
Al_2O_3/water	5.32
Copper/water	30.00
Poly(methylmethacrylate)/water	1.05
Poly(vinylchloride)/water	1.03
Poly(tetrafluoroethylene)/water	0.33

G_A decreases with increasing h as schematically shown in Fig. 4.19. As also shown in Fig. 4.19, G_A increases very sharply with h at small h values. A capture distance can be defined at which all the particles become strongly attracted to each other (coagulation). At very short distances, Born repulsion appears.

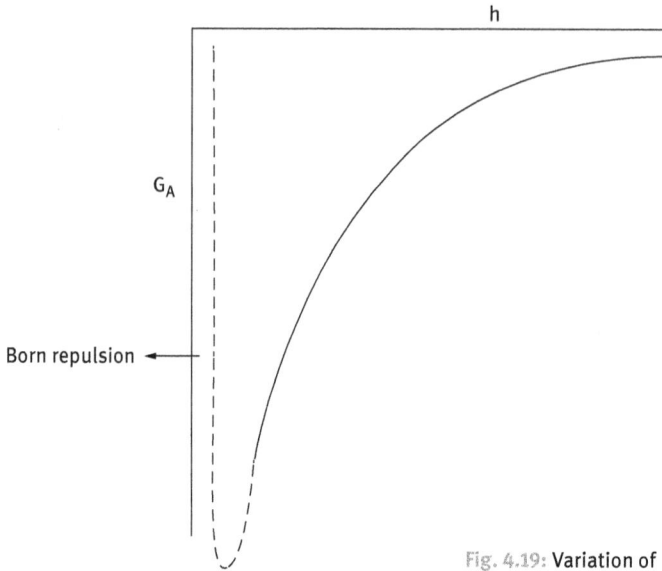

Fig. 4.19: Variation of G_A with h.

One of the problems with the microscopic theory of Hamaker [18] is the situation when one is dealing with particles in a medium such as an electrolyte solution. In this case the interaction of permanent dipoles is screened by the ionic environment and the low frequency interaction falls off more rapidly than would be the case in a vacuum. This problem can be solved by using the generalized macroscopic theory of Lifshitz [19] that is described in detail by Mahanty and Ninham [20]. It is based on the principle that the spontaneous electromagnetic fluctuations in two particles become correlated when they approach each other, causing a decrease in the free energy of the system. This theory treats the interacting bodies as continuous and ascribes the interactions to fluctuating electromagnetic fields arising from spontaneous electric and magnetic polarizations within the various media. One important result is that the interactions are described completely in terms of the complex dielectric constants of the media.

4.6 Total energy of interaction

4.6.1 Deryaguin–Landau–Verwey–Overbeek (DLVO) theory [8, 9]

Combining G_{el} and G_A results in the well-known theory of stability of colloids (DLVO theory) [8, 9],

$$G_T = G_{el} + G_A \qquad (4.49)$$

A plot of G_T versus h is shown in Fig. 4.20, which represents the case at low electrolyte concentrations, i.e. strong electrostatic repulsion between the particles. G_{el} decays

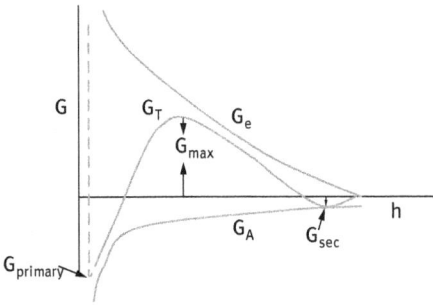

Fig. 4.20: Schematic representation of the variation of G_T with h according to DLVO theory.

exponentially with h, i.e. $G_{el} \rightarrow 0$ as h becomes large. G_A is $\propto 1/h$, i.e. G_A does not decay to 0 at large h.

At long distances of separation, $G_A > G_{el}$ resulting in a shallow minimum (secondary minimum). At very short distances, $G_A \gg G_{el}$ resulting in a deep primary minimum. At intermediate distances, $G_{el} > G_A$ resulting in energy maximum, G_{max}, whose height depends on ψ_0 (or ψ_d) and the electrolyte concentration and valency.

At low electrolyte concentrations ($< 10^{-2}$ mol dm^{-3} for a 1:1 electrolyte), G_{max} is high (> 25 kT) and this prevents particle aggregation into the primary minimum. The higher the electrolyte concentration (and the higher the valency of the ions), the lower the energy maximum.

Under some conditions (depending on electrolyte concentration and particle size), flocculation into the secondary minimum may occur. This flocculation is weak and reversible. By increasing the electrolyte concentration, G_{max} decreases until at a given concentration it vanishes and particle coagulation occurs. This is illustrated in Fig. 4.21 which shows the variation of G_T with h at various electrolyte concentrations.

Since approximate formulae are available for G_{el} and G_A, quantitative expressions for $G_T(h)$ can also be formulated. These can be used to derive expressions for the coagulation concentration, which is that concentration that causes every encounter between two colloidal particles to lead to destabilization. Verwey and Overbeek [9]

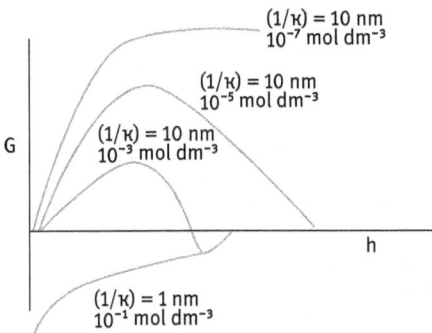

Fig. 4.21: Variation of G with h at various electrolyte concentrations.

introduced the following criteria for transition between stability and instability,

$$G_T(= G_{el} + G_A) = 0 \tag{4.50}$$

$$\frac{dG_T}{dh} = 0 \tag{4.51}$$

$$\frac{dG_{el}}{dh} = -\frac{dG_A}{dh} \tag{4.52}$$

Using the equations for G_{el} and G_A, the critical coagulation concentration, ccc, could be calculated. The theory predicts that the ccc is directly proportional to the surface potential ψ_0 and inversely proportional to the Hamaker constant A and the electrolyte valency Z. The ccc is inversely proportional to Z^6 at high surface potential and inversely proportional to Z^6 at low surface potential.

4.7 Criteria for stabilization of suspensions with double layer interaction

The two main criteria for stabilization are: (i) High surface or Stern potential (zeta potential), high surface charge. As shown in equation (4.5), the repulsive energy G_{el} is proportional to ψ_0^2. In practice, ψ_0 cannot be directly measured and, therefore, one instead uses the measurable zeta potential. (ii) Low electrolyte concentration and low valency of counter- and co-ions. As shown in Fig. 4.21, the energy maximum increases with decreasing electrolyte concentration. The latter should be lower than 10^{-2} mol dm^{-3} for 1 : 1 electrolyte and lower than 10^{-3} mol dm^{-3} for 2 : 2 electrolyte. One should ensure that an energy maximum in excess of 25 kT should exist in the energy-distance curve. When $G_{max} \gg kT$, the particles in the dispersion cannot overcome the energy barrier, thus preventing coagulation. In some cases, particularly with large and asymmetric particles, flocculation into the secondary minimum may occur. This flocculation is usually weak and reversible and may be advantageous for preventing the formation of hard sediments.

References

[1]   Lyklema, J., "Fundamentals of Interface and Colloid Science", Elsevier (Academic Press), Amsterdam (2005), Vol. IV.

[2]   Tadros, Th. F., "Suspensions", in "Encyclopedia of Colloid and Interface Science", Th. F. Tadros (ed.), Springer, Germany (2013).

[3]   Tadros, Th. F., "Interfacial Phenomena", De Gruyter, Germany (2015).

[4]   Gouy, G., J. Phys., **9**, 457 (1910); Ann. Phys., **7**, 129 (1917).

[5]   Chapman, D. L., Phil. Mag., **25**, 475 (1913)

[6]   Stern, O., Z. Elektrochem., **30**, 508 (1924).

[7]   Grahame, D. C., Chem. Rev., **41**, 44 (1947).

[8]   Deryaguin, B. V. and Landau, L., Acta Physicochem. USSR, **14**, 633 (1941).

[9]   Verwey, E. J. W. and Overbeek, J. Th. G., "Theory of Stability of Lyophobic Colloids", Elsevier, Amsterdam (1948).

[10] Goodwin, J., "Colloids and Interfaces with Surfactants and Polymers", John Wiley & Sons, London (2009).

[11] Deryaguin, B. V., Kolloid Z., **69**, 155 (1934).

[12] Russel, W. B., Saville, D. A. and Schowalter, W. R., "Colloidal Dispersions", Cambridge University Press, Cambridge (1989).

[13] Israelachvili, J., "Intermolecular and Surface Forces", Academic Press, London (1992).

[14] Keesom, W. H., Proc. Koninkl. Nederland. Wetenschap., **18**, 636 (1915); **23**, 939 (1920); Physik. Z., **22**, 129, 643 (1921).

[15] Debye, P., Physik. Z., **21**, 178 (1920); **22**, 302 (1921).

[16] London, F., Z. Physik., **63**, 245 (1930); Trans. Faraday Soc., **33**, 8 (1937).

[17] de Boer, J. H., Trans. Faraday Soc., **32**, 10 (1936).

[18] Hamaker, H. C., Physica, **4**, 1058 (1937).

[19] Lifshitz, E. M., Soviet Physics JETP **2**, 73 (1956).

[20] Mahanty, J. and Ninham, B. W., "Dispersion Forces", Academic Press, London (1976).

5 Steric stabilization of suspensions

5.1 Introduction

As mentioned in Chapter 4, stabilization of suspensions using electrostatic (double layer) repulsion requires low electrolyte concentration and valency. This may not be possible with many practical suspensions, in particular with those having high disperse phase (volume fraction) concentration. In this case, one must use nonionic surfactants or polymeric surfactants containing an "anchor" (strongly adsorbed) B chain that is insoluble in the dispersion medium and a stabilizing chain A that is highly soluble in the medium and strongly solvated by its molecules. Examples of B chains for hydrophobic solids in aqueous media are alkyl chain, polystyrene or polymethylmethacrylate. Examples for A chains in aqueous media are poly(ethylene oxide), poly(vinyl alcohol) or polysaccharide. These nonionic or polymeric surfactants adopt a special configuration at the solid/liquid interface with the B chains strongly attached to the hydrophobic surface leaving the A chains "dangling" in solution and giving an adsorbed polymer layer with thickness δ that increases with increasing molecular weight of the stabilizing chains. When two particles, each containing an adsorbed layer with thickness δ, approach to a separation distance h that is smaller than 2δ, strong repulsion occurs provided the A chains are in good solvent conditions. This repulsion is usually referred to as "steric" and is not strongly affected by moderate electrolyte concentration.

In this chapter I will start with two sections on the adsorption and conformation of surfactants and polymeric surfactants at the solid/liquid interface. This is key to understanding how these surfactants act as stabilizer. This is followed by a section on the interaction between particles containing adsorbed layers and the origin of steric stabilization.

5.2 Adsorption and orientation of nonionic surfactants at the solid/liquid interface

Surfactant adsorption is relatively simpler than polymeric surfactant adsorption. This stems from the fact that nonsurfactants consist of a small number of units and they mostly are reversibly adsorbed, allowing one to apply thermodynamic treatments. In this case, it is possible to describe the adsorption in terms of the various interaction parameters, namely chain-surface, chain-solvent and surface-solvent. Moreover, the conformation of the surfactant molecules at the interface can be deduced from these simple interactions parameters. However, in some cases the interaction parameters may involve ill-defined forces, such as hydrophobic bonding, solvation forces and chemisorption.

DOI 10.1515/9783110486872-006

Several types of nonionic surfactants exist, depending on the nature of the polar (hydrophilic) group [1]. The most common type is that based on a poly(oxyethylene) glycol group, i.e. $(CH_2-CH_2O)_n OH$ (where n can vary from as little as 2 units to as high as 100 or more units) linked either to an alkyl $(C_x H_{2x+1})$ or alkyl phenyl $(C_x H_{2x+1}-C_6 H_4-)$ group. These surfactants may be abbreviated as $C_x E_n$ or $C_x \phi E_n$ (where C refers to the number of C atoms in the alkyl chain, ϕ denotes $C_6 H_4$ and E denotes ethylene oxide). These ethoxylated surfactants are characterized by a relatively large head group compared to the alkyl chain (when n > 4). However, there are nonionic surfactants with a small head group such as amine oxides $(-N \to 0)$ head group, phosphate oxide $(-P \to 0)$ or sulphinyl-alkanol $(-SO-(CH_2)_n -OH)$ [1]. Most adsorption isotherms in the literature are based on the ethoxylated type of surfactants.

In comparing various nonionic surfactants, and in particular those based on poly(ethylene oxide) (PEO), it is useful to use the "hydrophilic-lipophilic-balance" (HLB) concept, which is simply gives the relative proportion of the hydrophilic (PEO and OH) and lipophilic (alkyl chain) components [1]. The HLB is simply given by the weight percent of PEO and OH divided by 5. For example, for a nonionic surfactant of $C_{12}H_{25}-O-(CH_2-CH_2-O)_4-H$, the HLB is 10.5.

The adsorbents used for studying nonionic surfactant adsorption range from apolar surfaces such as C black, organic pigments or polystyrene (low energy solids) to polar surfaces such as oxides or silicates (high energy solids).

The adsorption isotherm of nonionic surfactants are in many cases Langmuirian, like those of most other highly surface active solutes adsorbing from dilute solutions and adsorption is generally reversible. However, several other adsorption types are produced [1] and those are illustrated in Fig. 5.1. The steps in the isotherm may be explained in terms of the various solvent and adsorbate interactions. These adsorbate-adsorbate and adsorbate-adsorbent orientations are schematically illustrated in Fig. 5.2.

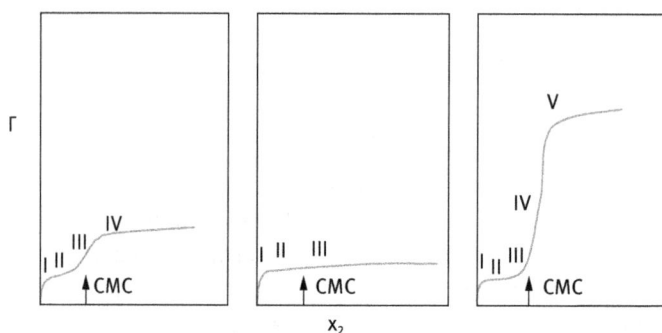

Fig. 5.1: Adsorption isotherms corresponding to the three adsorption sequences shown in Fig. 5.2.

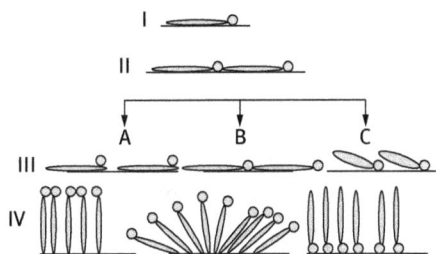

Fig. 5.2: Model for adsorption of nonionic surfactants.

In the first stage of adsorption (denoted by I in Fig. 5.1 and 5.2), surfactant-surfactant interaction is negligible (low coverage) and adsorption occurs mainly by van der Waals interaction. On a hydrophobic surface, the interaction is dominated by the hydrophobic portion of the surfactant molecule. This is mostly the case with pharmaceuticals and agrochemicals that have hydrophobic surfaces. Nevertheless, the polar groups of the surfactant may have some interaction with the surface and the hydrophilic EO groups can have a slight positive adsorption even on a nonpolar adsorbent. When the interaction is due to dispersion forces, the heat of adsorption is relatively small and corresponds to the heat liberated by replacing solvent molecules with surfactant. At this stage, the molecule tends to lie flat on the surface because its hydrophobic portion is positively adsorbed, as are also most types of hydrophilic head groups, especially large PEO chains. With the molecule lying parallel to the surface, the adsorption energy will increase in almost equal increments for each additional carbon atom in its alkyl chain, and the initial slope of the isotherm will increase accordingly according to Traube's rule. The same also happens with each additional EO group.

The approach to monolayer saturation with the molecules lying flat (Fig. 5.2 II) is accompanied by a gradual decrease in the slope of the adsorption isotherm as shown in Fig. 5.1. Although most of the "free" solvent molecules will have been displaced from the surface by the time the monolayer is complete, the surfactant molecules themselves will probably stay hydrated at this stage. An increase in the size of the surfactant molecule, e.g. by increasing the length of the alkyl chain or the PEO chain, will decrease the adsorption (expressed in $mol\,m^{-2}$). Increasing temperature should increase the adsorption because dehydration decreases the size of the adsorbate molecules. Increasing temperature reduces the solubility of the nonionic surfactant and this enhances its adsorption (higher surface activity).

The subsequent stages of adsorption are increasingly dominated by adsorbate-adsorbate interaction, although it is the adsorbate-adsorbent interaction that initially determines how the adsorption progresses when stage II is completed. The adsorbate-adsorbent interaction depends on the nature of the adsorbent and the HLB of the surfactant. When the hydrophilic group (e.g. PEO) is only weakly adsorbed it will be displaced from the surface by the alkyl chain of the surfactant molecule (Fig. 5.2 III A). This is particularly the case with nonpolar adsorbents such as C black or polystyrene when the surfactant has a short PEO chain (relatively low HLB number). However, if the interaction between the hydrophilic chain (PEO) and polar adsorbent such as silica or silicates is strong, the alkyl chain is displaced as is illustrated in Fig. 5.2 III C. The intermediate situation (Fig. 5.2 III B) occurs when neither type of displacement is favoured and the adsorbate molecules remain flat on the surface.

The change in the amount of adsorption in the third stage (stage III of Fig. 5.1) is unlikely to be large, but as the concentration of the surfactant in bulk solution approaches the critical micelle concentration (cmc) there will be a tendency for the alkyl chains of the adsorbed molecules to aggregate. This will cause the molecules to be vertically oriented and there will be a large increase in adsorption (stage IV). The lateral forces due to alkyl chain interactions will compress the head group, and for a PEO chain this will result in a less coiled, more extended conformation. The longer the alkyl chain, the greater will be the cohesion force and hence the smaller the surfactant cross-sectional area. This may explain the increase in saturation adsorption with increasing alkyl chain length and decreasing number of EO units in the PEO chain. With nonpolar adsorbents, the adsorption energy per methylene group is almost the same as the micellization energy, so surface aggregation can occur quite easily even at concentrations below the cmc. However, with polar adsorbents, the head group may be strongly bound to the surface, and partial displacement of a large PEO chain from the surface, needed for close packing, may not be achieved until the surfactant concentration is above the cmc. When the adsorption layer is like that shown in Fig. 5.2 IV C the surface becomes hydrophobic.

The interactions occurring in the adsorption layer during the fourth and subsequent stages of adsorption are similar to interactions in bulk solution where enthalpy changes caused by increasing alkyl-alkyl interactions balance those due to head group interactions and the dehydration process. For this reason, the heat of adsorption remains constant, although adsorption increases with increasing temperature due to dehydration of the head group and its more compact nature.

5.3 Polymeric surfactant adsorption

The process of polymer adsorption is fairly complicated. In addition to the usual adsorption considerations, such as polymer/surface, polymer/solvent and surface/solvent interactions, one of the principal problems to be resolved is the configuration (conformation) of the polymer at the solid/liquid interface [2, 3]. This was recognized by Jenkel and Rumbach in 1951 [2] who found that the amount of polymer adsorbed per unit area of the surface would correspond to a layer more than 10 molecules thick if all the segments of the chain are attached. They suggested a model in which each polymer molecule is attached in sequences separated by bridges which extend into solution. In other words not all the segments of a macromolecule are in contact with the surface. The segments which are in direct contact with the surface are termed "trains"; those in between and extended into solution are termed "loops"; the free ends of the macromolecule also extending in solution are termed "tails". This is illustrated in Fig. 5.3 (a) for a homopolymer. Examples of homopolymers that are formed from the same repeating units are poly(ethylene oxide) or poly(vinyl pyrrolidone). Such homopolymers may adsorb significantly at the S/L interface. Even if the adsorption energy per monomer segment to the surface is small (fraction of kT, where k is the Boltzmann constant and T is the absolute temperature), the total adsorption energy per molecule may be sufficient to overcome the unfavourable entropy loss of the molecule at the S/L interface. Clearly, homopolymers are not the most suitable dispersants. A small variant is to use polymers that contain specific groups that have high affinity to the surface. This is exemplified by partially hydrolysed poly(vinyl acetate) (PVAc), technically referred to as poly(vinyl alcohol) (PVA). The polymer is prepared by partial hydrolysis of PVAc, leaving some residual vinyl acetate groups. Most commercially available PVA molecules contain 4–12% acetate groups. These acetate groups, which are hydrophobic, give the molecule its amphipathic character. On a hydrophobic surface such as polystyrene, the polymer adsorbs with preferential attachment of the acetate groups on the surface, leaving the more hydrophilic vinyl alcohol segments dangling in the aqueous medium. The configuration of such "blocky" copolymers is illustrated in Fig. 5.3 (b). Clearly, if the molecule is made fully from hydrophobic segments, the chain will adopt a flat configuration as is illustrated in Fig. 5.3 (c). The most convenient polymeric surfactants are those of the block and graft copolymer type. A block copolymer is a linear arrangement of blocks of variable monomer composition. The nomenclature for a diblock is poly-A-block-poly-B and for a triblock is poly-A-block-poly-B-poly-A. An example of an A-B diblock is polystyrene block-polyethylene oxide and its conformation is represented in Fig. 5.3 (d). One of the most widely used triblock polymeric surfactants are the "Pluronics" (BASF, Germany) which consists of two poly-A blocks of poly(ethylene oxide) (PEO) and one block of poly(propylene oxide) (PPO). Several chain lengths of PEO and PPO are available. As will be discussed below, these polymeric triblocks can be applied as dispersants, whereby the assumption is made that the hydrophobic PPO chain resides at the hydrophobic surface, leaving the two PEO chains dangling in aqueous solution and hence providing steric repulsion. Several other triblock copolymers have been synthesized, although these are of limited commercial availability. Typical examples are triblocks of poly(methyl methacrylate)-block poly(ethylene oxide)-block poly(methyl methacrylate). The conformation of these triblock copolymers is illustrated in Fig. 5.3 (e). An alternative (and perhaps more efficient) polymeric surfactant is the amphipathic graft copolymer consisting of a polymeric backbone B (polystyrene or polymethyl methacrylate) and several A chains ("teeth") such as polyethylene oxide. This graft copolymer is sometimes referred to as a "comb" stabilizer. Its configuration is illustrated in Fig. 5.3 (f).

The polymer/surface interaction is described in terms of adsorption energy per segment χ^s. The polymer/solvent interaction is described in terms of the Flory–

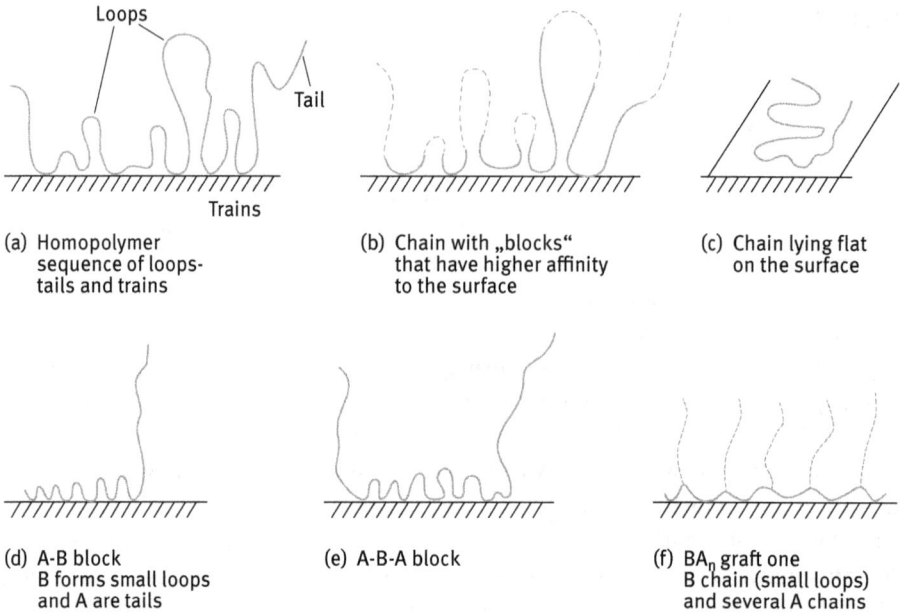

Fig. 5.3: Various conformations of macromolecules on a plane surface.

Huggins interaction parameter χ. For adsorption to occur, a minimum energy of adsorption per segment χ^s is required. When a polymer molecule adsorbs on a surface, it loses configurational entropy and this must be compensated by an adsorption energy χ^s per segment. This is schematically shown in Fig. 5.4, where the adsorbed amount Γ is plotted versus χ^s. The minimum value of χ^s can be very small (< 0.1 kT) since a large number of segments per molecule are adsorbed. For a polymer with say 100 segments and 10 % of these are in trains, the adsorption energy per molecule now reaches 1 kT (with $\chi^s = 0.1$ kT). For 1000 segments, the adsorption energy per molecule is now 10 kT.

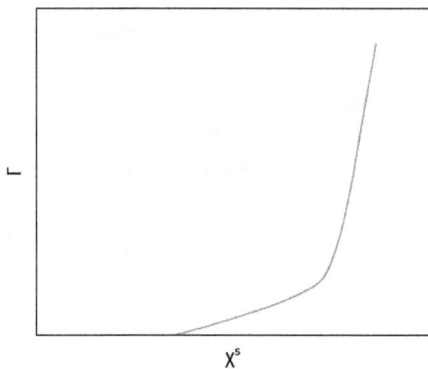

Fig. 5.4: Variation of adsorption amount Γ with adsorption energy per segment χ^s.

As mentioned above, homopolymers are not the most suitable for stabilization of dispersions. For strong adsorption, one needs the molecule to be "insoluble" in the medium and to have strong affinity ("anchoring") to the surface. For stabilization, one needs the molecule to be highly soluble in the medium and strongly solvated by its molecules; this requires a Flory–Huggins interaction parameter of less than 0.5. The above opposing effects can be resolved by introducing "short" blocks in the molecule which are insoluble in the medium and have a strong affinity to the surface, as for example partially hydrolysed polyvinyl acetate (88 % hydrolysed, i.e. with 12 % acetate groups), usually referred to as polyvinyl alcohol (PVA),

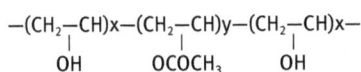

$$-(CH_2-CH)x-(CH_2-CH)y-(CH_2-CH)x-$$
$$\quad\ \ |\qquad\qquad |\qquad\qquad\ |$$
$$\quad\ OH\qquad\ OCOCH_3\qquad OH$$

As mentioned above, these requirements are better satisfied using A-B, A-B-A and BA_n graft copolymers. B is chosen to be highly insoluble in the medium and it should have high affinity to the surface. This is essential to ensure strong "anchoring" to the surface (irreversible adsorption). A is chosen to be highly soluble in the medium and strongly solvated by its molecules. The Flory–Huggins parameter χ can be applied in this case. For a polymer in a good solvent, χ has to be lower than 0.5; the smaller the χ value the better the solvent for the polymer chains. Examples of B for a hydrophobic particle in aqueous media are polystyrene and polymethylmethacrylate. Examples of A in aqueous media are polyethylene oxide, polyacrylic acid, polyvinyl pyrrolidone and polysaccharides. For nonaqueous media such as hydrocarbons, the A chain(s) could be poly(12-hydroxystearic acid).

For a full description of polymer adsorption one needs to obtain information on the following: (i) The amount of polymer adsorbed Γ (in mg or mol) per unit area of the particles. It is essential to know the surface area of the particles in the suspension. Nitrogen adsorption on the powder surface may give such information (by application of the BET equation) provided there will be no change in area on dispersing the particles in the medium. For many practical systems, a change in surface area may occur on dispersing the powder, in which case one has to use dye adsorption to measure the surface area (some assumptions have to be made in this case). (ii) The fraction of segments in direct contact with the surface, i.e. the fraction of segments in trains p (p = (number of segments in direct contact with the surface)/total number). (iii) The distribution of segments in loops and tails, $\rho(z)$, which extend in several layers from the surface. $\rho(z)$ is usually difficult to obtain experimentally, although recently application of small angle neutron scattering could obtain such information. An alternative and useful parameter for assessing "steric stabilization" is the hydrodynamic thickness, δ_h (thickness of the adsorbed or grafted polymer layer plus any contribution from the hydration layer). Several methods can be applied to measure δ_h as will be discussed below.

Measurement of the adsorption isotherm is by far the easiest to obtain. One measures the polymeric surfactant concentration before ($C_{initial}$, C_1) and after ($C_{equilibrium}$, C_2),

$$\Gamma = \frac{(C_1 - C_2)V}{A} , \tag{5.1}$$

where V is the total volume of the solution and A is the specific surface area ($m^2\ g^{-1}$). It is necessary in this case to separate the particles from the polymer solution after adsorption. This could be carried out by centrifugation and/or filtration. One should make sure that all particles are removed. To obtain this isotherm, one must develop a sensitive analytical technique for determining the polymeric surfactant concentration in the ppm range. It is essential to follow the adsorption as a function of time to determine the time required to reach equilibrium. For some polymer molecules, such as polyvinyl alcohol, PVA, and polyethylene oxide, PEO, (or blocks containing PEO), analytical methods based on complexation with iodine/potassium iodide or iodine/boric acid potassium iodide have been established. For some polymers with specific functional groups, spectroscopic methods may be applied, e.g. UV, IR or fluorescence spectroscopy. A possible method is to measure the change in refractive index of the polymer solution before and after adsorption. This requires very sensitive refractometers. High resolution NMR has been recently applied since the polymer molecules in the adsorbed state are in a different environment to those in the bulk. The chemical shift of functional groups within the chain is different in these two environments. This has the attraction of measuring the amount of adsorption without separating the particles.

The fraction of segments in direct contact with the surface can be directly measured using spectroscopic techniques: (i) IR if there is specific interaction between the segments in trains and the surface, e.g. polyethylene oxide on silica from non-aqueous solutions [4, 5]. (ii) Electron spin resonance (ESR); this requires labelling of the molecule. (iii) NMR, pulse gradient or spin ECO NMR. This method is based on the fact that the segments in trains are "immobilized" and hence they have lower mobility than those in loops and tails [6, 7].

An indirect method of determining p is to measure the heat of adsorption ΔH using microcalorimetry [7]. One should then determine the heat of adsorption of a monomer H_m (or molecule representing the monomer, e.g. ethylene glycol for PEO); p is then given by the equation,

$$p = \frac{\Delta H}{H_m n} , \tag{5.2}$$

where n is the total number of segments in the molecule.

The above indirect method is not very accurate and can only be used in a qualitative sense. It also requires very sensitive enthalpy measurements (e.g. using an LKB microcalorimeter).

The segment density distribution $\rho(z)$ is given by the number of segments parallel to the surface in the z-direction. Three direct methods can be applied for determining adsorbed layer thickness: ellipsometry, attenuated total reflection (ATR) and neutron scattering. Both ellipsometry and ATR [8] depend on the difference between refractive indices between the substrate, the adsorbed layer and bulk solution and they require a flat reflecting surface. Ellipsometry [8] is based on the principle that light undergoes a change in polarizability when it is reflected at a flat surface (whether covered or uncovered with a polymer layer).

The above limitations when using ellipsometry or ATR are overcome by the application technique of neutron scattering, which can be applied to both flat surfaces as well as particulate dispersions. The basic principle of neutron scattering is to measure the scattering due to the adsorbed layer, when the scattering length density of the particle is matched to that of the medium (the so-called "contrast-matching" method). Contrast matching of particles and medium can be achieved by changing the isotopic composition of the system (using deuterated particles and mixture of D_2O and H_2O). It was used for measurement of the adsorbed layer thickness of polymers, e.g. PVA or poly(ethylene oxide) (PEO) on polystyrene latex [9]. Apart from obtaining δ, one can also determine the segment density distribution $\rho(z)$.

The above technique of neutron scattering clearly gives a quantitative picture of the adsorbed polymer layer. However, its application in practice is limited since one needs to prepare deuterated particles or polymers for the contrast matching procedure. The practical methods for determination of the adsorbed layer thickness are mostly based on hydrodynamic methods. Several methods may be applied to determine the hydrodynamic thickness of adsorbed polymer layers of which viscosity, sedimentation coefficient (using an ultracentrifuge) and dynamic light scattering measurements are the most convenient. A less accurate method is from zeta potential measurements.

The most rapid technique for measuring δ_h is photon correlation spectroscopy (PCS) (sometime referred to as quasi-elastic light scattering), which allows one to obtain the diffusion coefficient of the particles with and without the adsorbed layer ($D\delta$ and D respectively). This is obtained from measurement of the intensity fluctuation of scattered light as the particles undergo Brownian diffusion [10]. When a light beam (e.g. monochromatic laser beam) passes through a dispersion, an oscillating dipole is induced in the particles, thus re-radiating the light. Due to the random arrangement of the particles (which are separated by a distance comparable to the wavelength of the light beam, i.e. the light is coherent with the interparticle distance, the intensity of the scattered light will, at any instant, appear as random diffraction or a "speckle" pattern. As the particles undergo Brownian motion, the random configuration of the speckle pattern changes. The intensity at any one point in the pattern will, therefore, fluctuate such that the time taken for an intensity maximum to become a minimum (i.e. the coherence time) corresponds approximately to the time required for a particle to move one wavelength. Using a photomultiplier of active area about the size of a diffraction maximum, i.e. approximately one coherence area, this intensity fluctuation can be measured. A digital correlator is used to measure the photocount or intensity correlation function of the scattered light. The photocount correlation function can be used to obtain the diffusion coefficient D of the particles. For monodisperse non-interacting particles (i.e. at sufficient dilution), the normalized correlation function [g(1)(τ)] of the scattered electric field is given by the equation,

$$[g^{(1)}(\tau)] = \exp-(\Gamma\tau),\tag{5.3}$$

where τ is the correlation delay time and Γ is the decay rate or inverse coherence time. Γ is related to D by the equation,

$$\Gamma = DK^2,\tag{5.4}$$

where K is the magnitude of the scattering vector that is given by,

$$K = \left(\frac{4n}{\lambda_0}\right)\sin\left(\frac{\theta}{2}\right),\tag{5.5}$$

where n is the refractive index of the solution, λ is the wavelength of light in vacuum and θ is the scattering angle.

From D, the particle radius R is calculated using the Stokes–Einstein equation,

$$D = \frac{kT}{6\pi\eta R},\tag{5.6}$$

where k is the Boltzmann constant and T is the absolute temperature. For a polymer coated particle, R is denoted Rδ, which is equal to R + δ_h. Thus, by measuring D_δ and D, one can obtain δ_h. It should be mentioned that the accuracy of the PCS method depends on the ratio of δ_h/R, since δ_h is determined by difference. Since the accuracy of the measurement is +1 %, δ_h should be at least 10 % of the particle radius. This method can only be used with small particles and reasonably thick adsorbed layers.

5.4 Interaction between particles containing adsorbed surfactant layers

When two particles, each with a radius R and containing an adsorbed polymer layer with a hydrodynamic thickness δh, approach each other to a surface-surface separation distance h that is smaller than 2δh, the polymer layers interact with each other resulting in two main situations [11]: (i) the polymer chains may overlap with each other; (ii) the polymer layer may undergo some compression. In both cases, there will be an increase in the local segment density of the polymer chains in the interaction region. This is schematically illustrated in Fig. 5.5. The real situation is perhaps in between the above two cases, i.e. the polymer chains may undergo some interpenetration and some compression.

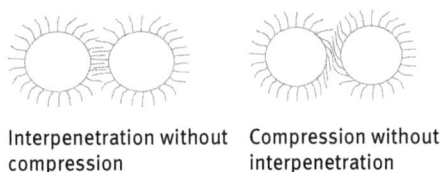

Interpenetration without compression   Compression without interpenetration

Fig. 5.5: Schematic representation of the interaction between particles containing adsorbed polymer layers.

Provided the dangling chains (the A chains in A-B, A-B-A block or BA_n graft copolymers) are in a good solvent, this local increase in segment density in the interaction zone will result in strong repulsion as a result of two main effects [11]: (i) An increase in the osmotic pressure in the overlap region as a result of the unfavourable mixing of the polymer chains, when these are in good solvent conditions. This is referred to as osmotic repulsion or mixing interaction and it is described by a free energy of interaction G_{mix}. (ii) Reduction of the configurational entropy of the chains in the interaction zone; this entropy reduction results from the decrease in the volume available for the chains when these are either overlapped or compressed. This is referred to as volume restriction interaction, entropic or elastic interaction and it is described by a free energy of interaction G_{elas}.

The combination of G_{mix} and G_{elas} is usually referred to as the steric interaction free energy, G_s, i.e.,

$$G_s = G_{mix} + G_{elas} . \tag{5.7}$$

The sign of G_{mix} depends on the solvency of the medium for the chains. If in a good solvent, i.e. the Flory–Huggins interaction parameter χ is less than 0.5, then G_{mix} is positive and the mixing interaction leads to repulsion (see below). In contrast, if $\chi > 0.5$ (i.e. the chains are in a poor solvent condition), G_{mix} is negative and the mixing interaction becomes attractive. G_{elas} is always positive and hence in some cases one can produce stable dispersions in a relatively poor solvent (enhanced steric stabilization).

As mentioned above, G_{mix} results from the unfavourable mixing of the polymer chains when these are in good solvent conditions. This is schematically shown in Fig. 5.6. Consider two spherical particles with the same radius and each containing an adsorbed polymer layer with thickness δ. Before overlap, one can define in each polymer layer a chemical potential for the solvent μ_i^α and a volume fraction for the polymer in the layer ϕ_2^α. In the overlap region (volume element dV), the chemical potential of the solvent is reduced to μ_i^β. This results from the increase in polymer segment concentration in this overlap region.

In the overlap region, the chemical potential of the polymer chains is now higher than in the rest of the layer (with no overlap). This amounts to an increase in the osmotic pressure in the overlap region; as a result solvent will diffuse from the bulk to the overlap region, thus separating the particles and hence a strong repulsive energy arises from this effect. The above repulsive energy can be calculated by considering the free energy of mixing of two polymer solutions, as for example treated by Flory and

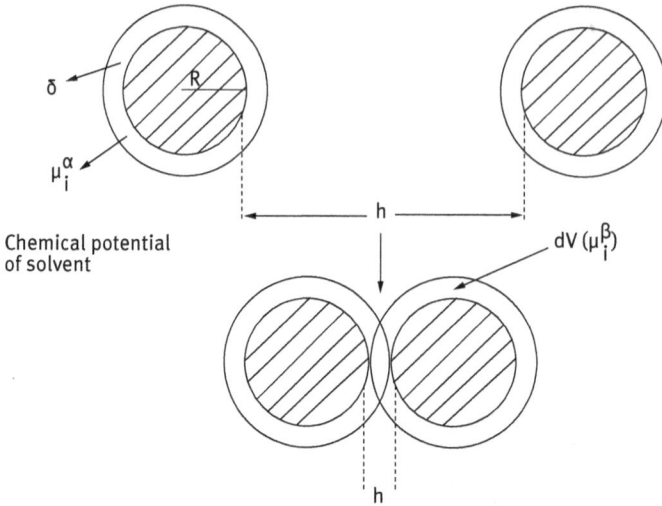

Fig. 5.6: Schematic representation of polymer layer overlap.

Krigbaum [12]. The free energy of mixing is given by two terms: (i) an entropy term that depends on the volume fraction of polymer and solvent and (ii) an energy term that is determined by the Flory–Huggins interaction parameter,

where n_1 and n_2 are the number of moles of solvent and polymer with volume fractions ϕ_1 and ϕ_2, k is the Boltzmann constant and T is the absolute temperature.

$$\delta(G_{mix}) = kT(n_1 \ln \phi_1 + n_2 \ln \phi_2 + \chi n_1 \phi_2). \tag{5.8}$$

The total change in free energy of mixing for the whole interaction zone, V, is obtained by summing over all the elements in V,

$$G_{mix} = \frac{2kTV_2^2}{V_1} v_2 \left(\frac{1}{2} - \chi\right) R_{mix}(h) \tag{5.9}$$

where V_1 and V_2 are the molar volumes of solvent and polymer respectively, v_2 is the number of chains per unit area and $R_{mix}(h)$ is a geometric function that depends on the form of the segment density distribution of the chain normal to the surface, $\rho(z)$. k is the Boltzmann constant and T is the absolute temperature.

Using the above theory one can derive an expression for the free energy of mixing of two polymer layers (assuming a uniform segment density distribution in each layer) surrounding two spherical particles as a function of the separation distance h between the particles [13].

The expression for G_{mix} is,

$$\frac{G_{mix}}{kT} = \left(\frac{2V_2^2}{V_1}\right) v_2 \left(\frac{1}{2} - \chi\right) \left(\delta - \frac{h}{2}\right)^2 \left(3R + 2\delta + \frac{h}{2}\right). \tag{5.10}$$

The sign of G_{mix} depends on the value of the Flory–Huggins interaction parameter χ: if $\chi < 0.5$, G_{mix} is positive and the interaction is repulsive; if $\chi > 0.5$, G_{mix} is negative and the interaction is attractive. The condition $\chi = 0.5$ and $G_{mix} = 0$ is termed the θ-condition. This corresponds to the case where the polymer mixing behaves as ideal, i.e. mixing of the chains does not lead to an increase or decrease in the free energy.

Elastic interaction results from the loss in configurational entropy of the chains on the approach of a second particle. As a result of this approach, the volume available for the chains becomes restricted, resulting in loss of the number of configurations. This can be illustrated by considering a simple molecule, represented by a rod that rotates freely in a hemisphere across a surface (Fig. 5.7). When the two surfaces are separated by an infinite distance ∞, the number of configurations of the rod is $\Omega(\infty)$, which is proportional to the volume of the hemisphere. When a second particle approaches to a distance h such that it cuts the hemisphere (losing some volume), the volume available to the chains is reduced and the number of configurations become $\Omega(h)$ which is less than $\Omega(\infty)$. For two flat plates, G_{elas} is given by the following expression [14],

$$\frac{G_{elas}}{kT} = -2v_2 \ln \left[\frac{\Omega(h)}{\Omega(\infty)} \right] = -2v_2 R_{elas}(h), \tag{5.11}$$

where $R_{elas}(h)$ is a geometric function whose form depends on the segment density distribution. It should be stressed that G_{elas} is always positive and could play a major role in steric stabilization. It becomes very strong when the separation distance between the particles becomes comparable to the adsorbed layer thickness δ.

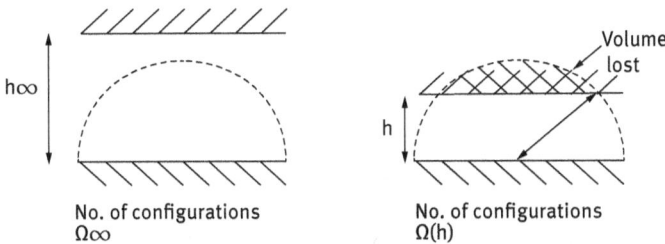

$h\infty$

No. of configurations $\Omega\infty$

Volume lost

h

No. of configurations $\Omega(h)$

Fig. 5.7: Schematic representation of configurational entropy loss on approach of a second particle.

Combining G_{mix} and G_{elas} with G_A gives the total energy of interaction G_T (assuming there is no contribution from any residual electrostatic interaction), i.e. [15],

$$G_T = G_{mix} + G_{elas} + G_A. \tag{5.12}$$

A schematic representation of the variation of G_{mix}, G_{elas}, G_A and G_T with surface-surface separation distance h is shown in Fig. 5.8. G_{mix} increases very sharply with decreasing h; when $h < 2\delta$, G_{elas} increases very sharply with decreasing h; when $h < \delta$, G_T versus h shows a minimum, G_{min}, at separation distances comparable to

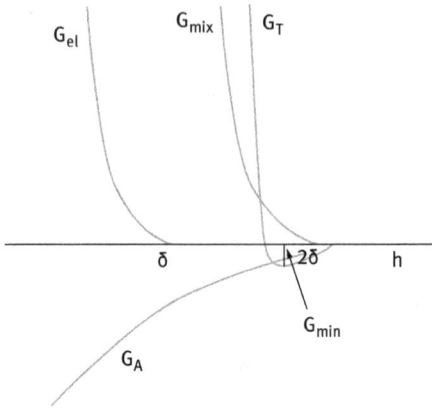

Fig. 5.8: Energy-distance curves for sterically stabilized systems.

2δ. When $h < 2\delta$, G_T shows a rapid increase with decreasing h. The depth of the minimum depends on the Hamaker constant A, the particle radius R and adsorbed layer thickness δ. G_{min} increases with increasing A and R. At a given A and R, G_{min} decreases with increasing δ (i.e. with increasing molecular weight, M_w, of the stabilizer). This is illustrated in Fig. 5.9 which shows the energy-distance curves as a function of δ/R. The larger the value of δ/R, the smaller the value of G_{min}. In this case the system may approach thermodynamic stability as is the case with nanodispersions.

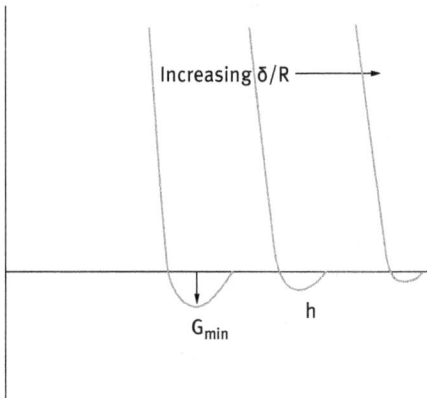

Fig. 5.9: Variation of G_{min} with δ/R.

For effective steric stabilization, the following criteria must be satisfied: (i) The particles should be completely covered by the polymer (the amount of polymer should correspond to the plateau value). Any bare patches may cause flocculation either by van der Waals attraction (between the bare patches) or by bridging flocculation (whereby a polymer molecule will become simultaneously adsorbed on two or more particles). (ii) The polymer should be strongly "anchored" to the particle surfaces, to prevent any displacement during particle approach. For this purpose A-B, A-B-A block and

BA$_n$ graft copolymers are the most suitable where the chain B is chosen to be highly insoluble in the medium and has a strong affinity to the surface. Examples of B groups for hydrophobic particles in aqueous media are polystyrene and polymethylmethacrylate. (iii) The stabilizing chain A should be highly soluble in the medium and strongly solvated by its molecules. Examples of A chains in aqueous media are poly(ethylene oxide) and poly(vinyl alcohol). (iv) δ should be sufficiently large (> 5 nm) to prevent weak flocculation.

References

[1]   Tadros, Th. F., "Applied Surfactants", Wiley-VCH, Germany (2005).
[2]   Tadros, Th. F., "Polymer Adsorption and Dispersion Stability", in "The Effect of Polymers on Dispersion Properties", Th. F. Tadros (ed.), Academic Press, London (1981).
[3]   Tadros, Th. F., "Polymeric Surfactants", in "Encyclopedia of Colloid and Interface Science", Th. F. Tadros (ed.), Springer, Germany (2013).
[4]   Killmann, E., Eisenlauer, E. and Horn, M. J., Polymer Sci. Polymer Symposium, **61**, 413 (1977).
[5]   Fontana, B. J. and Thomas, J. R., J. Phys. Chem., **65**, 480 (1961).
[6]   Robb, I. D. and Smith, R., Eur. Polym. J., **10**, 1005 (1974).
[7]   Cohen-Stuart, M. A., Fleer, G. J. and Bijesterbosch, B., J. Colloid Interface Sci., **90**, 321 (1982).
[8]   Abeles, F., in "Ellipsometry in the Measurement of Surfaces and Thin Films", E. Passaglia, R. R. Stromberg and J. Kruger (eds.), Nat. Bur. Stand. Misc. Publ., **256**, 41 (1964).
[9]   Barnett, K. G., Cosgrove, T., Vincent, B., Burgess, A., Crowley, T. L., Kims, J., Turner, J. D. and Tadros, Th. F., Disc. Faraday Soc. **22**, 283 (1981).
[10] Pusey, P. N., in "Industrial Polymers: Characterisation by Molecular Weights", J. H. S. Green and R. Dietz (eds.), London, Transcripta Books (1973).
[11] Napper, D. H., "Polymeric Stabilisation of Colloidal Dispersions", Academic Press, London (1983).
[12] Flory, P. J. and Krigbaum, W. R., J. Chem. Phys. **18**, 1086 (1950).
[13] Fischer, E. W., Kolloid Z. **160**, 120 (1958).
[14] Mackor, E. L. and van der Waals, J. H., J. Colloid Sci., **7**, 535 (1951).
[15] Hesselink, F. Th., Vrij, A. and Overbeek, J. Th. G., J. Phys. Chem. **75**, 2094 (1971).

6 Flocculation of suspensions

6.1 Introduction

As mentioned in Chapter 4, DLVO theory [1, 2] predicts the process of aggregation on addition of electrolytes with different valency. Addition of electrolyte reduces the range of the repulsive component (due to compression of the electrical double layer) and this results in reduction of the energy maximum, G_{max}. This was illustrated in Fig. 4.21 of Chapter 4, which shows the effect of addition of 1:1 electrolyte on the energy-distance curves. At very low electrolyte concentration of 10^{-7} mol dm^{-3} (corresponding to a double layer thickness of 1000 nm), the energy maximum is very high (much higher than 100 kT) and that prevents any close approach of the particles. In this case the particles remain dispersed for a very long period of time (some years). By increasing the electrolyte concentration to 10^{-5} mol dm^{-3} (corresponding to a double layer thickness of 100 nm), the energy maximum is still high (> 100 kT) and this prevents any aggregation of the particles. On increasing the electrolyte concentration to 10^{-3} mol dm^{-3} (corresponding to a double layer thickness of 10 nm), the energy maximum is reduced but still high enough (> 25 kT) to prevent aggregation. However, when the electrolyte concentration is increased to 10^{-1} mol dm^{-3} (corresponding to a double layer thickness of 1 nm), the energy maximum disappears and the energy-distance curve becomes attractive at all separation distances. In this case, the dispersion shows rapid coagulation and the particles in the aggregates are strongly bound to each other.

Another factor that affects the electrostatic repulsion is the magnitude of the surface or zeta potential. As discussed in Chapter 4, G_{elec} is proportional to the square of the surface or zeta potential. As an illustration, Fig. 6.1 shows calculations of the energy-distance curves for polystyrene latex particles of 500 nm radius at various NaCl concentrations and zeta potential [3].

It can be seen that a high energy maximum is obtained at 10^{-2} mol dm^{-3} NaCl and ζ-potential of -50 mV. When the ζ-potential is reduced to -20 mV while keeping the NaCl concentration the same, the maximum disappears. Also, at higher NaCl concentration of 4×10^{-1} mol dm^{-3} the maximum disappears even when the ζ-potential is increased to -30 mV.

Sterically stabilized suspensions can also undergo flocculation by four different mechanisms: (i) Weak flocculation when the thickness of the adsorbed layer δ becomes small enough to produce a deep minimum (few kT units, where k is the Boltzmann constant and T is the absolute temperature) that is sufficient to cause particle-particle aggregation. The flocs produced are weak and they can be broken by shaking. In addition, the process is reversible in the sense that on standing the floc structure is formed again. (ii) Incipient flocculation when the stabilizing chains are in poor solvent, with a Flory–Huggins interaction parameter > 0.5 (worse than a θ-solvent). This flocculation is catastrophic and in concentrated suspensions can be irreversible.

DOI 10.1515/9783110486872-007

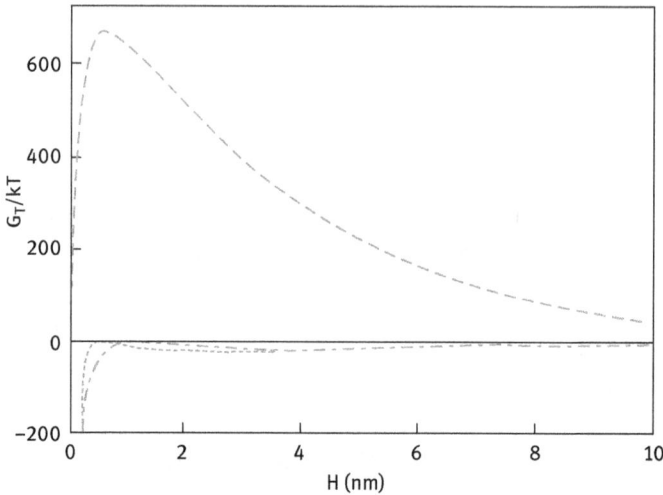

Fig. 6.1: Energy-distance curves for 500 nm radius polystyrene latex particles:
(—) 10^{-2} mol dm^{-3} NaCl and ζ-potential $= -50$ mV; ($\cdots\cdots$) 10^{-2} mol dm^{-3} NaCl and ζ-potential $= -20$ mV; ($\cdots$) 4×10^{-1} mol dm^{-3} NaCl and ζ-potential $= -30$ mV.

(iii) Bridging flocculation when there is insufficient polymer to saturate the surface. In this case the polymer chain can become simultaneously adsorbed on two or more particles. (iv) Depletion flocculation induced by addition of "free" (nonadsorbing) polymer. When the concentration of the free polymer exceeds a critical value, the free polymer chains become squeezed out from between the particles. In this case the osmotic pressure outside the particle surfaces becomes higher than that of the "free" polymer zone in between the particles. This results in attraction between the particles that form flocs.

Since the mechanisms of flocculation of electrostatically and sterically stabilized suspensions are very different, the processes will be described separately.

6.2 Kinetics of flocculation of electrostatically stabilized suspensions

6.2.1 Diffusion limited aggregation (fast flocculation kinetics)

Fast flocculation kinetics represents the case where no energy barrier exists and hence the process becomes diffusion controlled. This process was treated by Smoluchowki [4], who modelled the system as that of diffusing spherical particles that stick on collision, but the pair potential is zero up to this contact [5, 6]. If r is the centre-to-centre distance between a reference spherical particle with radius R and an approaching particle with the same radius R (for a monodisperse dispersion) then contact occurs when $r = 2R$. As both particles are diffusing, the net diffusion coefficient is equal to $2D \, m^2 \, s^{-1}$ [2]. The net velocity of the incoming particle is therefore $2D/R \, m \, s^{-1}$. The surface area of the "collision sphere" is $4\pi(2R)^2$. The flux resulting from Brownian diffusion J_B through the collision sphere, if there are initially no particles per unit volume, is

$$J_B = n_0 \frac{2D}{R} 4\pi(2R)^2 .$$ (6.1)

Since the above process is occurring with each particle, the collision frequency due to Brownian diffusion is,

$$c_B = \frac{n_0 J_B}{2} .$$ (6.2)

The factor of 2 is introduced to prevent double counting. The diffusion coefficient is given by the Stokes–Einstein equation,

$$D = \frac{kT}{6\pi\eta_0 R} ,$$ (6.3)

where η_0 is the viscosity of the medium.

Combining equations (6.1)–(6.3),

$$c_B = n_0^2 \frac{8kT}{3\eta_0} .$$ (6.4)

As each collision results in coagulation, the initial coagulation rate is given by,

$$-\frac{dn_0}{dt} = n_0^2 \frac{8kT}{3\eta_0} .$$ (6.5)

The half-life, $t_{1/2}$, for the rapid coagulation rate is determined for this second-order rate equation as,

$$t_{1/2} = \frac{3\eta_0}{4kT n_0} .$$ (6.6)

A simple analysis of fast flocculation kinetics is to consider the process to be represented by second-order kinetics and the process is simply diffusion controlled. The number of particles n at any time t may be related to the initial number (at $t = 0$) n_0 by the following expression,

$$n = \frac{n_0}{1 + k_0 n_0 t} ,$$ (6.7)

where k_0 is the rate constant for fast flocculation that is related to the diffusion coefficient of the particles D, i.e.,

$$k_0 = 8\pi DR .$$ (6.8)

D is given by the Stokes–Einstein equation (6.7).

Combining equations (6.3) and (6.8),

$$k_0 = \frac{4}{3} \frac{kT}{\eta_0} = 5.5 \times 10^{-18} \, m^3 \, s^{-1} \text{ for water at } 25 \, °C.$$ (6.9)

Equation (6.9) shows that the rate constant for flocculation is directly proportional to the temperature and inversely proportional to the viscosity of the medium. It should also be mentioned that the viscosity of the medium decreases with increasing temperature, which means that the overall effect of a temperature increase will be an increase in the rate constant.

The half-life $t_{1/2}$ ($n = (1/2)n_0$) can be calculated at various n_0 or volume fractions ϕ as given in Tab. 6.1.

Tab. 6.1: Half-life of suspension flocculation.

R / μm	ϕ			
	10^{-5}	10^{-2}	10^{-1}	5×10^{-1}
0.1	765 s	76 ms	7.6 ms	1.5 ms
1.0	21 h	76 s	7.6 s	1.5 s
10.0	4 months	21 h	2 h	25 min

6.2.2 Potential limited aggregation (slow flocculation kinetics)

Slow flocculation kinetics was treated by Fuchs [7] who considered the effect of the presence of an energy barrier. In this case, the pair potential slows the approach of two particles. At any distance, the fraction of particles with thermal energy in excess of the potential at that distance is given by the Boltzmann factor: $\exp(-G_T/kT)$. The flux through successive spherical shells as the particles approach is slowed from the simple collision case and only a fraction of the particles that encounter one another approach close enough to stick. The fraction of encounters that stick is $1/W$, where W is known as the stability ratio.

$$W = \frac{k_0}{k}.$$ (6.10)

W can be expressed as the ratio of the two fluxes,

$$W = 2R \int_{2R}^{\infty} \exp\left(\frac{G_T}{kT}\right) \frac{dr}{r^2}.$$ (6.11)

Reerink and Overbeek [8] pointed out that the maximum in the pair potential, G_{max}, is the dominant factor in restricting the approach of particles and they showed a useful approximation to the integral of equation (6.11),

$$W \approx \frac{1}{2\kappa R} \exp\left(\frac{G_{max}}{kT}\right).$$ (6.12)

Since G_{max} is determined by the salt concentration C and valency Z, one can derive an expression relating W to C and Z [8],

$$\log W = -2.06 \times 10^9 \left(\frac{R\gamma^2}{Z^2} \right) \log C, \tag{6.13}$$

where γ is a function that is determined by the surface potential ψ_0,

$$\gamma = \left[\frac{\exp(Ze\psi_0/kT) - 1}{\exp(Ze\psi_0/kT) + 1} \right]. \tag{6.14}$$

Plots of $\log W$ versus $\log C$ are shown in Fig. 6.2. The condition $\log W = 0$ ($W = 1$) is the onset of fast flocculation. The electrolyte concentration at this point defines the critical flocculation concentration, ccc. Above the ccc, $W < 1$ (due to the contribution of van der Waals attraction which accelerates the rate above the Smoluchowski value). Below the ccc, $W > 1$ and it increases with decreasing electrolyte concentration. Figure 6.2 also shows that the ccc decreases with increasing valency. At low surface potentials, ccc $\propto 1/Z^2$. This referred to as the Schultze–Hardy rule.

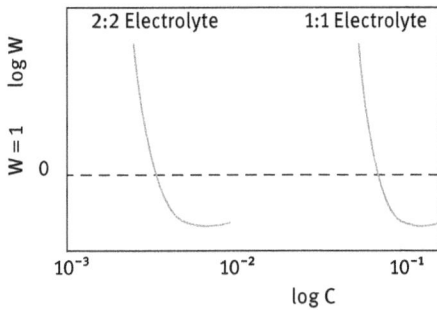

Fig. 6.2: log W–log C curves.

6.2.3 Weak (reversible) flocculation

Another mechanism of flocculation is that involving the secondary minimum (G_{min}) which is few kT units. In this case flocculation is weak and reversible and hence one must consider both the rate of flocculation (forward rate k_f) and deflocculation (backward rate k_b). In this case the rate or decrease of particle number with time is given by the expression,

$$-\frac{dn}{dt} = -k_f n^2 + k_b n. \tag{6.15}$$

The backward reaction (break-up of weak flocs) reduces the overall rate of flocculation.

6.2.4 Orthokinetic flocculation

This process of flocculation occurs under shearing conditions and is referred to as orthokinetic (to distinguish it from the diffusion controlled perikinetic process). The simplest analysis is for laminar flow, since for turbulent flow with chaotic vortices (as is the case in a high speed mixer) the particles are subjected to a wide and unpredictable range of hydrodynamic forces. For laminar flow, the particle will move at the velocity of the liquid at the plane coincident with the centre of the particle, v_p. In this case the total collision frequency due to flow, c_f, is given by the following expression,

$$c_f = \frac{16}{3} n_p^2 R^3 \left(\frac{dv}{dx}\right).$$

(6.16)

As the particles approach in the shear field, the hydrodynamic interactions cause the colliding pair to rotate and with the combination of the slowing approach due to liquid drainage (lubrication stress) and Brownian motion, not all collisions will lead to aggregation. Equation (6.16) must be reduced by a factor α (the collision frequency) to account for this,

$$c_f = \alpha \frac{16}{3} n_p^2 R^3 \left(\frac{dv}{dx}\right).$$

(6.17)

The collision frequency α is of the order 1 and a typical value would be $\alpha \approx 0.8$.
   (dv/dx) is the shear rate so that equation (6.17) can be written as,

$$c_f = \alpha \frac{16}{3} n_p^2 R^3 \dot{\gamma}.$$

(6.18)

And the rate of orthokinetic flocculation is given by,

$$-\frac{dn}{dt} = \alpha \frac{16}{3} n_p^2 R^3 \dot{\gamma}.$$

(6.19)

A comparison can be made between the collision frequency or rate of orthokinetic and perikinetic flocculation by comparing equations (6.19) and (6.8),

$$\frac{c_f}{c_B} = \frac{2\alpha \eta_0 R^3 \dot{\gamma}}{kT}.$$

(6.20)

If the particles are dispersed in water at a temperature of 25 °C, the ratio in equation (6.20) becomes,

$$\frac{c_f}{c_B} \approx 4 \times 10^{17} R^3 \dot{\gamma}.$$

(6.21)

When a liquid is stirred in a beaker using a rod, the velocity gradient r shear rate is in the range 1–10 s^{-1}, with a mechanical stirrer it is about 100 s^{-1} and at the tip of a turbine in a large reactor it can reach values as high as 1000–10 000 s^{-1}. This means that the particle radius R must be less than 1 µm if even slow mixing can be disregarded. This shows how the effect of shear can increase the rate of aggregation.
   It should be mentioned that the above analysis is for the case where there is no energy barrier, i.e. the Smoluchowski case [4]. In the presence of an energy barrier, i.e.

potential limited aggregation, one must consider the contribution due to the hydrodynamic forces acting on the colliding pair [6]. Figure 6.3 shows the forces acting on a
collision doublet in simple shear [3]. The figure shows the trajectory with the points at
which maximum compression and tension occurs, i.e. at an angle $\theta = 45°$ to the shear
plane. The particles have the same radius R and the reference particle is at $z = 0$.

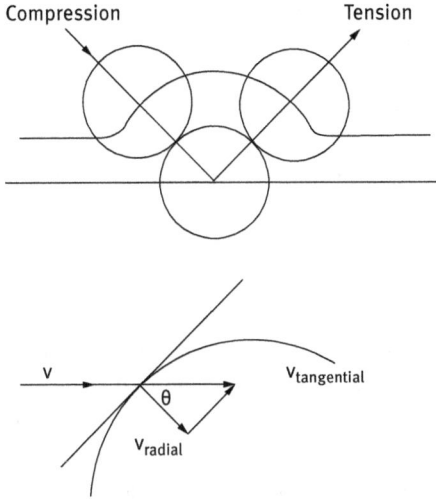

Fig. 6.3: Schematic representation of the geometry of a colliding pair of particles with
maximum compression and tension at $\theta = 45°$
to the shear plane.

The velocity of the streamline coincident with the centre of colliding particle, v, at the
orientation giving the maximum force is,

$$v = \dot{\gamma} 2R \sin(45). \tag{6.22}$$

The radial component of the Stokes drag force on the particle is given by,

$$F_h = 6\pi\eta_0 R v_{radial} = 6\pi\eta_0 R \cos(45). \tag{6.23}$$

F_h can be written as,

$$F_h = \pm\dot{\gamma} 6\pi\eta_0 2R^2 \sin(45) \cos(45), \tag{6.24a}$$
$$F_h = \pm\dot{\gamma} 6\pi\eta_0 2R^2, \tag{6.24b}$$

where the sign $\pm$ indicates compression (+) or tension (−). The trajectory would be
altered by the colloidal forces on approach, i.e. whether there is net repulsion or attraction. Equation (6.24b) can be used to indicate where the stability or instability
boundaries are for a particular dispersion. This requires calculation of the interparticle force at the maximum and minimum points on the force-distance curve. Some
calculations were made by Goodwin [3] for polystyrene latex particles with a radius of
500 nm in the presence of 50 mM 1 : 1 electrolyte and a zeta potential of −40 mV. F_T is

given by,

$$F_T = 2\pi\varepsilon_r\varepsilon_0\kappa R\psi_d \frac{\exp(-\kappa H)}{1 + \exp(-\kappa H)} - \frac{A_{11}R}{12H^2} . \tag{6.25}$$

The value of κR is 368 and therefore the interparticle forces change in the region very close to the particle surface. This indicates that hydrodynamics control the trajectories until the particles are very close to each other. The force-distance curve is shown in Fig. 6.4, where ζ is assumed to be equal to ψ_d.

Figure 6.4 shows a force maximum F_{max} of 3.96×10^{-10} N and a force minimum F_{min} of -3.8×10^{-11} N. The stability boundaries are calculated as a function of shear rate as illustrated in Fig. 6.5. This clearly shows the change in the aggregation state at different values of ζ-potential [3].

Several features can be identified from the stability map shown in Fig. 6.5. At ζ-potentials less than -20 mV, the dispersion is coagulated at all shear rates. With a small increase in the ζ-potential above -20 mV, the dispersion shows weak flocculation (secondary minimum aggregation) at low shear rates, but at high shear rates of the order of 10^5 s^{-1}, the hydrodynamic forces are sufficient to cause the dispersion to form doublets which are coagulated. This shear-induced coagulation is referred to as orthokinetic flocculation as discussed above. It is interesting to note that the shear forces on this particle size, ionic strength and diffusion potential combination will only break down the doublets flocculated in the secondary minimum when the shear rates exceed 10^3 s^{-1}. Although such high shear rates are readily attainable in a viscometer, they would require a high stirrer speed when using a paddle stirrer. Clearly

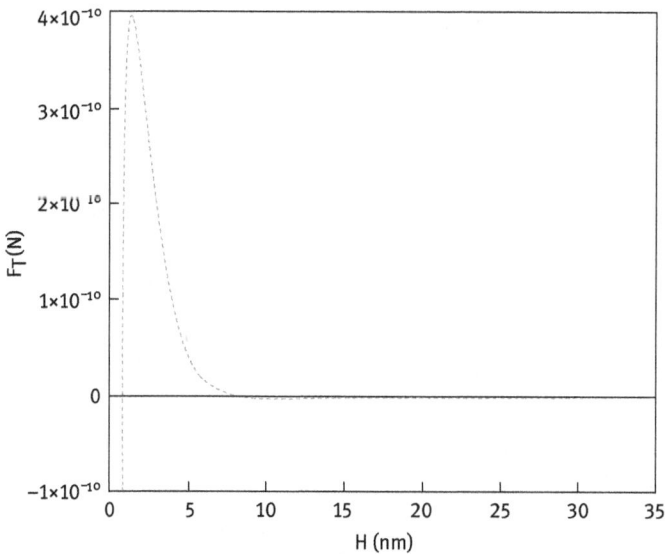

Fig. 6.4: Force-distance curve for 500 nm polystyrene particles at 50 mM 1:1 electrolyte and ζ-potential of -40 mV [3].

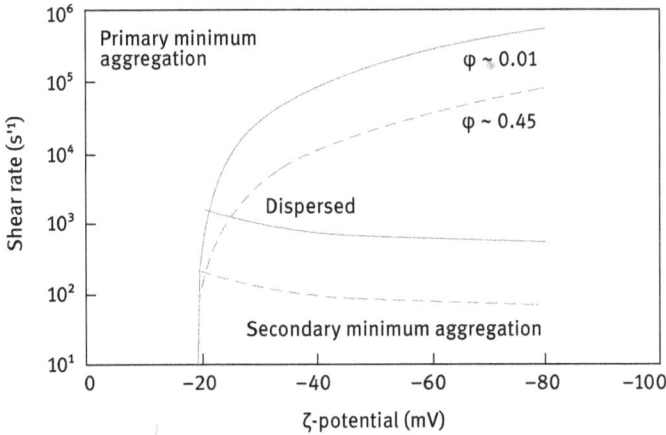

Fig. 6.5: Stability map for polystyrene latex dispersions (R = 500 nm and 1:1 electrolyte concentration of 50 mM) as a function of ζ-potential and shear rate.

such shear rates are easily achieved in pumps and large reactors with turbine mixers. Such shear rates can also be achieved when using rotor-stator mixers (such as Ultra-Turrax and Silverson mixers). Equations (6.24a) and (6.24b) show that the shear forces increase as the square of the particle radius and so the stability boundaries drop rapidly with increasing particle size as the colloidal forces change more slowly with radius than the hydrodynamic forces. Thus particles with a radius of 3–4 μm are much more sensitive to shear-induced aggregation than particles with an order of magnitude lower in radius (0.3–0.4 μm).

The effect of increasing the volume fraction of the dispersion, ϕ, as is the case with most practical systems, has a large impact on the shear-induced aggregation as clearly illustrated in Fig. 6.5, which shows the stability map as ϕ is increased from 0.01 to 0.45. When dealing with concentrated dispersions, one must consider the "multibody" hydrodynamic forces. Using a mean field approximation, Krieger and Dougherty [9, 10] related the viscosity of the suspension η to that of the medium η_0 by the following semi-empirical equation,

$$\eta = \eta_0 \left(1 - \frac{\phi}{\phi_p} \right)^{-[\eta]\phi_p}, \tag{6.26}$$

where ϕ_p is the maximum packing fraction, that is ≈ 0.605 (for random packing), and $[\eta]$ is the intrinsic viscosity that is equal to 2.5 for hard spheres. Equation (6.26) becomes,

$$\eta = \eta_0 \left(1 - \frac{\phi}{0.605} \right)^{-1.513}. \tag{6.27}$$

The stability boundaries for $\phi = 0.45$ are shown in Fig. 6.5 (dashed line) and this clearly shows the drop resulting from the increase in the viscous forces at high volume fraction as predicted by equation (6.27). The boundaries drop in proportion to the viscous forces as expected, so it is easier to break up flocculated pairs with an applied shear field. This means that a larger fraction of the stable area that is occupied by the particles that can be considered "dispersed" occurs at lower shear rates when compared with the case of dilute dispersions. The range of stability under shear at moderate ζ-potentials is reduced with increasing the solid volume fraction.

6.2.5 Aggregate structure

When dealing with aggregated systems, the floc structure plays an important role in applications. For example, the rheological properties change dramatically so that handling can become very difficult. A good example is the case of "dewatering" of suspensions by filtration. Filtration may start as an easy separation process once the system is aggregated, but the final "dewatering" is limited, so subsequent drying can be a slow and expensive process. In this case, a weakly aggregated structure would be a preferable situation so that collapse of the filter cake to high solids density can be achieved. In the case of ceramic pastes, the rheology of the weakly aggregated open structures is excellent for shape formation with minimum elastic recovery after yield has occurred at moderate to high stress. However, this open structure results in considerable shrinkage on drying and firing.

The mode of aggregation that occurs in the absence diffusion limited aggregation of a barrier can result in the formation of an open-dendritic or fractal type of structure. In this case, the particles collide and stick as they diffuse. Computer models generate this type of open branched structure and some careful experiments have confirmed these models [6]. As these aggregates grow by accretion of "stick" particles, they grow into each other and span the available space [3]. This point is referred to as the "percolation threshold". At higher concentrations, denser structures result and these are more difficult to define by a single parameter such as the "fractal dimension". These structures are modified in practice by addition of coagulants and application of shear during mixing. The shear forces on these large and fragile structures compact them to relatively high densities [11]. In some cases, systems of monodisperse particles can be compacted by shearing the coagulating system to random packing densities $\phi \approx 0.64$. These strongly aggregated systems are "metastable" structures. The lowest energy configuration would be a very dense unit with the maximum number and/or area of contacts. However, the fractal structure of a dilute strongly aggregated system would be very long lived in the absence of external forces since $(G_{min})_{primary} \gg kT$ and densification purely by diffusive motion would be imperceptibly slow.

The structures obtained depend on the processing and strength of the attractive interaction. The latter can be controlled by addition of materials to the surface prior to coagulation, e.g. by adding nonionic surfactants or polymers which provide a steric barrier thus limiting the aggregation to weak flocculation [12].

6.3 Flocculation of sterically stabilized dispersions

6.3.1 Weak flocculation

Weak flocculation occurs when the thickness of the adsorbed layer is small (usually < 5 nm), particularly when the particle radius and Hamaker constant are large. The minimum depth required for causing weak flocculation depends on the volume fraction of the suspension. The higher the volume fraction, the lower the minimum depth required for weak flocculation. This can be understood if one considers the free energy of flocculation that consists of two terms, an energy term determined by the depth of the minimum (G_{min}) and an entropy term that is determined by reduction in configurational entropy on aggregation of particles,

$$\Delta G_{flocc} = \Delta H_{flocc} - T\Delta S_{flocc} . \tag{6.28}$$

With dilute suspensions, the entropy loss on flocculation is larger than with concentrated suspensions. Hence for flocculation of a dilute suspension, a higher energy minimum is required when compared to the case with concentrated suspensions.

The above flocculation is weak and reversible, i.e. on shaking the container redispersion of the suspension occurs. On standing, the dispersed particles aggregate to form a weak "gel". This process (referred to as sol ↔ gel transformation) leads to reversible time dependence of viscosity (thixotropy). On shearing the suspension, the viscosity decreases and when the shear is removed, the viscosity is recovered. This phenomenon is applied in paints. On application of the paint (by a brush or roller), the gel is fluidized, allowing uniform coating of the paint. When shearing is stopped, the paint film recovers its viscosity and this avoids any dripping.

6.3.2 Incipient flocculation

Incipient flocculation occurs when the solvency of the medium is reduced to become worse than θ-solvent (i.e. $\chi > 0.5$). This reduction in solvency can be induced by temperature changes [12–15] or addition of a nonsolvent [12–15] for the stabilizing chain. When the solvency is reduced, the dispersion often exhibits a sharp transition from long-term stability to fast flocculation. This process of incipient flocculation is, for example, observed when a dispersion stabilized by poly(ethylene oxide) moieties is heated. Over a few degrees temperature rise, the turbidity of the dispersion rises sharply, indicating excessive flocculation. Flocculation can also occur by the addition of a nonsolvent, e.g. by addition of ethanol to polymethylmethacrylate dispersion stabilized by poly(hydroxystearic) acid in a hydrocarbon solvent [12–15]. The critical point at which flocculation is first observed is referred to as the critical flocculation temperature (CFT) or critical flocculation concentration of the added nonsolvent (CFV).

An illustration of incipient flocculation is given in Fig. 6.6, where χ was increased from < 0.5 (good solvent) to > 0.5 (poor solvent). One of the characteristic features of sterically stabilized systems, which distinguish them from electrostatically stabilized dispersions, is the temperature dependence of stability. Indeed, some dispersions flocculate on heating [12–15]; others flocculate on cooling [12–15]. Furthermore, in some cases dispersions can be produced which do not flocculate at any accessible temperature, whilst some sterically stabilized systems have been found to flocculate both on heating and cooling. This temperature dependence led Napper [13] to describe stability in terms of the thermodynamic process that governs stabilization. Thus, the temperature dependence of the Gibbs free energy of interaction (ΔG_R) for two sterically stabilized particles or droplets is given by,

$$\frac{\partial \Delta G_R}{\partial T} = -\Delta S_R \,, \tag{6.29}$$

where ΔS_R is the corresponding entropy change.

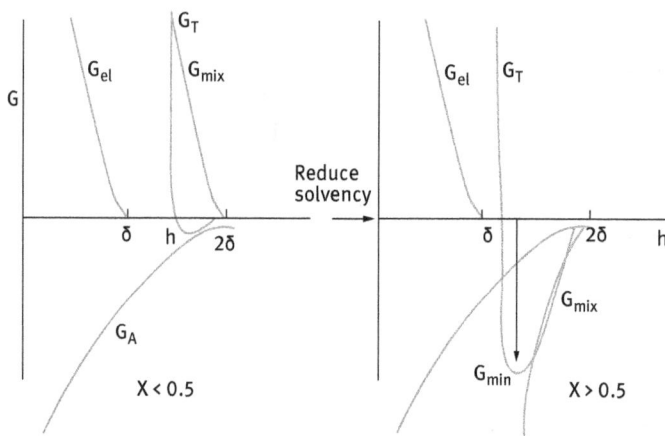

Fig. 6.6: Influence of reduction in solvency on the energy-distance curve.

In passing from the stability to the instability domain, ΔG_R must change sign, i.e. from being positive to being negative. It is convenient to split ΔG_R into its enthalpy and entropy contributions,

$$\Delta G_R = \Delta H_R - T\Delta S_R \,. \tag{6.30}$$

Thus the sign of ΔG_R will depend on the signs and relative magnitudes of ΔH_R and ΔS_R as summarized in Tab. 6.2.

As a result of extensive investigations on model sterically stabilized dispersions, it has been demonstrated that a strong correlation exists between the critical flocculation point and the θ-point of the stabilizing moieties in free solution. As mentioned before, the θ-point is that at which $\chi = 0.5$, i.e. the point at which the second virial

Tab. 6.2: Types of steric stabilization.

ΔH_R	ΔS_R	$\Delta H_R/T\Delta S_R$	ΔG_R	Type	Flocculation
+	+	> 1	+	Enthalpic	On Heating
−	−	< 1	+	Entropic	On cooling
+	−	≥	+	Combined	Not accessible
		<		Enthalpic = Entropic	

coefficient of the polymer chain is equal to zero. The absolute methods for the determination of the θ-point include light scattering and osmotic pressure measurements. Less sound methods for determining the θ-point depend on establishing the phase diagrams of polymer solutions.

Using lattices with terminally-anchored polymer chains of various kinds, it has been established that the CFT is independent of the molar mass of the chain, the size of the particle core and the nature of the disperse phase [13]. The CFT correlates strongly with the θ-temperature. Similar correlations have been found between the CFV and the θ-point [13]. However, such correlations are only obtained if the surface is fully covered by the polymer chains. Under conditions of incomplete coverage, flocculation occurs in dispersion media that are better than θ-solvents. This may be due to lateral movement of the stabilizer, desorption or even bridging flocculation (see below).

The correlation between the critical flocculation point and the θ-point implies that G_{mix} dominates the steric interaction. It has been argued that the contribution from G_{el} can be neglected until $h < \delta$, i.e. the polymer layer from one particle comes into direct contact with the second interface. With the high molar mass chains, the contribution from G_A to the total interaction is also negligible. This means that G_T is approximately equal to G_{mix}, which shows that χ is the main parameter controlling the stability. This is clearly illustrated in Fig. 6.6, which shows a significant value of G_{min} when $\chi > 0.5$.

Thus, by measuring the θ-point (CFT or CFV) for the polymer chains (A) in the medium under investigation (which could be obtained from light scattering or viscosity measurements) one can establish the stability conditions for a dispersion before its preparation. This procedure also helps in designing effective steric stabilizers, such as block and graft copolymers.

6.3.3 Depletion flocculation

Depletion flocculation is produced by addition of "free" nonadsorbing polymer [15]. In this case, the polymer coils cannot approach the particles to a distance Δ (that is determined by the radius of gyration of free polymer R_G), since the reduction of entropy on close approach of the polymer coils is not compensated by an adsorption energy. The suspension particles will be surrounded by a depletion zone with thickness Δ. Above a critical volume fraction of the free polymer, ϕ_p^+, the polymer coils are "squeezed out" from between the particles and the depletion zones begin to interact. The interstices between the particles are now free from polymer coils and hence an osmotic pressure is exerted outside the particle surface (the osmotic pressure outside is higher than in between the particles) resulting in weak flocculation [15]. A schematic representation of depletion flocculation is shown in Fig. 6.7.

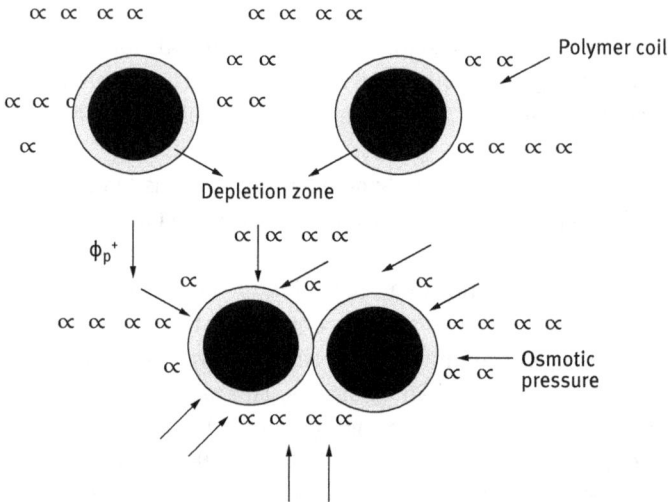

Fig. 6.7: Schematic representation of depletion flocculation.

The magnitude of the depletion attraction free energy, G_{dep}, is proportional to the osmotic pressure of the polymer solution, which in turn is determined by ϕp and molecular weight M. The range of depletion attraction is proportional to the thickness of the depletion zone, Δ, which is roughly equal to the radius of gyration, R_G, of the free polymer. A simple expression for G_{dep} is [15],

$$G_{dep} = \frac{2\pi R\Delta^2}{V_1}(\mu_1 - \mu_1^0)\left(1 + \frac{2\Delta}{R}\right),$$ (6.31)

where V_1 is the molar volume of the solvent, μ_1 is the chemical potential of the solvent in the presence of free polymer with volume fraction ϕp and μ_1^0 is the chemical potential of the solvent in the absence of free polymer. $(\mu_1 - \mu_1^0)$ is proportional to the osmotic pressure of the polymer solution.

6.3.4 Bridging flocculation by polymers and polyelectrolytes

Certain long-chain polymers may adsorb in such a way that different segments of the same polymer chain are adsorbed on different particles, thus binding or "bridging" the particles together, despite the electrical repulsion [16, 17]. With polyelectrolytes of opposite charge to the particles, another possibility exists; the particle charge may be partly or completely neutralized by the adsorbed polyelectrolyte, thus reducing or eliminating the electrical repulsion and destabilizing the particles.

Effective flocculants are usually linear polymers, often of high molecular weight, which may be nonionic, anionic or cationic in character. Ionic polymers should be strictly referred to as polyelectrolytes. The most important properties are molecular weight and charge density. There are several polymeric flocculants that are based on natural products, e.g. starch and alginates, but the most commonly used flocculants are synthetic polymers and polyelectrolytes, e.g. polyacrylamide and copolymers of acrylamide and a suitable cationic monomer such as dimethylaminoethyl acrylate or methacrylate. Other synthetic polymeric flocculants are poly(vinyl alcohol), poly(ethylene oxide) (nonionic), sodium polystyrene sulphonate (anionic) and polyethyleneimine (cationic).

As mentioned above, bridging flocculation occurs because segments of a polymer chain adsorb simultaneously on different particles thus linking them together. Adsorption is an essential step and this requires favourable interaction between the polymer segments and the particles. Several types of interactions are responsible for adsorption that is irreversible in nature: (i) electrostatic interaction when a polyelectrolyte adsorbs on a surface bearing oppositely charged ionic groups, e.g. adsorption of a cationic polyelectrolyte on a negative oxide surface such as silica; (ii) hydrophobic bonding that is responsible for adsorption of nonpolar segments on a hydrophobic surface, e.g. partially hydrolyzed poly(vinyl acetate) (PVA) on a hydrophobic surface such as polystyrene; (iii) hydrogen bonding, for example interaction of the amide group of polyacrylamide with hydroxyl groups on an oxide surface; (iv) ion binding as is the case with adsorption of anionic polyacrylamide on a negatively charged surface in the presence of Ca^{2+}.

Effective bridging flocculation requires that the adsorbed polymer extends far enough from the particle surface to attach to other particles and that there is sufficient free surface available for adsorption of these segments of extended chains. When excess polymer is adsorbed, the particles can be restabilized, either because of surface saturation or by steric stabilization as discussed before. This is one explanation of the fact that an "optimum dosage" of flocculant is often found; at low concentration there is insufficient polymer to provide adequate links and with larger amounts restabilization may occur. A schematic picture of bridging flocculation and restabilization by adsorbed polymer is given in Fig. 6.8.

If the fraction of particle surface covered by polymer is θ, then the fraction of uncovered surface is $(1 - \theta)$ and the successful bridging encounters between the par-

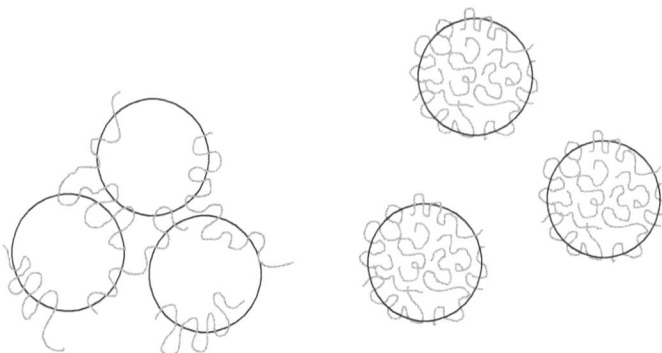

Fig. 6.8: Schematic illustration of bridging flocculation (left) and restabilization (right) by adsorbed polymer.

ticles should be proportional to $\theta(1 - \theta)$, which has its maximum when $\theta = 0.5$. This is the well-known condition of "half-surface-coverage" that has been suggested as giving the optimum flocculation.

An important condition for bridging flocculation with charged particles is the role of electrolyte concentration. This determines the extension ("thickness") of the double layer which can reach values as high as 100 nm (in 10^{-5} mol dm^{-3} 1 : 1 electrolyte such as NaCl). For bridging flocculation to occur, the adsorbed polymer must extend far enough from the surface to a distance over which electrostatic repulsion occurs (> 100 nm in the above example).

This means that at low electrolyte concentrations quite high molecular weight polymers are needed for bridging to occur. As the ionic strength is increased, the range of electrical repulsion is reduced and lower molecular weight polymers should be effective.

In many practical applications, it has been found that the most effective flocculants are polyelectrolytes with a charge opposite to that of the particles. In aqueous media most particles are negatively charged, and cationic polyelectrolytes such as polyethyleneimine are often necessary. With oppositely charged polyelectrolytes it is likely that adsorption occurs to give a rather flat configuration of the adsorbed chain, due to the strong electrostatic attraction between the positive ionic groups on the polymer and the negatively charged sites on the particle surface. This would probably reduce the probability of bridging contacts with other particles, especially with fairly low molecular weight polyelectrolytes with high charge density. However, the adsorption of a cationic polyelectrolyte on a negatively charged particle will reduce the surface charge of the latter, and this charge neutralization could be an important factor in destabilizing the particles. Another mechanism for destabilization has been suggested by Gregory [8] who proposed an "electrostatic-patch" model. This applies to cases where the particles have a fairly low density of immobile charges and the polyelectrolyte has a fairly high charge density. Under these conditions, it is not physically possible for each surface site to be neutralized by a charged segment on the polymer chain, even though the particle may have sufficient adsorbed polyelectrolyte to achieve overall neutrality. There are then "patches" of excess positive charge, corresponding to the adsorbed polyelectrolyte chains (probably in a rather flat configuration), surrounded by areas of negative charge, representing the original particle surface. Particles which have this "patchy" or "mosaic" type of surface charge distribution may interact in such a way that the positive and negative "patches" come into contact, giving quite strong attraction (although not as strong as in the case of bridging flocculation). A schematic illustration of this type of interaction is given in Fig. 6.9. The electrostatic patch concept (which can be regarded as another form of "bridging") can explain a number of features of flocculation of negatively charged particles with positive polyelectrolytes. These include the rather small effect of increasing the molecular weight and the effect of ionic strength on the breadth of the flocculation dosage range and the rate of flocculation at optimum dosage.

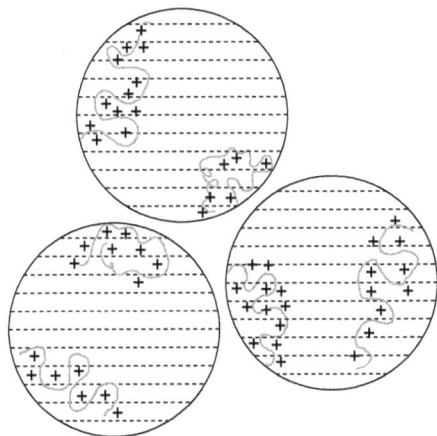

Fig. 6.9: "Electrostatic patch" model for the interaction of negatively charged particles with adsorbed cationic polyelectrolytes.

References

[1] Deryaguin, B. V. and Landau, L., Acta Physicochem. USSR, **14**, 633 (1941).
[2] Verwey, E. J. W. and Overbeek, J. Th. G., "Theory of Stability of Lyophobic Colloids", Elsevier, Amsterdam (1948).
[3] Goodwin, J., "Colloids and Interfaces with Surfactants and Polymers", John Wiley and Sons, United Kingdom (2009).
[4] Smoluchowski, M. V., Physik. Z., **17**, 557, 585 (1916).
[5] Hunter, R. J., "Foundation of Colloid Science", Vol. 1, Oxford University Press, Oxford (1987).
[6] Russel, W. B., Saville, D. A. and Schowalter, W., "Colloidal Dispersions", Cambridge University Press, Cambridge (1989).
[7] Fuchs, N., Z. Physik, **89**, 736 (1936).

[8]   Reerink, H. and Overbeek, J. Th. G., Discussion Faraday Soc., **18**, 74 (1954).

[9]   Krieger, I. M., and Dougherty, T. J., Trans. Soc. Rheol, **3**, 137 (1959).

[10] Krieger, I. M., Advances Colloid and Interface Sci., **3**, 111 (1972).

[11] Goodwin, J. W. and Mercer-Chalmers, J., "Flow Induced Aggregation of Colloidal Dispersions", in "Modern Aspects of Colloidal Dispersions", R. H. Ottewill and A.R Rennie (eds.), Kluwer, Dordrecht (1998), pp. 61–75.

[12] Tadros, Th. F., "Polymer Adsorption and Dispersion Stability", in "The Effect of Polymers on Dispersion Properties", Th. F. Tadros (ed.), Academic Press, London (1981).

[13] Napper, D. H., "Polymeric Stabilisation of Colloidal Dispersions", Academic Press, London (1983).

[14] Tadros, Th. F., "Polymeric Surfactants", in "Encyclopedia of Colloid and Interface Science", Th. F. Tadros (ed.), Springer, Germany (2013).

[15] Asakura, S. and Oosawa, F., J. Phys. Chem., **22**, 1255 (1954); Asakura, S. and Oosawa, F., J. Polym. Sci., **33**, 183 (1958).

[16] Gregory, J., in "Solid/Liquid Dispersions", Th. F. Tadros (ed.), Academic Press, London (1987).

[17] Gregory, J., "Flocculation Fundamentals", in "Encyclopedia of Colloid and Interface Science", Th. F. Tadros (ed.), Springer, Germany (2013).

7 Ostwald ripening in suspensions

7.1 Driving force for Ostwald ripening

The driving force of Ostwald ripening is the difference in solubility between smaller and larger particles [1–5]. Small particles with radius r_1 will have higher solubility than larger particle with radius r_2. This can be easily recognized from the Kelvin equation [6], which relates the solubility of a particle or droplet $S(r)$ to that of a particle or droplet with infinite radius $S(\infty)$,

$$S(r) = S(\infty) \exp\left(\frac{2\sigma V_m}{rRT}\right), \qquad (7.1)$$

where σ is the solid/liquid or liquid/liquid interfacial tension, V_m is the molar volume of the disperse phase, R is the gas constant and T is the absolute temperature. The quantity $2\sigma V_m/(RT)$ has the dimension of length and is termed the characteristic length with an order of $\approx 1\,\mathrm{nm}$.

A schematic representation of the enhancement the solubility $c(r)/c(0)$ with decreasing particle size according to the Kelvin equation is shown in Fig. 7.1.

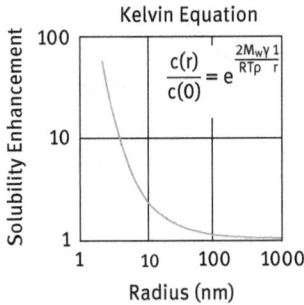

Fig. 7.1: Solubility enhancement with decreasing particle or droplet radius.

It can be seen from Fig. 7.1 that the solubility of suspension particles increases very rapidly with decreasing radius, particularly when r < 100 nm. This means that a particle with a radius of say 4 nm will have about 10 times solubility enhancement compared say with a particle with 10 nm radius which has a solubility enhancement of only 2 times. Thus, with time, molecular diffusion will occur between the smaller and larger particle or droplet, with the ultimate disappearance of most of the small particles. This results in a shift in the particle size distribution to larger values on storage of the suspension. This could lead to the formation of a suspension with average particle size > 2 µm. This instability can cause severe problems, such as sedimentation, flocculation and even flocculation of the suspension.

DOI 10.1515/9783110486872-008

For two particles with radii r_1 and r_2 ($r_1 < r_2$),

$$\frac{RT}{V_m} \ln \left[\frac{S(r_1)}{S(r_2)} \right] = 2\sigma \left[\frac{1}{r_1} - \frac{1}{r_2} \right] . \tag{7.2}$$

Equation (7.2) is sometimes referred to as the Ostwald equation and it shows that the greater the difference between r_1 and r_2, the higher the rate of Ostwald ripening. That is why in preparation of suspensions, one aims at producing a narrow size distribution.

A second driving force for Ostwald ripening in suspensions is due to polymorphic changes. If, for example, a drug has two polymorphs A and B, the more soluble polymorph, say A (which may be more amorphous) will have higher solubility than the less soluble (more stable) polymorph B. During storage, polymorph A will dissolve and recrystallize as polymorph B. This can have a detrimental effect on bioefficacy, since the more soluble polymorph may be more active.

7.2 Kinetics of Ostwald ripening

The kinetics of Ostwald ripening is described in terms of the theory developed by Lifshitz and Slesov [7] and by Wagner [8] (referred to as LSW theory). The LSW theory assumes that: (i) mass transport is due to molecular diffusion through the continuous phase; (ii) the dispersed phase particles are spherical and fixed in space; (iii) there are no interactions between neighbouring particles (the particles are separated by a distance much larger than the diameter of the particles) and (iv) the concentration of the molecularly dissolved species is constant except adjacent to the particle boundaries.

The rate of Ostwald ripening ω is given by,

$$\omega = \frac{d}{dr}(r_c^3) = \left(\frac{8\sigma DS(\infty)V_m}{9RT} \right) f(\phi) = \left(\frac{4DS(\infty)\alpha}{9} \right) f(\phi) , \tag{7.3}$$

where r_c is the radius of a particle that is neither growing nor decreasing in size, D is the diffusion coefficient of the disperse phase in the continuous phase, $f(\phi)$ is a factor that reflects the dependence of ω on the disperse volume fraction and a is the characteristic length scale ($= 2gV_m/(RT)$).

Particles with $r > r_c$ grow at the expense of smaller ones, while particles with $r < r_c$ tend to disappear. The validity of the LSW theory was tested by Kabalanov et al. [9, 10] using emulsions of 1,2 dichloroethane-in-water where the droplets were fixed to the surface of a microscope slide to prevent their coalescence. The evolution of the droplet size distribution was followed as a function of time by microscopic investigations.

The LSW theory predicts that droplet growth over time will be proportional to r_c^3. This is illustrated in Fig. 7.2 for dichloroethane-in-water emulsions.

Another consequence of LSW theory is the prediction that the size distribution function $g(u)$ for the normalized droplet radius $u = r/r_c$ adopts a time-independent

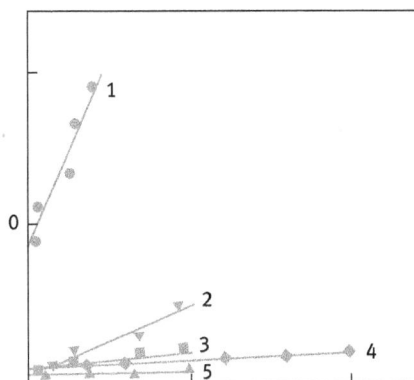

Fig. 7.2: Variation of average cube radius with time during Ostwald ripening in emulsions of: (1) 1,2 dichloroethane; (2) benzene; (3) nitrobenzene; (4) toluene; (5) p-xylene.

form given by:

$$g(u) = \frac{81eu^2 \exp[1/(2u/3 - 1)]}{32^{1/3}(u + 3)^{7/3}(1.5 - u)^{11/3}} \quad \text{for } 0 < u \le 1.5 \tag{7.4}$$

and

$$g(u) = 0 \quad \text{for } u > 1.5. \tag{7.5}$$

A characteristic feature of the size distribution is the cut-off at $u > 1.5$.

A comparison of the experimentally determined size distribution (dichloroethane-in-water emulsions) with the theoretical calculations based on LSW theory is shown in Fig. 7.3.

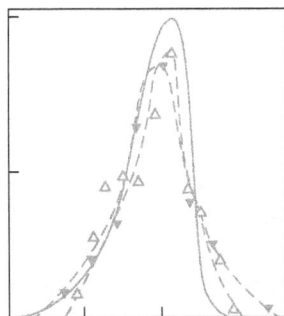

Fig. 7.3: Comparison between theoretical function g(u) (full line) and experimentally determined functions obtained for 1,2 dichloroethane droplets at time 0 (open triangles) and 300 s (inverted solid triangles).

The influence of the alkyl chain length of the hydrocarbon on the Ostwald ripening rate of emulsions was systematically investigated by Kabalanov et al. [9, 10]. Increasing the alkyl chain length of the hydrocarbon used for the emulsion results in a decrease of the oil solubility. According to LSW theory, this reduction in solubility should result in a decrease of the Ostwald ripening rate. This was confirmed by the results of Kabalanov et al. [9, 10] who showed that the Ostwald ripening rate decreases with increasing alkyl chain length from C_9–C_{16}. The same applies to suspensions, i.e. reduction of particle solubility results in a reduction of the Ostwald ripening rate. Although the results showed the linear dependence of the cube of the droplet radius with time in accordance with LSW theory, the experimental rates were ≈ 2–3 times higher than the theoretical values. The deviation between theory and experiment has been ascribed to the effect of Brownian motion [9, 10]. LSW theory assumes that the particles or droplets are fixed in space and molecular diffusion is the only mechanism of mass transfer. For particles or droplets undergoing Brownian motion, one must take into account the contributions of molecular and convective diffusion as predicted by the Peclet number,

$$Pe = \frac{rv}{D},\qquad(7.6)$$

where v is the velocity of the droplets that is approximately given by,

$$v = \left(\frac{3kT}{M}\right)^{1/2},\qquad(7.7)$$

where k is the Boltzmann constant, T is the absolute temperature and M is the mass of the droplet. For $r = 100$ nm, $Pe = 8$, indicating that mass transfer will be accelerated with respect to that predicted by LSW theory.

LSW theory assumes that there are no interactions between the particles or droplets and it is limited to low particle or oil volume fractions. At higher volume fractions the rate of ripening depends on the interaction between diffusion spheres of neighbouring particles. It is expected that suspensions with higher volume fractions of solid will have broader particle size distribution and faster absolute growth rates than those predicted by LSW theory. However, experimental results using high surfactant concentrations (5 %) showed the rate to be independent of the volume fraction in the range $0.01 \leq \phi \leq 0.3$. It has been suggested that the particles may have been screened from one another by surfactant micelles [11]. A strong dependence on volume fraction has been observed for fluorocarbon-in-water emulsions [12]. A threefold increase in ω was found when φ was increased from 0.08 to 0.52.

It has been suggested that micelles play a role in facilitating mass transfer between particles or droplets by acting as carriers of solute molecules [13–16]. Three mechanisms were suggested: (i) molecules are transferred via direct particle/micelle collisions; (ii) molecules exit the particle and are trapped by micelles in the immediate vicinity of the particle; (iii) molecules exit the particles collectively with a large number of surfactant molecules to form a micelle.

In mechanism (i) the micellar contribution to the rate of mass transfer is directly proportional to the number of particle/micelle collisions, i.e. to the volume fraction of micelles in solution. In this case the molecular solubility of the particle in the LSW equation is replaced by the micellar solubility, which is much higher. Large increases in the rate of mass transfer would be expected with increasing micelle concentration. Numerous studies indicate, however, that the presence of micelles affects mass transfer to only a small extent [16]. Results were obtained for decane-in-water emulsions using sodium dodecyl sulphate (SDS) as emulsifier at concentrations above the critical micelle concentration (cmc). This is illustrated in Fig. 7.4 which shows plots of

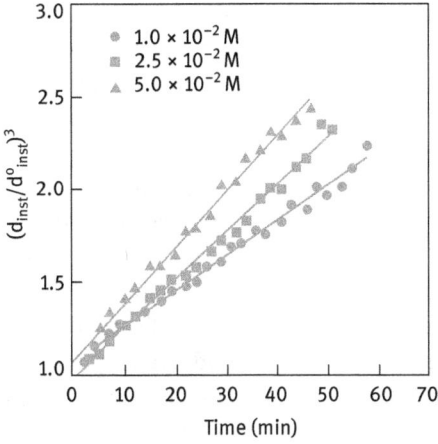

Fig. 7.4: Variation of $(d_{inst}/d_{inst}^0)^3$ with time for decane-in-water emulsions for different SDS concentrations above the cmc.

$(d_{inst}/d_{inst}^0)^3$, where d_{inst} is the diameter at time t and d_{inst}^0 is the diameter at time 0, as a function of time.

The results showed only a two-fold increase in ω above the cmc. This result is consistent with many other studies which showed an increase in mass transfer of only 2–5 times with increasing micelle concentration. The lack of strong dependence of mass transfer on micelle concentration for ionic surfactants may result from electrostatic repulsion between the emulsion droplets and micelles, which provides a high energy barrier preventing droplet/micelle collision.

In mechanism (ii), a micelle in the vicinity of a particle rapidly takes up dissolved molecules from the continuous phase. This "swollen" micelle diffuses to another particle, where the molecule is redeposited. Such a mechanism would be expected to result in an increase in mass transfer over and above that expected from LSW theory by a factor ϕ given by the following equation,

$$\phi = 1 + \frac{\phi_s \Gamma D_m}{D} = 1 + \frac{\chi^{eq} D_m}{c^{eq} D}, \tag{7.8}$$

where ϕ_s is the volume fraction of micelles in solution, $\chi^{eq} = \phi_s c_m^{eq}$ is the net solubility in the micelle per unit volume of micellar solution reduced by the density of the solute, $\Gamma = c_m^{eq}/c^{eq} \approx 10^6 - 10^{11}$ is the partition coefficient for the molecule between the micelle and bulk aqueous phase at the saturation point, D_m is the micellar diffusivity ($\approx 10^{-6} - 10^{-7}$ cm^2 s^{-1}). For a decane-water nanoemulsion in the presence of 0.1 mol dm^{-3} SDS, equation (7.8) predicts an increase in the rate of ripening by three orders of magnitude, in sharp contrast to the experimental results.

To account for the discrepancy between theory and experiment in the presence of surfactant micelles, Kabalanov [15] considered the kinetics of micellar solubilization and he proposed that the rate of oil monomer exchange between the oil droplets and the micelles is slow, and rate determining. Thus at low micellar concentration, only a small proportion of the micelles are able to rapidly solubilize the oil. This leads to a small, but measurable increase in the Ostwald ripening rate with micellar concentration. Taylor and Ottewill [16] proposed that micellar dynamics may also be important. According to Aniansson et al. [17], micellar growth occurs in a stepwise fashion and is characterized by two relaxation times, τ_1 and τ_2. The short relaxation time τ_1 is related to the transfer of monomers in and out of the micelles, while the long relaxation time τ_2 is the time required for break-up and reformation of the micelle. At low SDS (0.05 mol dm^{-3}) concentration, $\tau_2 \approx 0.01$ s, whereas at higher SDS concentration (0.2 mol dm^{-3}), $\tau_2 \approx 6$ s. Taylor and Ottewill [16] suggested that, at low SDS concentration, τ_2 may be fast enough to have an effect on the Ostwald ripening rate, but at 5 % SDS τ_2 may be as long as 1000 s (taking into account the effect of solubilization on τ_2) which is too long to have a significant effect on the Ostwald ripening rate.

When using nonionic surfactant micelles, larger increases in the Ostwald ripening rate might be expected due to the larger solubilization capacities of the nonionic surfactant micelles and absence of electrostatic repulsion between the nanoemulsion droplets and the uncharged micelles. This was confirmed by Weiss et al. [18] who found a large increase in the Ostwald ripening rate in tetradecane-in-water emulsions in the presence of Tween 20 micelles.

7.3 Thermodynamic theory of crystal growth

This theory developed by Gibbs [19] and Volmer [20, 21] is based on the assumption of a stepwise process in which the crystal grows layer by layer. Volmer [20, 21] originally assumed that a thin adsorption layer exists at the phase boundary where the atoms or molecules of the growing substance lose some of their energy as they approach the surface of the crystal without losing all their degrees of freedom. These particles are able to migrate along the surface of the crystal like molecules in a two-dimensional gas. Equilibrium between the adsorption layer and the solution is established immediately. The growth rate is determined by the capture of particles from the adsorption layer by the crystal lattice. The collision of particles in the adsorption layer results in the formation of two-dimensional nuclei, which grow to form a new crystalline layer. The time taken for a layer to form a nucleus is considerably shorter than the time necessary for the formation of a two-dimensional crystal. Thus, the formation of two-dimensional nuclei is the rare determining step.

The formation of a two-dimensional nucleus is similar to the formation of a three-dimensional nucleus as discussed in Chapter 2. According to Gibbs [19], the free energy of formation of a spherical nucleus, ΔG, is given by the sum of two contributions: a positive surface energy term ΔG_s, which increases with increasing radius r of the nucleus, and a negative contribution, ΔG_v, due to the appearance of a new phase, which also increases with increasing r,

$$\Delta G = \Delta G_s + \Delta G_v. \tag{7.9}$$

Thus, the sign of ΔG depends on the relative magnitudes of ΔG_s and ΔG_v, which in turn depend on the size of the two-dimensional nucleus of the new phase, l. ΔG_s for a two-dimensional nucleus is given by,

$$\Delta G_s = jl\kappa , \tag{7.10}$$

where κ is the specific linear energy and j is a shape factor.

ΔG_v is given by,

$$\Delta G_v = \frac{jl^2\kappa}{M} RT \ln \frac{S}{S_0} , \tag{7.11}$$

where (S/S_0) is the relative supersaturation.

Therefore,

$$\Delta G = jl\kappa - \left(\frac{jl^2\kappa}{M} \right) RT \ln(\frac{S}{S_0}) . \tag{7.12}$$

In the initial stages of nucleation, ΔG_s increases faster with increasing l when compared to ΔG_v and ΔG remains positive, reaching a maximum at a critical size l^*, after which it decreases and eventually becomes negative. This occurs since the second term in equation (7.12) rises faster with increasing l than the first term (l^2 versus l). When ΔG becomes negative, growth becomes spontaneous and the cluster grows very fast. This is illustrated in Fig. 7.5. This figure shows the critical size of the nucleus l^* above which growth becomes spontaneous. The free energy maximum ΔG^* at the critical radius represents the barrier that has to be overcome before growth becomes spontaneous. Both l^* and ΔG^* can be obtained by differentiating equation (7.12) with respect to l and equating the result to zero. This gives the following expressions,

$$l^* = \frac{\kappa M}{2\rho RT \ln(S/S_0)} , \tag{7.13}$$

$$\Delta G^* = \frac{j\kappa^2 M}{4\rho RT \ln(S/S_0)} . \tag{7.14}$$

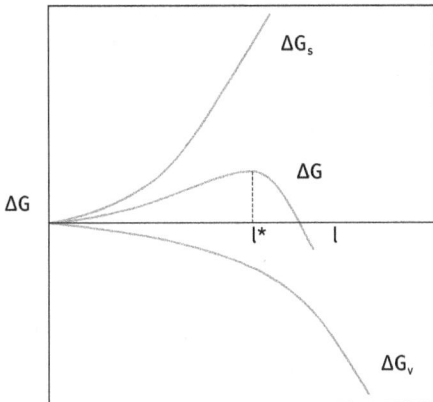

Fig. 7.5: Variation of free energy of formation of a nucleus with size l.

It is clear from equations (7.10)–(7.14) that the free energy of formation of a nucleus and the critical size l* above which the cluster formation grows spontaneously depend on two main parameters, κ and (S/S_0), both of which are influenced by the presence of surfactants. κ is influenced in a direct way by adsorption of surfactant on the surface of the nucleus; this adsorption lowers κ and this reduces l* and ΔG^*. In other words, spontaneous formation of clusters occurs at smaller critical radius. In addition, surfactant adsorption stabilizes the nuclei against any flocculation. The presence of micelles in solution also affects the process of nucleation and growth directly and indirectly. The micelles can act as "nuclei" on which growth may occur. In addition, the micelles may solubilize the molecules of the material, thus affecting the relative supersaturation and this can have an effect on nucleation and growth.

The probability W of the formation of two-dimensional nuclei can be described by an Arrhenius type relationship,

$$W = K' \exp\left(-\frac{\Delta G_{max}}{RT}\right) \tag{7.15}$$

and the linear growth rate (dl/dt) is given by,

$$\frac{dl}{dt} = K'' \exp\left(-\frac{\Delta G_{max}}{RT}\right) = K'' \exp\left(-\frac{j\kappa^2 M}{\rho R^2 T^2 \ln(S/S_0)}\right). \tag{7.16}$$

When a two-dimensional crystal grows on the face of a crystal, it evidently produces a layer of thickness r_0. The time interval t between the formation of the nuclei is inversely proportional to the formation of a two-dimensional nucleus, i.e.,

$$\frac{dl}{dt} = \frac{r_0}{t} = r_0 AW' = r_0 3\kappa l^2 K'' \exp\left(-\frac{j\kappa^2 M}{\rho R^2 T^2 \ln(S/S_0)}\right), \tag{7.17}$$

where A is the surface area of a growing crystal and W is the probability of formation of a nucleus per unit surface area.

When a small crystal grows to a sufficient size, several nuclei may form on the surface. The time of formation of a monolayer then reduces to: $t_0 = (\lambda/a_f)$, where λ is the rate of growth of nuclei on the surface of a crystal and a_f is the average final size of two-dimensional nuclei that is given by,

$$a_f = \left(\frac{\lambda}{W_1}\right)^{1/3} \tag{7.18}$$

and,

$$\frac{dl}{dt} = \frac{r_0}{t_0} = r_0(W\lambda^2)^{1/3} = K \exp\left(-\frac{j\kappa^2 M}{3\rho R^2 T^2 \ln(S/S_0)}\right). \tag{7.19}$$

Thus, the rate of growth of small crystals depends on the degree of supersaturation and on their surface area. With large crystals (dl/dt) it depends only on the degree of supersaturation.

7.4 Molecular-kinetic theory of crystal growth

Kossel [22] developed a model for the growth of crystals in which he considered the energy of interaction of various particles as they are deposited on the surface of a crystal, at its corners, edges, etc., in a manner similar to a "brick laying" process. Various energies will be encountered in attaching the particles to the various regions of the crystal. On the other hand, Stranskii and Kaishev [23, 24] considered the growth of crystals from a different point of view. Namely, the work necessary to detach the particles from the surface of the crystal to infinity. According to Stranskii and Kaishev [24], the rate of crystal growth, which is proportional to the nucleation rate, is given by,

$$v = K' a_2 \exp\left(-\frac{Wa_2^*}{RT}\right) \exp\left(-\frac{L\kappa}{2kT}\right),$$ (7.20)

where Wa_2^* is the work of detachment of a two-dimensional nucleus from the surface of a crystal and L is the perimeter of a two-dimensional nucleus.

7.5 The influence of dislocations on crystal growth

According to the above theories, the growth of crystals can take place only under conditions of appreciable supersaturation, mostly $> 1.5\%$, which ensures the necessary work of formation of two-dimensional nuclei. However, experiments on the growth of various crystals have shown that crystal growth can take place at extremely low supersaturation. The existence of a critical finite supersaturation for the growth of crystals has only been established for a few materials and then for individual faces of crystals, being different from case to case; at the most it is about 1%. However, this discrepancy is not too surprising [25] since the crystals do not have a complete perfect surface needing fresh two-dimensional nucleation in order to grow. This discrepancy may be attributed to crystal dislocations and structural defects. The latter include cracks, surface kinks and surface roughness.

According to Cabrera and Burton [26] and Frank [25], such defects result in the formation of steps at which crystals can grow without the need of formation of nuclei. Screw dislocations are of special importance in the growth of real crystals. If just one dislocation of this type emerges at the centre of the face, that crystal face can grow perpetually up a "spiral staircase". The general importance of dislocations for crystal growth accounts for many observations, such as the individual behaviour of each crystal face, particularly on the microscopic scale.

The growth of a crystal in steps is due to three processes: (i) exchange of particles between adsorption layer and solution; (ii) diffusion of adsorbed particles to the steps, as well as exchange of particles between steps and adsorption layer; (iii) diffusion of particles adsorbed by steps in the direction of kinks and exchange with these kinks. Using these assumptions, Burton, Cabrera and Frank [25, 26] arrived at the following expression for the growth rate R of a crystal from solution,

$$R = \frac{D N_0 \Omega a \beta(x_0)}{2 \kappa_0 \rho_c}, \tag{7.21}$$

where x_0 is the average distance between kinks in the steps, $\beta(x_0)$ is the supersaturation of the solution at distance x_0, D is the diffusion coefficient, N_0 is the equilibrium or steady state concentration of the crystallizing substance in solution, a is the distance between two equilibrium configurations on the surface of a crystal and Ω is the volume of one molecule.

The critical size of a nucleus ρ_c is given by,

$$2\rho_c = 2\gamma \frac{a}{kT} \beta(x_0), \tag{7.22}$$

where γ is a constant (Euler constant). $\beta(x_0)$ is given by,

$$\frac{\beta(x_0)}{\beta} = \left[1 + \frac{2\pi a (\delta - Y_0)}{x_0 Y_0} + \frac{2a}{x_0} \ln \left(\frac{Y_0}{x_0} \right) \right]^{-1}. \tag{7.23}$$

Y_0 is the distance between successive steps, δ is the thickness of the unsaturated layer at the surface of the crystal and β is the supersaturation elsewhere in the solution.

At low supersaturation, the third term on the right-hand side of equation (7.23) is the largest and R depends on $\beta(x_0)$ in a parabolic manner. At higher supersaturations, the second term in equation (7.23) becomes more important and $R = f(\beta)$ becomes linear. In this case equation (7.23) takes the form,

$$R_1 = \frac{D N_0 \Omega \beta}{\delta}. \tag{7.24}$$

Thus, the linear growth rate should be observed at $\beta \geq 10^{-3}$.

7.6 Influence of impurities on crystal growth and habit

It has long been known that trace concentrations of certain additives can have pronounced effects on crystal growth and habit. These effects are of great importance in many fields of science and technology, but the mechanism by which these additives affect crystal growth is not clear. It is generally agreed that additives must adsorb on a crystal surface in order to affect the growth on that face.

Sufficient effects on growth behaviour with very small impurity additives are usually produced by large organic molecules on colloidal materials. One part in 10^4 or 10^5 of such materials may be sufficient to completely alter the growth. The effects of large molecules are usually nonspecific, presumably due to their adsorption on almost any point of the crystal.

Assuming growth to be governed by creation and subsequent lateral motion of steps in the crystal surface [25, 26], it is possible to derive an expression for the effect of impurities on the flow of these steps [27]. Consider the surface of a growing crystal which contains a uniform average concentration of steps n. Suppose there is a constant flux J_i of impurity molecules deposited on the crystal surface per unit time. Assuming that the impurity molecules are immobile on the surface; as the step moves along it will be stopped by a pair of molecules that are less than $2\rho_c$ apart (where ρ_c is the medium radius of curvature of the step corresponding to the supersaturation) and will squeeze itself between pair o impurities that are more than $2\rho_c$ apart. Since the steps are curved, their average velocity v will be smaller than v_0, the velocity in the absence of impurities. A rough estimate of this reduction in velocity is given by the following approximate equation,

$$v = v_0(1 - 2\rho_c d^{1/2})^{1/2} , \tag{7.25}$$

where d is the average density of impurities just ahead of the step.

Assuming, for simplicity, that once a step has passed beyond a certain point on the crystal the impurities adsorbed then become occluded in the crystal and do not offer a significant obstacle to the advance of the following step, it is clear that the density d ahead of any step will be given by,

$$d = \frac{J_i}{nv} . \tag{7.26}$$

This expression automatically makes the flux of impurities being adsorbed in the crystal equal J_i. Substituting equation (7.26) into (7.25) and rearranging, the following equation, which may be solved for v, is obtained,

$$v^2 = v + a = 0 , \tag{7.27}$$

$$v^2 = \frac{v}{v_0}; \quad \alpha^2 = \frac{4\rho_c^2 J_i}{v_0} , \tag{7.28}$$

where $v = nv$ is the flow rate of step in the presence of impurity and $v_0 = nv_0$ is the corresponding rate in the absence of impurities.

7.7 Polymorphic changes

As mentioned in the introduction, when the compound used for formulation of a suspension concentrate exists in two polymorphs, crystal growth may take place as a result of reversion of the thermodynamically less stable form to the more stable form. If this is the case, crystal growth is virtually unaffected by temperature, i.e. it is an isothermal process, which is solvent mediated. Crystal growth involving such polymorphic changes has been carried out by various investigators [28–30]. A thermodynamic analysis based on Gibbs' theory to account for the polymorphic changes can be

made. If a crystal exists in two polymorphic forms, α and β, the Gibbs free energy is
given by the expressions,

$$\Delta G^{\alpha} = \Delta G_v^{\alpha} \sum_i^N A_i \Delta G_s^{\alpha}, \tag{7.29}$$

$$\Delta G^{\beta} = \Delta G_v^{\beta} + \sum_i^N A_i \Delta G_s^{\beta}, \tag{7.30}$$

where V is the crystal volume and A_i is the area. If $\Delta G^{\alpha} \neq \Delta G^{\beta}$, there exists a thermodynamic potential (driving force) to establish equilibrium by an appropriate change of
phase or crystal habit. By this mechanism the less soluble phase grows at the expense
of the more soluble phase. The different polymorphs can be characterized by X-ray
diffraction.

7.8 Crystal growth inhibition

It is clear from the above discussion on crystal growth in suspension concentrates that
solid particles with substantial solubility or those that exist in various polymorphs are
the rule rather than the exception. The task of the formulation scientist is to reduce
crystal growth to an acceptable level depending on the application. This is particularly the case with pharmaceutical and agrochemical suspensions, where crystal
growth leads to a shift of particle size distribution to larger values. Apart from reducing
the physical stability of the suspension, e.g. increased sedimentation, the increase in
particle size of the active ingredient reduces its bioavailability (reduction of disease
control). Unfortunately, crystal growth inhibition is still an "art", rather than a "science", in view of the lack of adequate fundamental understanding of the process at a
molecular level.

Since suspension concentrates are prepared by using a wetting/dispersing agent
(see Chapter 3), it is important to discuss how these agents can affect the growth rate.
In the first place, the presence of wetting/dispersing agents influences the process of
diffusion of the molecules from the surface of the crystal to the bulk solution. The
wetting/dispersing agent may affect the rate of dissolution by affecting the rate of
transport away from the boundary layer [1], although their addition is not likely to
affect the rate of dissolution proper (passage from the solid to the dissolved state in the
immediate adjacent layer). If the wetting/dispersing agent forms micelles which can
solubilize the solute, the diffusion coefficient of the solute in the micelles is greatly reduced. However, as a result of solubilization, the concentration gradient of the solute
is increased to an extent depending on the extent of solubilization. The overall effect
may be an increase in the crystal growth rate as a result of solubilization. In contrast,
if the diffusion rate of the molecules of the wetting/dispersing agent is sufficiently
rapid, their presence will lower the flux of the solute molecules compared to that in

the absence of the wetting/dispersing agent. In this case, the wetting/dispersing agent will lower the rate of crystal growth.

Secondly, wetting/dispersing agents are expected to influence growth when the rate is controlled by surface nucleation [1]. Adsorption of wetting/dispersing agents on the surface of the crystal can drastically change the specific surface energy and make it inaccessible to the solute molecules. In addition, if the wetting/dispersing agent is preferentially adsorbed at one or more of the faces of the crystal (for example by electrostatic attraction between a highly negative face of the crystal and cationic surfactant), surface nucleation is no longer possible at this particular face (or faces). Growth will then take place at the remaining faces, which are either bare or incompletely covered by the wetting/dispersing agent. This will result in a change in crystal habit.

The role of surfactants in modifying the crystal habit of adipic acid has been systematically studied by Michaels and collaborators [31–33]. These authors investigated the effect of various surfactants, of the anionic and cationic type, on the growth of adipic acid crystals from aqueous solution. Microscopic measurements of the crystals permitted calculation of the individual growth rates of the (001), (010) and (110) faces. The growth rate is governed by the rate at which solute is supplied to the individual steps on the crystal faces and the spacing between them. In other words, the growth rate is proportional to the step velocity and the distance between steps. Surfactants may alter the growth rate by changing either of these. At constant step velocity, the spacing between steps may be altered, with a corresponding modification in the growth rate, by a variation in the rate of step generation. With a constant step spacing, an alteration of step velocity will likewise modify the growth rate. Sodium dodecylbenzene sulphonate (NaDBS) retards the growth on the (010) and (110) faces more than on the (001) face, thus favouring the formation of a prismatic or needle crystals. Cationic surfactants, such as cetyltrimethyl ammonium chloride, have the opposite effect, thus favouring growth of the micaceous faces. Michaels et al. [31–33] concluded that the anionic surfactants are physically adsorbed on the faces of adipic acid crystals, while the cationics appear to be chemisorbed. In all cases, the surfactants retarded crystal growth by adsorption on the crystal faces, thus reducing the area on which nucleation would occur. In fact with relatively large crystals, the influence of surfactants on crystal growth can be correlated satisfactorily with the Langmuir adsorption isotherm. Surfactants, in general, exhibit a far greater retarding influence on the crystal growth of very small crystals than on the growth of larger ones.

From the above discussion, it can be seen that surfactants (wetting/dispersing agents), if properly chosen, may be used for crystal growth inhibition and control of habit formation. Inhibition of crystal growth can also be achieved by polymeric surfactants and other additives. For example, Simon Elli et al. [34] found that the crystal growth of the drug sulphathiazole can be inhibited by the addition of poly(vinylpyrrolidone) (PVP). The inhibition effect depends on the concentration and molecular weight of PVP. A minimum concentration (expressed as grams PVP/100 ml) of polymer is required for inhibition, which increases with increasing molecular weight of the polymer. However, if the concentration is expressed in $mol\,dm^{-3}$, the reverse is true, i.e. the higher the molar mass of PVP the lower the number of moles required for inhibition. This led Simonelli et al. [34] to conclude that inhibition must involve kinetic effects, i.e. the rate of diffusion of PVP to the surfaces. If the rate of deposition of PVP is relatively slow as compared to that of sulphathiazole molecules, it is buried by the "avalanche" of the precipitating sulphathiazole molecules. If, on the other hand, its rate is rapid, it in turn can bury the precipitating sulphathiazole molecules and sufficiently cover the crystal surface to cause inhibition of crystal growth. Clearly, a higher PVP concentration would be needed at higher supersaturation rates to cause inhibition. This is due to the increase in diffusion rate at higher supersaturation [34].

Carless et al. [35] reported that the crystal growth of cortisone acetate in aqueous suspensions can be inhibited by addition of cortisone alcohol. Crystal growth in this system is mainly inhibited by polymorphic transformation [35]. The authors assumed that cortisone alcohol is adsorbed onto the particles of the stable form and this prevents the arrival of new cortisone acetate molecules which would result in crystal growth. The authors also noticed that the particles change their shape, growing to long needles. This means that the cortisone alcohol fits into the most dense lattice plane of the cortisone acetate crystal, thus preventing preferential growth on that face.

Many block ABA and graft BA_n copolymers (with B being the "anchor" part and A the stabilizing chain) are very effective in inhibiting crystal growth. The B chain adsorbs very strongly on the surface of the crystal and sites become unavailable for deposition. This has the effect of reducing the rate of crystal growth. Apart from their influence on crystal growth, the above copolymers also provide excellent steric stabilization, providing the A chain is chosen to be strongly solvated by the molecules of the medium.

References

[1] Tadros, Th. F., Advances in Colloid and Interface Science, **12**, 141 (1980).
[2] Tadros, Th. F., "Applied Surfactants", Wiley-VCH, Germany (2005).
[3] Tadros, Th. F., "Dispersion of Powders in Liquids and Stabilisation of Suspensions", Wiley-VCH, Germany (2012).
[4] Tadros, Th. F., "Suspensions", in "Encyclopedia of Colloid and Interface Science", Th. F. Tadros (ed.), Springer, Germany (2013).
[5] Tadros, Th. F., "Nanodispersions", De Gruyter, Germany (2016).
[6] Thomson, W., (Lord Kelvin), Phil. Mag., **42**, 448 (1871).
[7] Lifshitz I. M. and Slesov, V. V., Sov. Phys. JETP, **35**, 331 (1959).
[8] Wagner, C., Z. Electrochem., **35**, 581 (1961).
[9] Kabalnov, A. S. and Schukin, E. D., Adv. Colloid Interface Sci., **38**, 69 (1992).
[10] Kabalnov, A. S., Makarov, K. N. , Pertsov, A. V. and Schukin, E. D., J. Colloid Interface Sci, **138**, 98 (1990).

[11] Taylor, P., Colloids and Surfaces A, **99**, 175 (1995).

[12] Ni, Y., Pelura, T. J., Sklenar, T. A., Kinner, R. A. and Song, D., Art. Cells Blood Subs. Immob. Biorech., **22**, 1307 (1994).

[13] Karaboni, S., van Os, N. M., Esselink, K. and Hilbers, P. A. J., Langmuir, **9**, 1175 (1993).

[14] Soma, J. and Papadadopoulos, K. D., J. Colloid Interface Sci., **181**, 225 (1996).

[15] Kabalnov, A. S., Langmuir, **10**, 680 (1994).

[16] Taylor, P. and Ottewill, R. H., Colloids and Surfaces A, **88**, 303 (1994).

[17] Aniansson, E. A. G., Wall, S. N., Almagren, M., Hoffmann, H., Ulbricht, W., Zana, R., Lang, J. and Tondre, C., J. Phys. Chem., **80**, 905 (1976).

[18] Weiss, J., Coupland, J. N., Brathwaite, D. and McClemments, D. J., Colloids and Surfaces A, **121**, 53 (1997).

[19] Gibbs, J. W., "Scientific Papers", Longman Green, London (1906).

[20] Volmer, M., "Kinetic der Phasenbildung", Stemkopf, Dresden (1939).

[21] Volmer, M., Trans. Faraday Soc., **28**, 359 (1932).

[22] Kossel, W., Nachr. Ges. Wiss., Gottingen, **123**, 348 (1927).

[23] Stranskii, I. N., Z. Phys. Chem., **136**, 259 (1928).

[24] Stranskii, I. N. and Kaishev, R., Z. Phys. Chem., **26B** 100 (1934); **26**, 317 (1934); **36**, 393 (1934).

[25] Frank, F. C., Disc. Faraday Soc., **5**, 48, 67 (1949).

[26] Cabrera, N. and Burton, W., Disc. Faraday Soc., **5**, 33, 40 (1949).

[27] Cabrera, N. and Vermilyea, D. A., Proceedings International Conference on Crystal Growth, John Wiley and Sons, London (1958) p. 393.

[28] Pearson, J. T. and Varney, G., J. Pharm. Pharmac. Suppl., **21**, 60 (1969).

[29] Pearson, J. T. and Varney, G., J. Pharm. Pharmac. Suppl., **25**, 62 (1973).

[30] Pfeiffer, P. R., J. Pharm. Pharmac., **23**, 75 (1971).

[31] Michaels, A. S. and Golville, A. Jr., J. Phys. Chem., **64**, 13 (1960).

[32] Michaels, A. S. and Tausch, F. W. Jr., J. Phys. Chem., **65**, 1730 (1961).

[33] Michaels, A. S., Brian, P. L. T. and Bech, W. F., Chem. Phys. Appl. Surface Active Substances, Proceedings 4th Int. Congress, **2**, 1053 (1967).

[34] Simonelli, P. A., Mehta, S. C. and Higuchi, W. I., J. Pharm. Sci., **59**, 633 (1970).

[35] Carless, J. E., Moustafa, M. A. and Rapson, H. D. C., J. Pharm. Pharmac., **20**, 630 (1968).

8 Sedimentation of suspensions and prevention of formation of dilatant sediments

8.1 Introduction

Most suspensions undergo separation on standing as a result of the density difference between the particles and the medium, unless the particles are small enough for Brownian motion to overcome gravity [1–4]. This is illustrated in Fig. 8.1 for three cases of suspensions.

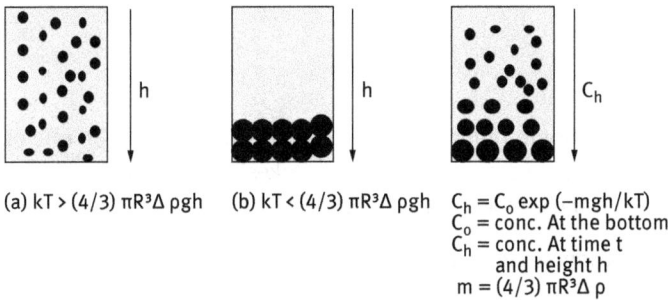

(a) $kT > (4/3) \pi R^3 \Delta \rho g h$ (b) $kT < (4/3) \pi R^3 \Delta \rho g h$ $C_h = C_o \exp(-mgh/kT)$
C_o = conc. At the bottom
C_h = conc. At time t
    and height h
$m = (4/3) \pi R^3 \Delta \rho$

Fig. 8.1: Schematic representation of sediment suspensions.

Case (a) represents the situation when the Brownian diffusion energy (which is in the region of kT, where k is the Boltzmann constant and T is the absolute temperature) is much larger than the gravitational potential energy (which is equal to $(4/3)\pi R^3 \Delta \rho g h$, where R is the particle radius, $\Delta \rho$ is the density difference between the particles and medium, g the acceleration due to gravity and h is the height of the container). Under these conditions, the particles become randomly distributed throughout the whole system, and no separation occurs. This situation may occur with nanosuspensions with radii less than 100 nm, particularly if $\Delta \rho$ is not large, say less than 0.1. In contrast, when $(4/3)\pi R^3 \Delta \rho g h \gg kT$, complete sedimentation occurs as illustrated in Fig. 8.1 (b) with suspensions of uniform particles. In such a case, the repulsive force necessary to ensure colloid stability enables the particles to move past each other to form a compact layer [1–4]. As a consequence of the dense packing and small spaces between the particles, such compact sediments (which are technically referred to as "clays" or "cakes") are difficult to redisperse. In rheological terms (see Chapter 9), the close packed sediment is shear thickening that is referred to as dilatancy, i.e. rapid increase in the viscosity with increasing shear rate.

DOI 10.1515/9783110486872-009

The most practical situation is that represented by Fig. 8.1 (c), where a concentration gradient of the particles occurs across the container. The concentration of particles C can be related to that at the bottom of the container C_0 by the following equation,

$$C = C_0 \exp\left(-\frac{mgh}{kT}\right),$$
(8.1)

where m is the mass of the particles that is given by $(4/3)\pi R^3 \Delta\rho$ (R is the particle radius and $\Delta\rho$ is the density difference between particle and medium), g is the acceleration due to gravity and h is the height of the container.

8.2 Sedimentation rate of suspensions

8.2.1 Very dilute suspensions

For a very dilute suspension of rigid noninteracting particles ($\phi \leq 0.01$), the rate of sedimentation, v_0, can be calculated by application of Stokes' law, where the hydrodynamic force is balanced by the gravitational force,

$$\text{Hydrodynamic Force} = 6\pi\eta R v_0,$$
(8.2)

$$\text{Gravity Force} = (4/3)\pi R^3 \Delta\rho g,$$
(8.3)

$$v_0 = \frac{2}{9}\frac{R^2 \Delta\rho g}{\eta},$$
(8.4)

where η is the viscosity of the medium (water).

v_0 calculated for three particle sizes (0.1, 1 and 10 µm) for a suspension with density difference $\Delta\rho = 0.2$ is 4.4×10^{-9}, 4.4×10^{-7} and 4.4×10^{-5} m s^{-1} respectively. The time needed for complete sedimentation in a 0.1 m container is 250 days, 60 hours and 40 minutes respectively.

8.2.2 Moderately concentrated suspensions

For moderately concentrated suspensions, $0.2 > \phi > 0.01$, sedimentation is reduced as a result of hydrodynamic interaction between the particles, which no longer sediment independently of each other [5, 6]. Several contributions to the change in sedimentation rate, as a result of increasing particle number concentration, have been considered. The first of these contributions arises from the upward flux of fluid volume that accompanies the downward flux of volume of solid material in order to maintain a zero mean volume flux at each point in a homogeneous suspension. This change in fluid environment for one sphere causes the mean sedimentation rate to differ from its value at infinite dilution by an amount $-\phi v_0$. The second, and the largest, contribution arises from the drag down of the fluid that adheres to the spherical particles.

This downward flux of fluid in the inaccessible shells surrounding the rigid spheres is accompanied by an equal upward flux of volume in the part of the fluid that is accessible to the centre of the test sphere. In other words, the reduction in sedimentation rate arises from the diffuse upward current, which compensates for the downward flux in fluid volume in the inaccessible shells surrounding the rigid sphere. This contributes $-4.5\phi v_0$ in the mean sedimentation velocity. The third contribution to the change in sedimentation velocity arises from the motion of the spheres, which collectively generates a velocity distribution in the fluid such that the second derivative of velocity ∇^2 has a nonzero mean. This property of the environment for a particular sphere changes its mean velocity by $0.5\phi v_0$. The fourth contribution arises from the interaction between the spheres. When the test sphere whose velocity is being averaged is near one of the other spheres in the suspension, the interaction between these two spheres gives the test sphere a translational velocity that is significantly different from that which is estimated from the velocity distribution in the absence of the second sphere. This gives a further change in the mean sedimentation rate of $-1.5\phi v_0$. Therefore, the average velocity v can be related to that at infinite dilution v_0 by taking into account the above four contributions, i.e.,

$$v = v_0 + [-\phi v_0 - 4.5\phi v_0 + 0.5\phi v_0 - 1.55\phi v_0]$$
$$= v_0(1 - 6.55\phi) \tag{8.5}$$

This means that for a suspension with $\phi = 0.1$, $v = 0.345 v_0$, i.e. the rate is reduced by a factor of ≈ 3.

8.2.3 Concentrated suspensions

For more concentrated suspensions ($\phi > 0.2$), the sedimentation velocity becomes a complex function of ϕ. An increase in the concentration of suspension leads to a considerable increase in the complexity of the dependence of sedimentation rate on particle size. This is because there is a decrease in the distance between the particles in the disperse phase and also interactions between them occur (either directly or indirectly through the dispersion medium). In addition, an increase in the concentration of the solid phase in the suspension brings an increase in the density and viscosity of the whole disperse system. At high values of the volume fraction of the solid phase ($\phi > 0.1$) displacement of the dispersion medium occurs and of small particles sedimenting out originally by larger particles. At even higher volume fraction ($\phi > 0.4$), the particles tend to sediment in what is known as "hindered sedimentation" mode, whereby all particles sediment at the same rate independent of their size. The closeness of packing prevents differential movement of any large particles through the suspension and the observed sedimentation rate becomes very much less than the Stokes sedimentation rate for any single particle. For a suspension sedimenting in this way, the solid appears to "condense" slowly to a larger volume fraction leaving a clear supernatant liquid separated from the solid sediment by a sharp interface.

The above phenomenon of "hindered sedimentation" was theoretically analysed by Kynch [7] who considered the case of sedimentation of monodisperse particles. He assumed that the velocity v of any particle is a function only of the local concentration n of the particles in its immediate vicinity. The particle flux S, i.e. the number of particles passing a horizontal section, per unit area, per unit time, is given by,

$$S = nv.\tag{8.6}$$

It is assumed that the concentration is the same everywhere across any horizontal layer. The concentration n varies from zero at the top of the sedimentation vessel to some maximum value n_m at the bottom and presumably the velocity v of fall decreases from a finite value u to zero. By considering the flux of particles at various levels in a sedimenting suspension, Kynch [7] derived expressions for the decrease of the height of the suspension with time, when the initial concentration n remains constant and when the initial concentration increases towards the bottom with increasing n in the concentration n range covered during sedimentation. The x versus t diagram obtained by Kynch [7] is shown in Fig. 8.2.

The curve in Fig. 8.2 is characterized by three sections: AOB, where the concentration n is the same as the initial concentration; OCD, where the concentration is a maximum n_m; and OBC, where there is a continuous but extremely rapid increase in concentration from n_B to the maximum concentration n_m. Thus the suspension falls

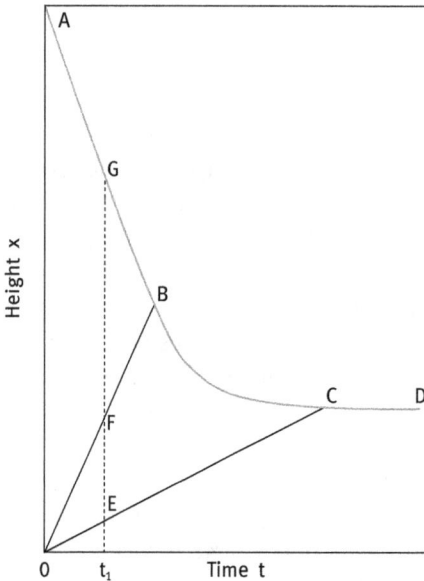

Fig. 8.2: Fall of surface of a suspension according to Kynch [7].

like a "plug" which has the initial density of the suspension (section AOB), depositing a "plug" of maximum density at the bottom (section OCD). At time t_1, there would be a suspension having the initial density above F, and a "plug" of maximum density below E. Between E and F, the density would decrease (in the upward direction) from the maximum to the initial density. Thus, for a suspension which is not flocculated, the sedimentation is constant with time in the first stage, becoming logarithmic in the third stage and the second stage is transition between the two. Unfortunately, Kynch's analysis [7] did not take into account the obvious change of sedimentation velocity with concentration, arising from hydrodynamic interactions [5, 6].

Buscall et al. [8] attempted to relate the decrease of sedimentation velocity with the increase in particle volume fraction ϕ to the reduction in relative viscosity with increasing ϕ. A schematic representation for the variation of v with ϕ is shown in Fig. 8.3, which also shows the variation in relative viscosity with ϕ. It can be seen that v decreases exponentially with increasing ϕ and ultimately it approaches zero when ϕ approaches a critical value ϕ_p (the maximum packing fraction). The relative viscosity shows a gradual increase with increasing ϕ and when $\phi = \phi_p$, the relative viscosity approaches infinity.

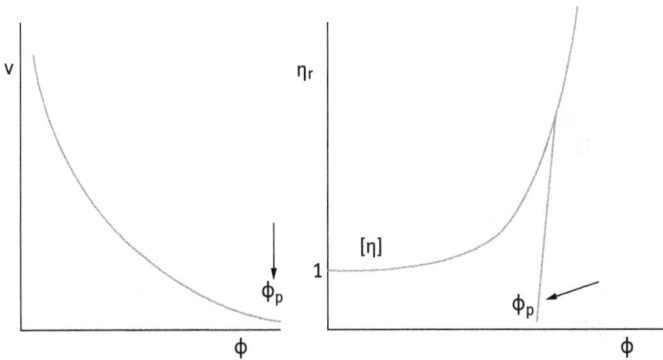

Fig. 8.3: Variation of v and η_r with ϕ.

The maximum packing fraction, ϕ_p, can be easily calculated for monodisperse rigid spheres. For hexagonal packing, $\phi_p = 0.74$, whereas for random packing, $\phi_p = 0.64$. The maximum packing fraction increases with polydisperse suspensions. For example, for a bimodal particle size distribution (with a ratio of $\approx 10:1$), $\phi_p > 0.8$.

It is possible to relate the relative sedimentation rate (v/v_0) to the relative viscosity η/η_0,

$$\left(\frac{v}{v_0}\right) = \left(\frac{\eta_0}{\eta}\right)^\alpha .$$

(8.7)

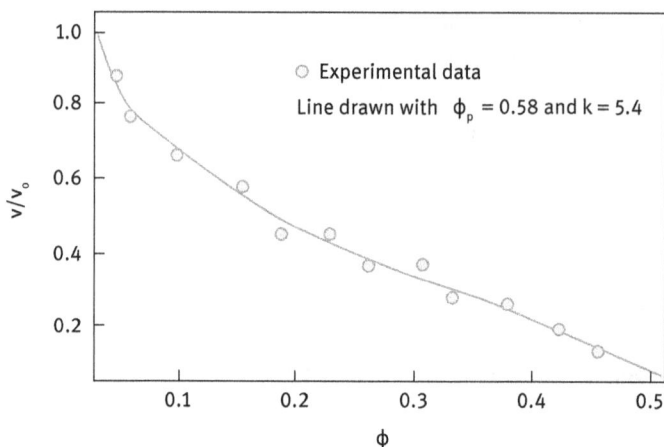

Fig. 8.4: Variation of sedimentation rate with volume fraction for polystyrene dispersions.

The relative viscosity is related to the volume fraction ϕ by the Dougherty–Krieger equation for hard spheres [9],

$$\frac{\eta}{\eta_0} = \left(1 - \frac{\phi}{\phi_p}\right)^{-[\eta]\phi_p}, \tag{8.8}$$

where $[\eta]$ is the intrinsic viscosity (= 2.5 for hard spheres).
Combining equations (8.7) and (8.8),

$$\frac{v}{v_0} = \left(1 - \frac{\phi}{\phi_p}\right)^{\alpha[\eta]\phi_p} = \left(1 - \frac{\phi}{\phi_p}\right)^{k\phi_p}. \tag{8.9}$$

The above empirical relationship was tested for sedimentation of polystyrene latex suspensions with R = 1.55 μm in 10^{-3} mol dm^{-3} NaCl [8]. The results are shown in Fig. 8.4.

The circles are the experimental points and the solid line is calculated using equation (8.9) with $\phi_p = 0.58$ and $k = 5.4$.

8.2.4 Sedimentation of flocculated suspensions

Michaels and Bolger [10] used a different model to describe the sedimentation of flocculated kaolin suspensions. Three types of sedimentation curves were considered, depending on the concentration range of the suspension. These are illustrated in Fig. 8.5. Curve a is the case of a dilute suspension in which the aggregates are considered to be spherical and sedimenting individually, thus producing a sharp interface. The sedimentation curve starts with a linear part, with the rate v_0 being a function of ϕ_A (the volume fraction of the aggregates). Michaels and Bolger [10] used Richardson and

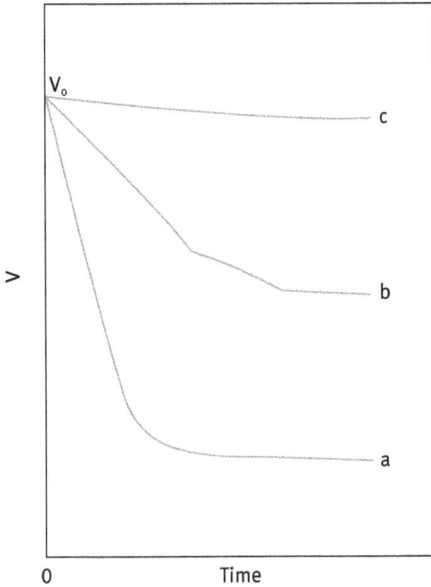

Fig. 8.5: Three types of sedimentation curves according to Michaels and Bolger [10].

Zaki's formula [11] for v_0 and they obtained the following equation,

$$v_0 = \frac{g(\rho - \rho_0)d_A^2}{18\eta C_A}(1 - C_A\phi)^{4.56}, \tag{8.10}$$

where d_A is the mean diameter of aggregates, $\phi = (1 - \varepsilon)$ is the volume fraction of the solids with ε being a measure of porosity, i.e. $(1 - \varepsilon)$ is the volumetric density, ρ is the density of the particles and ρ_0 that of the medium. C_A $(= \phi_A/\phi)$ is a factor that characterizes the "looseness" of the aggregates (the ratio of immobilized liquid to the total volume).

In the intermediate concentration range (curve b), the sedimentation curve was accounted for by Michaels and Bolger [10] by considering a network model for the aggregates. The maximum sedimentation rate is given by,

$$v_1' = \frac{g(\rho - \rho_0)d_p^2}{32\eta}(1 - C_{AF}\phi_F), \tag{8.11}$$

where d_p is the mean pore diameter in the network, C_{AF} $(= \phi_A/\phi_F)$, where ϕ_F is the ratio of the volume of the flocs forming the aggregates and the volume of suspension. The characteristics of the flocs can be determined by studying the sediment volume in detail.

The sedimentation of highly concentrated flocculated suspension (curve c) shows a slow decrease in the sediment volume with time and in some cases a clear liquid layer is "squeezed out" of the liquid bound to the particle surfaces to the top of the container. This process is sometimes referred to as syneresis and the compaction of the solid aggregates is referred to as consolidation.

8.2.5 Sedimentation in non-Newtonian fluids

The sedimentation of particles in non-Newtonian fluids, such as aqueous solutions containing high molecular weight compounds (e.g. hydroxyethyl cellulose or xanthan gum) usually referred to as "thickeners", is not simple since these non-Newtonian solutions are shear thinning with their viscosity decreasing with increasing shear rate [12]. These solutions show a Newtonian region at low shear rates or shear stresses, usually referred to as the residual or zero shear viscosity $\eta(0)$. This is illustrated in Fig. 8.6, which shows the variation of stress σ and viscosity η with shear rate $\dot{\gamma}$.

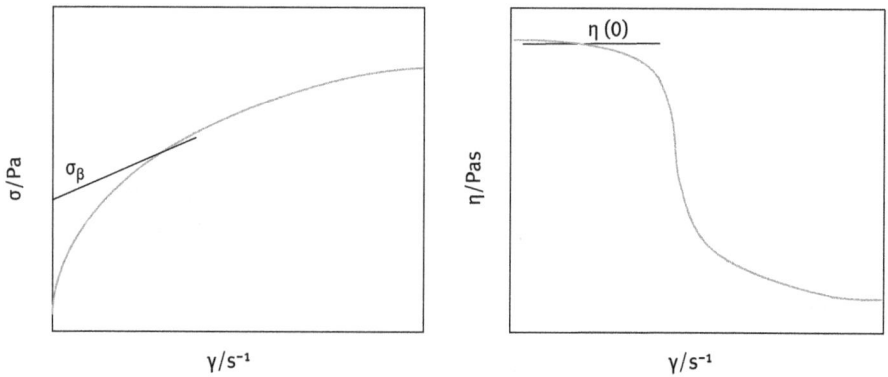

Fig. 8.6: Flow behaviour of "thickeners".

The viscosity of a polymer solution increases gradually with increasing concentration and at a critical concentration, C^*, the polymer coils with a radius of gyration R_G and a hydrodynamic radius R_h (which is higher than R_G due to solvation of the polymer chains) begin to overlap and viscosity increases rapidly. This is illustrated in Fig. 8.7, which shows the variation of $\log \eta$ with $\log C$.

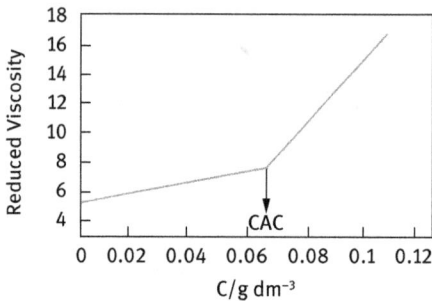

Fig. 8.7: Variation of reduced viscosity with HMHEC concentration.

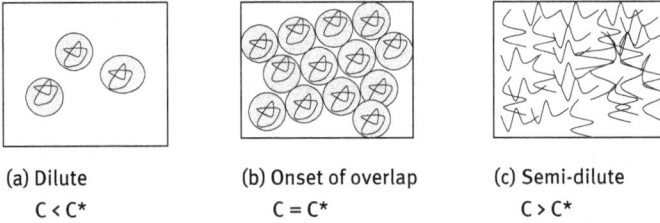

(a) Dilute
$C < C^*$

(b) Onset of overlap
$C = C^*$

(c) Semi-dilute
$C > C^*$

Fig. 8.8: Crossover between dilute and semi-dilute solutions.

In the first part of the curve, $\eta \propto C$, whereas in the second part (above C^*), $\eta \propto C^{3.4}$. A schematic representation of polymer coil overlap is shown in Fig. 8.8, which shows the effect of gradually increasing the polymer concentration. The polymer concentration above C^* is referred to as the semi-dilute range [13].

C^* is related to R_G and the polymer molecular weight M by,

$$C^* = \frac{3M}{4\pi R_G^3 N_{av}}.$$ (8.12)

N_{av} is the Avogadro number. As M increases, C^* becomes progressively lower. This shows that to produce physical gels at low concentrations by simple polymer coil overlap, one has to use high molecular weight polymers.

Another method to reduce the polymer concentration at which chain overlap occurs is to use polymers that form extended chains such as xanthan gum which produces conformation in the form of a helical structure with a large axial ratio. These polymers give much higher intrinsic viscosities and they show both rotational and translational diffusion. The relaxation time for the polymer chain is much higher than a corresponding polymer with the same molecular weight but produces random coil conformation.

The above polymers interact at very low concentrations and the overlap concentration can be very low ($< 0.01\%$). These polysaccharides are used in many formulations to produce physical gels at very low concentrations, thus reducing sedimentation.

The shear stress, σ_p, exerted by a particle (force/area) can be simply calculated [12],

$$\sigma_p = \frac{(4/3)\pi R^3 \Delta\rho g}{4\pi R^2} = \frac{\Delta\rho R g}{3}.$$ (8.13)

For a 10 µm radius particle with a density difference $\Delta\rho$ of $0.2\,\text{g cm}^{-3}$, the stress is equal to,

$$\sigma_p = \frac{0.2 \times 10^3 \cdot 10 \times 10^{-6} \cdot 9.8}{3} \approx 6 \times 10^{-3}\,\text{Pa}.$$ (8.14)

For smaller particles smaller stresses are exerted.

Thus, to predict sedimentation, one has to measure the viscosity at very low stresses (or shear rates). These measurements can be carried out using a constant stress rheometer (Carrimed, Bohlin, Rheometrics, Haake or Physica) as will be discussed in Chapter 9.

Usually, one obtains a good correlation between the rate of sedimentation, v, and the residual viscosity, η(0). Above a certain value of η(0), v becomes equal to 0. Clearly to minimize creaming or sedimentation one has to increase η(0); an acceptable level for the high shear viscosity η_∞ must be achieved, depending on the application. In some cases, a high η(0) may be accompanied by a high η_∞ (which may not be acceptable for applications).

As discussed above, the stress exerted by the particles is very small, in the region of $10^{-3} - 10^{-1}$ Pa depending on the particle size and the density of the particles. Clearly to predict sedimentation, one needs to measure the viscosity at this low stresses [12]. This is illustrated for solutions of ethyl hydroxyethyl cellulose (EHEC) in Fig. 8.9.

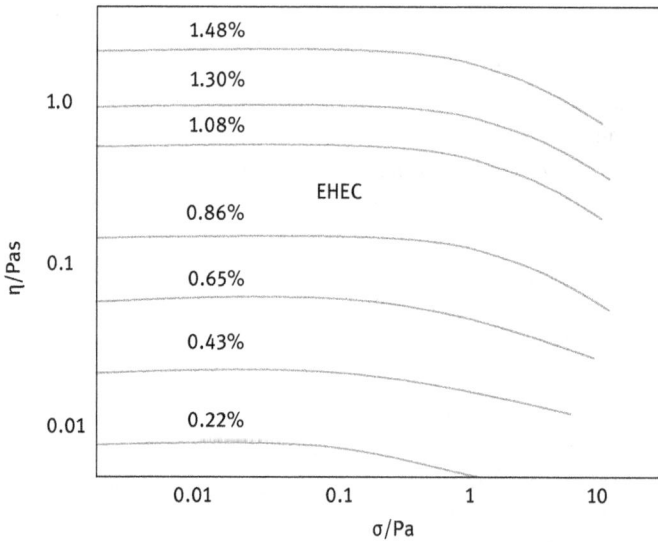

Fig. 8.9: Constant stress (creep) measurements for PS latex dispersions as a function of EHEC concentration.

The results in Fig. 8.9 show that below a certain critical value of shear stress the viscous behaviour is Newtonian and this critical stress value is in the region of 0.1 Pa. Above this stress, viscosity decreases with increasing shear stress, indicating shear thinning behaviour. The plateau viscosity values at low shear stress (< 0.1 Pa) give the limiting and residual viscosity η(0), i.e. the viscosity at near zero shear rate.

The settling rate of a dispersion of polystyrene latex with radius 1.55 μm and at 5 % w/v was measured as a function of ethyl hydroxyethyl cellulose (EHEC) concentration, using the same range as in Fig. 8.10. The settling rate, expressed as v/R^2 where R is the particle radius, is plotted versus $\eta(0)$ in Fig. 8.10 (on a log-log scale). As is clear, a linear relationship between $\log(v/R^2)$ and $\log \eta(0)$ is obtained, with a slope of -1, over three decades of viscosity, indicating that the rate of settling is proportional to $[\eta(0)]^{-1}$.

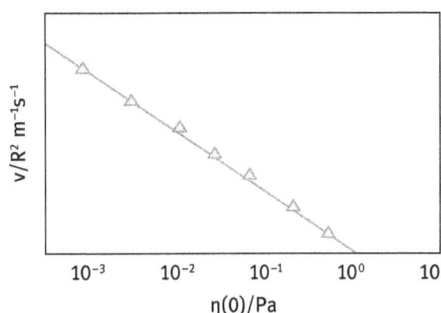

Fig. 8.10: Sedimentation rate versus $\eta(0)$.

The maximum shear stress developed by an isolated spherical particle as it settles through a medium of viscosity η is given by the expression [14],

$$\text{shear stress} = \frac{3v\eta}{2R}. \tag{8.15}$$

For particles at the coarse end of the colloidal range, the magnitude of this quantity will be in the range $10^{-2} - 10^{-5}$ Pa. From the data obtained on EHEC solutions and given in Fig. 8.9, it can be seen that in this range of shear stresses the solutions behave as Newtonian fluids with zero shear viscosity $\eta(0)$. Hence isolated spheres should obey equation (8.4) with η_0 replaced by $\eta(0)$. Consequently, it can be concluded that the rate of sedimentation of a particle is determined by the zero shear rate behaviour of the medium in which it is suspended. With the present system of polystyrene latex, no sedimentation occurred when $\eta(0)$ was greater than 10 Pa s.

The situation with more practical dispersions is more complex due to the interaction between the thickener and the particles. Most practical suspensions show some weak flocculation and the "gel" produced between the particles and thickener may undergo some contraction as a result of the gravity force exerted on the whole network. A useful method to describe separation in these concentrated suspensions is to follow the relative sediment volume V_t/V_0 or relative sediment height h_t/h_0 (where the subscripts t and o refers to time t and zero time respectively) with storage time. For good physical stability, the values of V_t/V_0 or h_t/h_0 should be as close as possible to unity (i.e. minimum separation). This can be achieved by balancing the gravitational force exerted by the gel network with the bulk "elastic" modulus of the suspension. The latter is related to the high frequency modulus G' (see Chapter 12 on rheology).

8.3 Prevention of sedimentation and formation of dilatant sediments

As mentioned before, dilatant sediments are produced with suspensions that are colloidally stable. These dilatant sediments are difficult to redisperse and hence they must be prevented from forming on standing. Several methods may be applied to prevent sedimentation and formation of clays or cakes in a suspension and these are summarized below.

8.3.1 Balance of the density of the disperse phase and medium

It is clear from Stokes law that if $\Delta\rho = 0$, $v_0 = 0$. This method can be applied only when the density of the particles is not much larger than that of the medium (e.g. $\Delta\rho \approx 0.1$). For example, with many organic solids having densities in the region of 1.1–$1.2\,\text{g cm}^{-3}$, by dissolving an inert substance in the continuous phase, such as sugar or glycerol, one may achieve density matching. However, apart from its limitation to particles with density not much larger than the medium, the method is not very practical since density matching can only occur at one temperature. Liquids usually have large thermal expansion, whereas densities of solids vary comparatively little with temperature.

8.3.2 Reduction of particle size

As mentioned above, if R is significantly reduced (to values below $0.1\,\mu\text{m}$), Brownian diffusion can overcome the gravity force and no sedimentation occurs. This is the principle of formation of nanosuspensions. In this case, the Brownian diffusion kT can overcome the gravity force $(4/3)\pi R^3 \Delta\rho gh$ as shown in Fig. 8.1.

8.3.3 Use of high molecular weight thickeners

As discussed above, high molecular weight materials, such as hydroxyethyl cellulose or xanthan gum, when added above a critical concentration (at which polymer coil overlap occurs) will produce very high viscosity at low stresses or shear rates (usually in excess of several hundred Pa s) and this will prevent sedimentation of the particles. In relatively concentrated suspensions, the situation becomes more complex, since the polymer molecules may lead to flocculation of the suspension, by bridging, depletion (see below), etc. Moreover, the polymer chains at high concentrations tend to interact with each other above a critical concentration C* (the so-called "semi-dilute" region discussed above). Such interaction leads to viscoelasticity (see Chapter 9 on rheology), whereby the flow behaviour shows an elastic component characterized by an elastic modulus G' (energy elastically stored during deformation) and a viscous component G'' (loss modulus resulting from energy dissipation during flow). The elastic behaviour of such relatively concentrated polymer solutions plays a major role in reducing settling and prevention of formation of dilatant clays. A good example of such viscoelastic polymer solution is that of xanthan gum, a high molecular weight polymer (molecular weight in excess of 10^6 Daltons). This polymer shows viscoelasticity at relatively low concentration ($< 0.1\%$) as a result of the interaction of the polymer chains, which are very long. This polymer is very effective in reducing settling of coarse suspensions at low concentrations (in the region of 0.1–0.4% depending on the volume fraction of the suspension).

It should be mentioned that to arrive at the optimum concentration of polymer required to prevent settling and claying of a suspension concentrate, one needs to evaluate the rheological characteristics of the polymer solution, on one hand, and the whole system (suspension and polymer) on the other (see Chapter 9). This will provide the formulator with the necessary information on the interaction of polymer coils with each other and with the suspended particles. Moreover, one should be careful in applying high molecular weight materials to prevent settling, depending on the system. For example, with suspensions used as coatings, such as paints, time effects in flow are very important. In this case, the polymer used for preventing settling must show reversible time dependence of viscosity (i.e. thixotropy). In other words, the polymer used has to be shear thinning on application to ensure uniform coating, but once the shearing force is removed, the viscosity has to build up quickly in order to prevent undesirable flow. On the other hand, with suspensions that need to be diluted on application, such as agrochemical suspension concentrates, it is necessary to choose a polymer that disperses readily into water, without the need for vigorous agitation. One should also consider the temperature variation of the rheology of the polymer solution. If the rheology undergoes considerable change with temperature, the suspension may clay at high temperatures. One should also consider the ageing of the polymer, which may result from chemical or microbiological degradation. This would result in reduction of viscosity with time, and settling or claying may occur on prolonged storage.

It is worth mentioning that the use of thickeners to reduce sedimentation suffers from some serious drawbacks. Firstly, for suspensions that require dilution before application, such as agrochemical suspension concentrates, the high viscosity of the system may require vigorous agitation. Secondly, this high viscosity may prevent spontaneity of dispersion on dilution. Thirdly, with most thickened suspensions the viscosity may show a reduction with increasing temperature. This means that on storage of the suspension at high temperature (e.g. 50 °C), sedimentation may occur resulting in separation of the formulation. Such a problem may be overcome by using a mixture of high molecular weight polymer (e.g. xanthan gum) and inert finely divided solids such as clays or oxides as will be discussed below.

8.3.4 Use of "inert" fine particles

Several fine particulate inorganic material produce "gels" when dispersed in aqueous media, e.g. sodium montmorillonite or silica. These particulate materials produce three-dimensional structures in the continuous phase as a result of interparticle interaction. For example, sodium montmorillonite (referred to as swellable clays) form gels at low and intermediate electrolyte concentrations. This can be understood from the knowledge of the structure of the clay particles. These consist of plate-like particles consisting of one octahedral alumina sheet sandwiched between two tetrahedral silica sheets. This is shown schematically in Fig. 8.11, which also shows the change in the spacing of these sheets. In the tetrahedral sheet, tetravalent Si is sometimes replaced by trivalent Al. In the octahedral sheet, there may be replacement of trivalent Al by divalent Mg, Fe, Cr or Zn. The small size of these atoms allows them to take the place of small Si and Al.

This replacement is usually referred to as isomorphic substitution whereby an atom of lower positive valence replaces one of higher valence, resulting in a deficit of positive charge or excess of negative charge. This excess of negative layer charge is compensated by adsorption at the layer surfaces of cations that are too big to be accommodated in the crystal. In aqueous solution, the compensation cations on the layer surfaces may be exchanged by other cations in solution, and hence may be re-

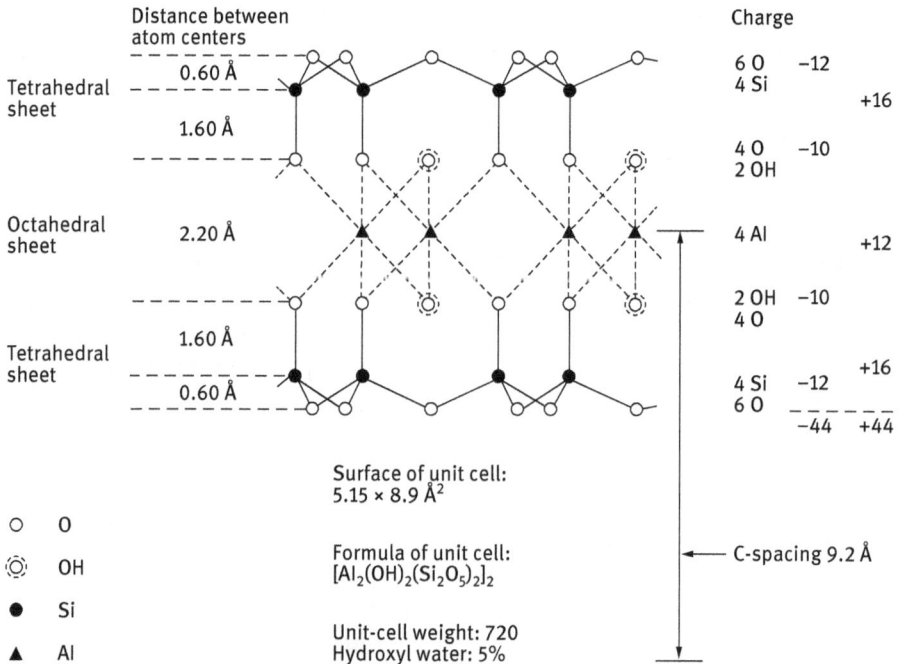

Fig. 8.11: Atom arrangement in one unit cell of 2 : 1 layer mineral.

ferred to as exchangeable cations. With montmorillonite, the exchangeable cations are located on each side of the layer in the stack, i.e. they are present in the external surfaces as well as between the layers. This causes a slight increase of the local spacing from about 9.13 Å to about 9.6 Å; the difference depends on the nature of the counterion. When montmorillonite clays are placed in contact with water or water vapour, the water molecules penetrate between the layers, causing interlayer swelling or (intra)crystalline swelling. This leads to a further increase in the basal spacing to 12.5–20 Å, depending on the type of clay and cation. This interlayer swelling leads, at most, to doubling of the volume of dry clay where four layers of water are adsorbed. The much larger degree of swelling, which is the driving force for "gel" formation (at low electrolyte concentration) is due to osmotic swelling. It has been suggested that swelling of montmorillonite clays is due to the electrostatic double layers that are produced between the charge layers and cations. This is certainly the case at low electrolyte concentration where the double layer extension (thickness) is large.

As discussed above, the clay particles carry a negative charge as a result of isomorphic substitution of certain electropositive elements by elements of lower valency. The negative charge is compensated by cations, which in aqueous solution form a diffuse layer, i.e. an electric double layer is formed at the clay plate/solution interface. This double layer has a constant charge, which is determined by the type and degree of isomorphic substitution. However, the flat surfaces are not the only surfaces of the plate-like clay particles, they also expose an edge surface. The atomic structure of the edge surfaces is entirely different from that of the flat-layer surfaces. At the edges, the tetrahedral silica sheets and the octahedral alumina sheets are disrupted, and the primary bonds are broken. The situation is analogous to that of the surface of silica and alumina particles in aqueous solution. On such edges, therefore, an electric double layer is created by adsorption of potential determining ions (H^+ and OH^-) and one may, therefore identify an isoelectric point (IEP) as the point of zero charge (pzc) for these edges. With broken octahedral sheets at the edge, the surface behaves as Al-OH with an IEP in the region of pH 7–9. Thus in most cases the edges become negatively charged above pH 9 and positively charged below pH 9.

Van Olphen [15] suggested a mechanism of gel formation of montmorillonite involving interaction of the oppositely charged double layers at the faces and edges of the clay particles. This structure, which is usually referred to as a "card-house" structure, was considered to be the reason for the formation of the voluminous clay gel. However, Norrish suggested that the voluminous gel is the result of the extended double layers, particularly at low electrolyte concentrations. A schematic picture of gel formation produced by double layer expansion and "card-house" structure is shown in Fig. 8.12.

Evidence for the picture illustrated in Fig. 8.12 was obtained by Van Olphen [15] who measured the yield value of 3.22% montmorillonite dispersions as a function of NaCl concentration as shown in Fig. 8.13. When C = 0, the double layers are extended and gel formation is due to double layer overlap (Fig. 8.12 (a)). First addition

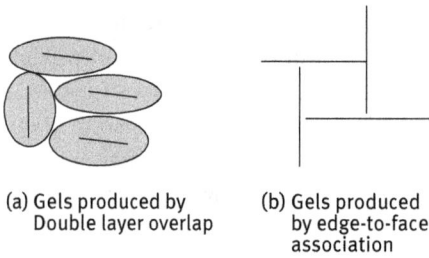

(a) Gels produced by   (b) Gels produced
    Double layer overlap   by edge-to-face   Fig. 8.12: Schematic representation of gel for-
                                  association   mation in aqueous clay dispersions.

of NaCl causes compression of the double layers and hence the yield value decreases very rapidly. At intermediate NaCl concentrations, gel formation occurs as a result of face-to-edge association (house of cards structure) (Fig. 8.12 (b)) and the yield value increases very rapidly with increasing NaCl concentration. If the NaCl concentration is increased further, face-to-face association may occur and the yield value decreases (the gel is destroyed).

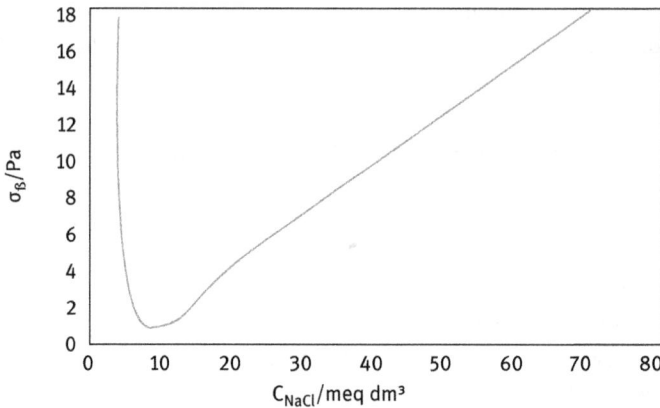

Fig. 8.13: Variation of yield value with NaCl concentration for 3.22 % sodium montmorillonite dispersions.

Finely divided silica, such as Aerosil 200 (produced by Degussa), produces gel structures by simple association (by van der Waals attraction) of the particles into chains and cross chains. When incorporated in the continuous phase of a suspension, these gels prevent sedimentation. In aqueous media, the gel strength depends on the pH and electrolyte concentration. As an illustration, Fig. 8.14 shows the variation of viscosity and yield value with Aerosil silica (which has been dispersed by sonication) concentration at three different pH values. In all cases, the viscosity and yield value show a rapid increase above a certain silica concentration that depends on the pH of the system.

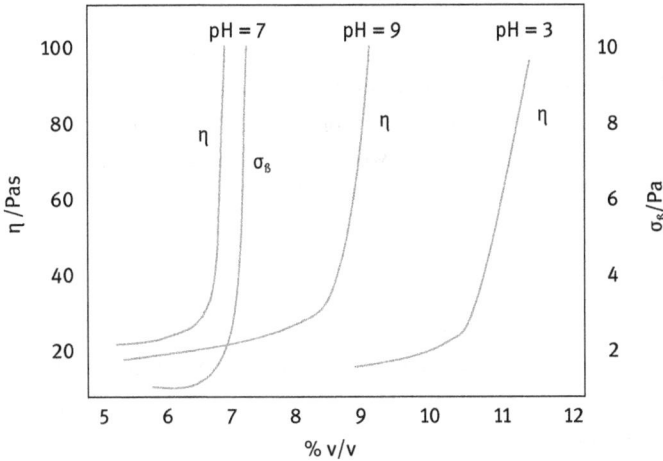

Fig. 8.14: Variation of viscosity η and yield value σ_β with Aerosil 200 concentration at three pH values.

At pH = 3 (near the isoelectric point of silica), the particles are aggregated (forming flocs) and the increase in viscosity occurs at relatively high silica concentration (> 11 % v/v). At pH = 7, the silica becomes negatively charged and the double layers stabilize the silica particles against aggregation. In this case the particles remain as small units and the viscosity and yield value increases sharply above 7 % v/v. At pH = 9, some aggregation occurs as a result of the electrolyte released on adjusting the pH; in this case the viscosity increases at higher concentration (> 9 % v/v) when compared with the results at pH = 7.

The above results clearly indicate the importance of pH and electrolyte concentration in gelation of silica. It seems that the optimum gel formation occurs at neutral pH.

8.3.5 Use of mixtures of polymers and finely divided particulate solids

By combining thickeners such as hydroxyethyl cellulose or xanthan gum with particulate solids such as sodium montmorillonite, a more robust gel structure can be produced. By using such mixtures, the concentration of the polymer can be reduced, thus overcoming the problem of dispersion on dilution (e.g. with many agrochemical suspension concentrates). This gel structure may be less temperature dependent and could be optimized by controlling the ratio of the polymer and the particles. If these combinations of say sodium montmorillonite and a polymer such as hydroxyethyl cellulose, polyvinyl alcohol (PVA) or xanthan gum are balanced properly, they can provide a "three-dimensional structure", which entraps all the particles and stops settling and formation of dilatant clays. The mechanism of gelation of such combined systems depends to a large extent on the nature of the solid particles, the polymer and the conditions. If the polymer adsorbs on the particle surface (e.g. PVA on sodium montmorillonite or silica) a three-dimensional network may be formed by polymer bridging. Under conditions of incomplete coverage of the particles by the polymer, the latter becomes simultaneously adsorbed on two or more particles. In other words, the polymer chains act as "bridges" or "links" between the particles.

8.3.6 Controlled flocculation ("self-structured" systems)

For systems where the stabilization mechanism is electrostatic in nature, for example those stabilized by ionic surfactants or polyelectrolytes, the Deryaguin–Landau–Verwey–Overbeek (DLVO) theory [16, 17] predicts the appearance of a secondary attractive minimum at large particle separations (see Chapter 4). This attractive minimum can reach sufficient values, in particular for large (> 1 μm) and asymmetric particles, for weak flocculation to occur. The depth of this minimum not only depends on particle size but also on the Hamaker constant, the surface (or zeta) potential and electrolyte concentration and valency. Thus by careful control of zeta potential and electrolyte concentration, it is possible to arrive at a secondary minimum of sufficient depth for weak flocculation. This results in the formation of a weakly structured "gel" throughout the suspension. This self-structured gel can prevent sedimentation and formation of dilatant clays. As an illustration, Fig. 8.15 shows energy distance curves for a suspension stabilized by naphthalene formaldehyde sulphonated condensate (an anionic polyelectrolyte with a modest molecular weight of ≈ 1000) at various NaCl concentrations [18].

It can be seen that at low NaCl concentration, the secondary minimum is absent and hence this suspension will show sedimentation and formation of a dilatant clay. This can be illustrated by measuring the sediment height and number of turns required to redisperse the suspension, as illustrated in Fig. 8.16. At 10^{-2} and 5×10^{-2} mol dm^{-3} NaCl, a secondary minimum appears in the energy-distance curve which is sufficiently deep for weak flocculation to occur. This is particularly the case with 5×10^{-2} mol dm^{-3} NaCl where the minimum depth reaches ≈ 50 kT. At such concentration the sediment height increases and the number of turns required for redispersion reaches a small value as is illustrated in Fig. 8.16.

The higher the valency of the electrolyte, the lower the concentration required to reach a sufficiently deep minimum for weak flocculation to occur. This is clearly shown in Fig. 8.16 where the sediment height and number of turns for redispersion are plotted as a function of Na$_2$SO$_4$ and AlCl$_3$ concentrations. It can be seen that for Na$_2$SO$_4$ a rapid increase in sediment height and decrease in the number of turns for redispersion occurs at ≈ 5×10^{-3} mol dm^{-3}, whereas for AlCl$_3$ this occurs at ≈ 5×10^{-4} mol dm^{-3}. Clearly, the electrolyte concentration (which depends on the electrolyte valency) must be chosen carefully to induce sufficient flocculation to prevent the formation of a dila-

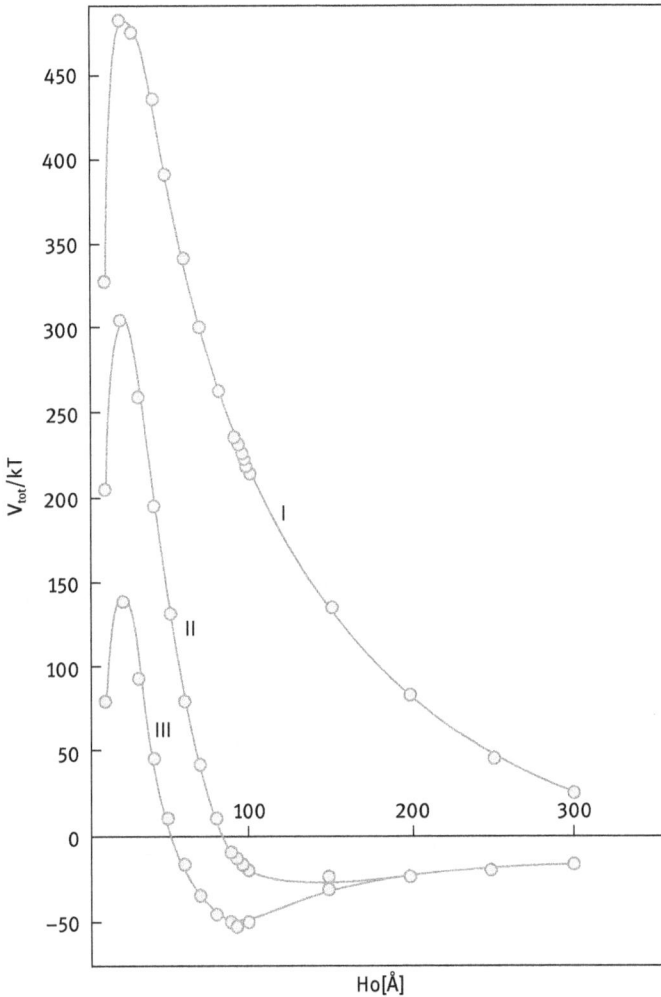

Fig. 8.15: Energy-distance curves for electrostatically stabilized suspension at various NaCl concentrations: (I) 10^{-3} mol dm^{-3}; (II) 10^{-2} mol dm^{-3}; (III) 5×10^{-2} mol dm^{-3}.

tant sediment, but this concentration should not result in irreversible coagulation (i.e. primary minimum flocculation). With polyelectrolytes, irreversible coagulation is prevented as a result of the contribution of steric interaction (which operates at moderate concentrations of electrolyte).

The application of the concept of controlled flocculation to pharmaceutical suspensions has been discussed by various authors [19–21] who demonstrated how flocculation of a sulphaguanidine suspension, by AlCl$_3$, could be used for the preparation of readily dispersible suspensions, which after prolonged storage retain satisfactory physical properties. The work was extended to other types of drug suspensions such

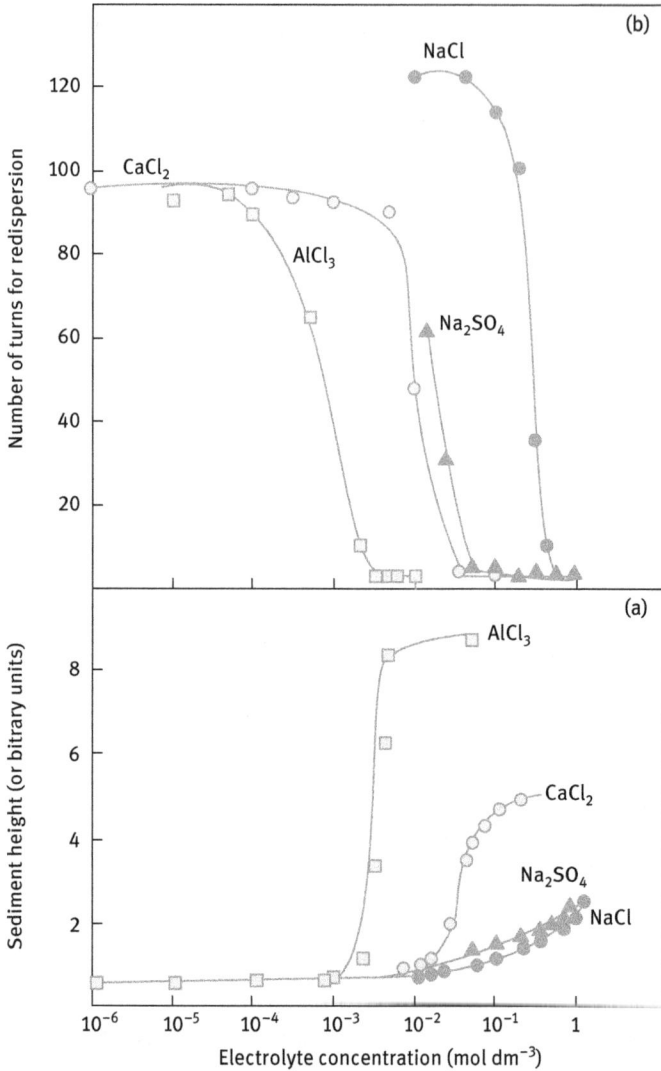

Fig. 8.16: Sediment height and redispersion as a function of electrolyte concentration.

as grisofulvin, hydrocortisone and sulphamerazine [22, 23]. The flocculation observed was interpreted by DLVO theory. It was suggested that flocculation occurs in a minimum whose depth is restricted owing to steric stabilization by the surfactant film. It should be mentioned that when using $AlCl_3$ in controlling flocculation, hydrolysable species are produced above pH 4 and these species play a major role in controlling the flocculation.

For systems stabilized with nonionic surfactants or macromolecules, the energy-distance curve also shows a minimum whose depth depends on particle size and

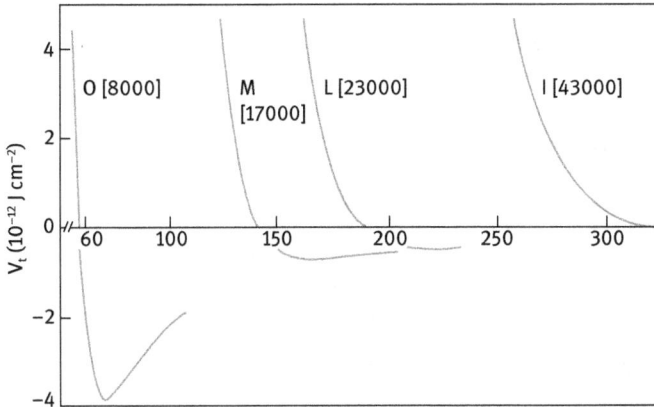

Fig. 8.17: Total energy of interaction versus separation distance for polystyrene latex with adsorbed layers of PVA of various thicknesses.

shape, Hamaker constant and adsorbed layer thickness δ. Thus for a given particulate system, having a given particle size and shape and Hamaker constant, the minimum depth can be controlled by varying the adsorbed layer thickness δ. This is illustrated in Fig. 8.17, where the energy-distance curves for polystyrene latex particles containing adsorbed polyvinyl alcohol (PVA) layers of various molecular weights, i.e. various δ values, are shown. These calculations were made using the theory of Hesselink et al. [24] together with the experimentally measured parameters of δ [25]. It is clear from Fig. 8.17 that with the high molecular weight PVA fractions δ is large and the energy minimum is too small for flocculation to occur. This is certainly the case with M = 43 000 (δ = 19.7 nm), M = 28 000 (δ = 14.0 nm) and M = 17 000 (δ = 9.8 nm). However with M = 8000 (δ = 3.3 nm), an appreciable attraction prevails at a separation distance in the region of 2δ. In this case a weakly flocculated open structure could be produced for preventing formation of dilatant sediments. To illustrate this point, dispersions stabilized with PVA of various M were slowly centrifuged at 50 g, and the sediment was freeze dried and examined by electron microscopy. The results are shown in Fig. 8.18, which clearly illustrates the close-packed sediment obtained with the high molecular weight PVA fractions and the weakly flocculated open structure obtained with the fraction with low molecular weight.

It should be mentioned that the minimum depth needed to induce flocculation depends on the volume fraction of the suspension. This can be understood if one considers the balance between the interaction energy and entropy terms in the free energy of flocculation, i.e.,

$$\Delta G_{floc} = \Delta G_h - T\Delta S_h ,\qquad(8.16)$$

where ΔG_h is the interaction energy term (which is negative), which is determined by the depth of the minimum in the energy-distance curve and ΔS_h is the entropy loss on flocculation. On flocculation, configurational entropy is lost and ΔS_h is therefore

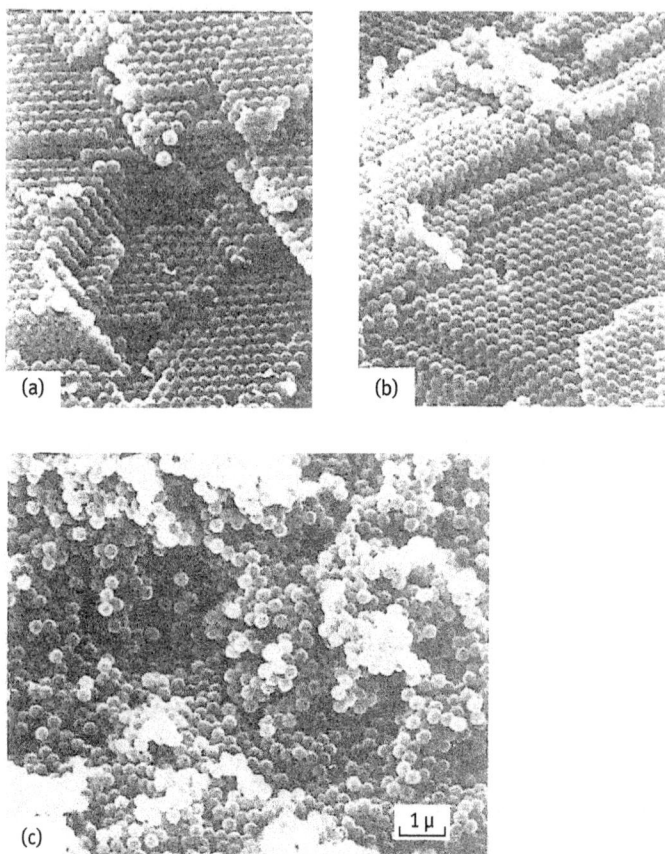

Fig. 8.18: Scanning electron micrographs of polystyrene latex sediments stabilized with PVA fractions: (C) M = 43 000 (δ = 19.7 nm); (L) M = 28 000 (δ = 14.0 nm); (O) M = 8000 (δ = 3.3 nm).

negative. This means that the term $T\Delta S_{li}$ is positive (i.e. it opposes flocculation). The condition for flocculation is $\Delta G \leq 0$, and therefore the ΔG_h required for flocculation depends on the magnitude of the $T\Delta S_h$ term. The latter term decreases with increasing volume fraction ϕ, and hence the higher the ϕ value, the smaller the ΔG_h required for flocculation. This means that with more concentrated suspensions weak flocculation of a sterically stabilized suspension occurs at lower minimum depth.

8.3.7 Depletion flocculation

As discussed in Chapter 6, addition of free nonadsorbing polymer can produce weak flocculation above a critical volume fraction of the free polymer, ϕ_p^+, which depends on its molecular weight and the volume fraction of suspension. This weak flocculation produces a "gel" structure that reduces sedimentation. According to Asakura and Oosawa [26, 27], when two particles approach each other within a distance of separation that is smaller than the diameter of the free polymer coil, exclusion of the polymer molecules from the interstices between the particles takes place, leading to the formation of a polymer-free zone. This is illustrated in Fig. 8.19, which shows the situation below and above ϕ_p^+. As a result of this process, an attractive energy, associated with the lower osmotic pressure in the region between the particles, is produced. Fleer et al. [28] derived the following expression for the interaction of hard spheres in the presence of nonadsorbing polymer, i.e. the decreases in free energy resulting from the transfer of solvent molecules from the depletion zone to bulk solution,

$$G_{dep} = \frac{2\pi R}{V_0}(\mu_1 - \mu_1^0)\Delta^2 \left(1 + \frac{2\Delta}{3R}\right), \tag{8.17}$$

where V_0 is the molecular volume of the solvent, μ_1 is the solvent chemical potential at ϕ_p and μ_1^0 is the chemical potential of the pure solvent. Since $\mu_1 < \mu_1^0$, $(\mu_1 - \mu_1^0)$ is negative and G_{dep} is negative, resulting in flocculation.

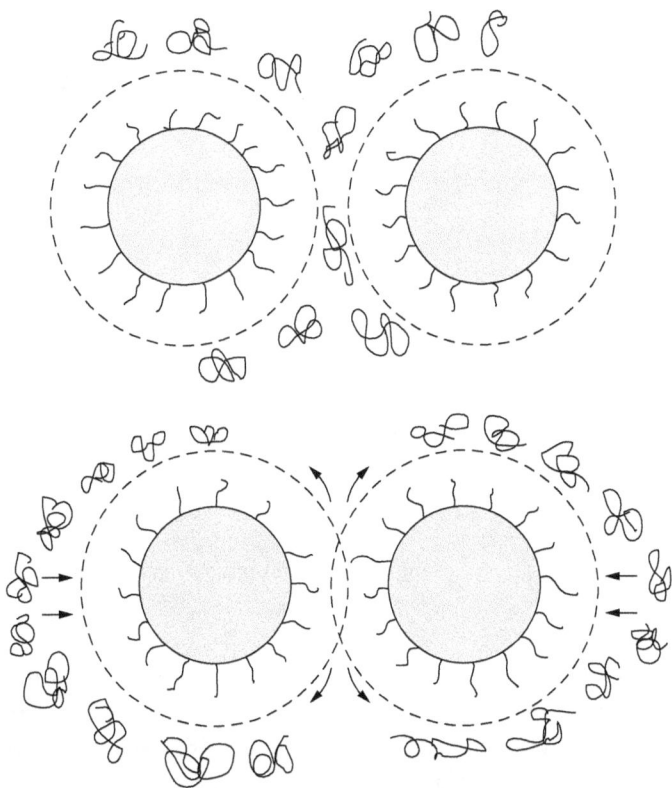

Fig. 8.19: Schematic representation of depletion flocculation.

The above phenomenon of flocculation can be applied to the prevention of settling and claying by forming an open structure that, under some conditions, can fill the whole volume of the suspension. Above ϕ_p^+, the suspension becomes weakly flocculated, and the extent of flocculation increases with a further increase in the concentration of free nonadsorbing polymer. This is illustrated for a suspension of an agrochemical (ethirimol, a fungicide) that is stabilized by a graft copolymer of poly(methyl methacrylate)/methacrylic acid with poly(ethylene oxide) side chains to which poly(ethylene oxide) (PEO) with M = 20 000, 35 000 or 90 000 is added for flocculation [29]. Rheological measurements showed that above ϕ_p^+ a rapid increase in the yield value is produced. This is shown in Fig. 8.20. Above ϕ_p^+ one would expect significant reduction in formation of dilatant sediments. This is illustrated in Fig. 8.21, which shows a plot of sediment height and number of revolutions for redispersion as a function of ϕ_p (for PEO 20 000). It is clear that addition of PEO results in weak flocculation and redispersion becomes easier. This redispersion is maintained up to $\phi_p = 0.04$ above which it becomes more difficult due to the increase in viscosity.

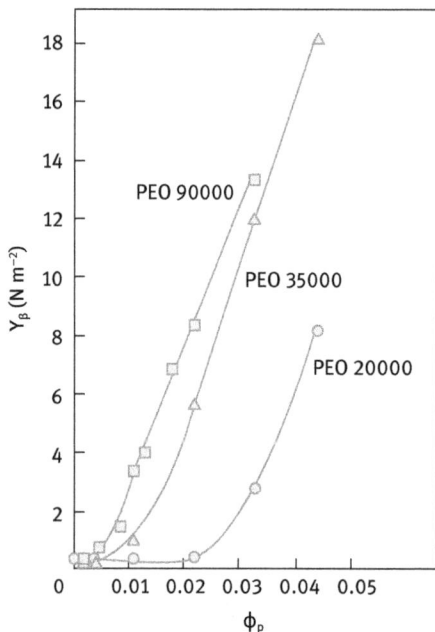

Fig. 8.20: Variation of yield value with PEO concentration.

Another example for the application of depletion flocculation was obtained for the same suspension but using hydroxyethyl cellulose (HEC) with various molecular weights. The weak flocculation was studied using oscillatory measurements. Figure 8.22 shows the variation of the complex modulus G^* with ϕ_p. Above a critical ϕ_p value (that depends on the molecular weight of HEC), G^* increases very rapidly with a

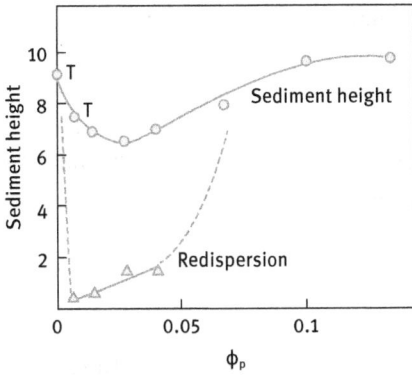

Fig. 8.21: Variation of sediment height and redispersion with ϕ_p for PEO 20 000.

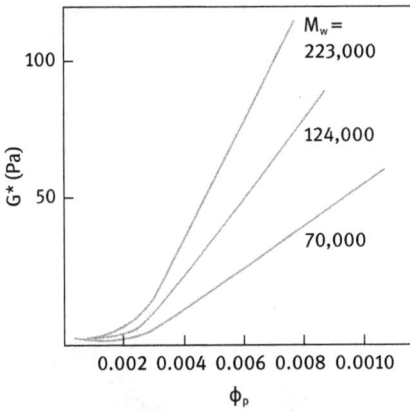

Fig. 8.22: Variation of G^* with ϕ_p for HEC with various molecular weights.

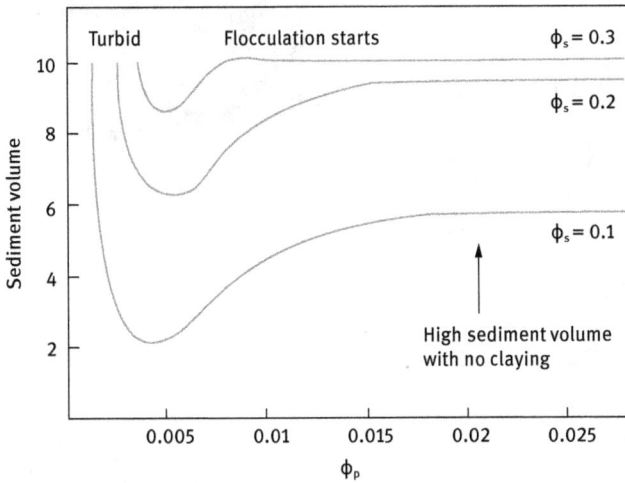

Fig. 8.23: Variation of sediment volume with ϕ_p for HEC (M = 70 000).

further increase in ϕ_p. When ϕ_p reaches an optimum concentration, sedimentation is prevented. This is illustrated in Fig. 8.23, which shows the sediment volume in 10 cm cylinders as a function ϕ_p for various volume fractions of the suspension ϕ_s. At sufficiently high volume fraction of the suspensions ϕ_s and high volume fraction of free polymer ϕ_p, a 100 % sediment volume is reached and this is effective in eliminating sedimentation and formation of dilatant sediments.

8.3.8 Use of liquid crystalline phases

Surfactants produce liquid crystalline phases at high concentrations [30]. Three main types of liquid crystals can be identified as illustrated in Fig. 8.24: hexagonal phase (sometimes referred to as middle phase), cubic phase and lamellar phase (neat phase). All these structures are highly viscous and they also show elastic response. If produced in the continuous phase of suspensions, they can eliminate sedimentation of the particles. These liquid crystalline phases are particularly useful for application in liquid detergents which contain high surfactant concentrations. Their presence reduces sedimentation of the coarse builder particles (phosphates and silicates).

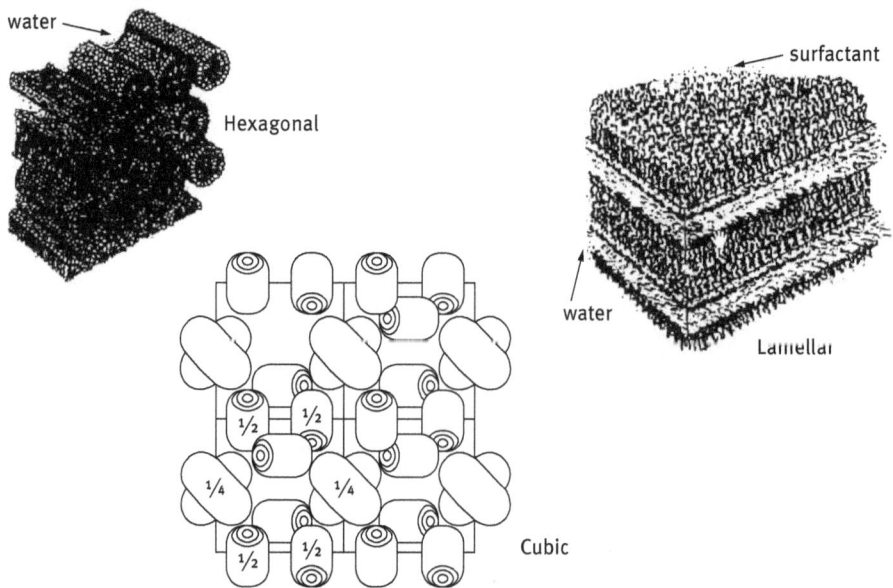

Fig. 8.24: Schematic picture of liquid crystalline phases.

References

[1]   Tadros, Th. F., Advances in Colloid and Interface Science, **12**, 141 (1980).
[2]   Tadros, Th. F., (ed.), "Solid/Liquid Dispersions", Academic Press, London (1987) Chapter 11.
[3]   Tadros, Th. F., "Dispersion of Powders in Liquids and Stabilisation of Suspensions", Wiley-VCH, Germany (2012).
[4]   Tadros, Th. F., "Suspensions", in: "Encyclopedia of Colloid and Interface Science", Th. F. Tadros (ed.), Springer, Germany (2013).
[5]   Bachelor, G. K., J. Fluid Mech., **52**, 245 (1972).
[6]   Bachelor, G. K., J. Fluid Mech., **79**, 1 (1976).
[7]   Kynch, G. J., Trans. Faraday Soc., **48**, 166 (1952).
[8]   Buscall, R., Goodwin, J. W., Ottewill, R. H. and Tadros, Th. F., J. Colloid Interface Sci., **85**, 78 (1982).
[9]   Krieger, I. M., Advances Colloid Interface Sci., **3**, 45 (1971).
[10]   Michaels, A. S. and Bolger, J. C., Ind. Eng. Chem., **1**, 24 (1962).
[11]   Richardson, J. F. and Zaki, W. N., Trans. Inst. Chem. Engrs., **32**, 35 (1954).
[12]   Tadros, Th. F., "Rheology of Dispersions", Wiley-VCH, Germany (2010).
[13]   de Gennes, P. G., "Scaling Concepts in Polymer Physics", Cornell University Press, Ithaca, London (1979).
[14]   Happel, J. and Brenner, H., "Low Reynolds Number Hydrodynamics", Prentice Hall, London (1965).
[15]   van Olphen, H., "Clay Colloid Chemistry", Wiley, New York (1963).
[16]   Deryaguin, B. V. and Landau, L., Acta Physicochem. USSR, **14**, 633 (1941).
[17]   Verwey, E. J. W. and Overbeek, J. Th. G., "Theory of Stability of Lyophobic Colloids", Elsevier, Amsterdam (1948).
[18]   Tadros, Th. F., Colloids and Surfaces, **18**, 427 (1986).
[19]   Haines, B. S. and Martin, A. N., J. Pharm. Sci., **50**, 228, 723, 756 (1961).
[20]   Wilson, R. G. and Canow, B. E., J. Pharm. Sci., **50**, 757 (1963).
[21]   Mathews, B. A. and Rhodes, C. T., J. Pharm. Sci., **57**, 557, 569 (1986).
[22]   Jones, R. D. C., Mathews, B. A. and Rhodes, C. T., J. Pharm. Sci., **59**, 529 (1971).
[23]   Mathews, B. A. and Rhodes, C. T., J. Pharm. Sci., **59**, 529 (1971).
[24]   Hesselink, F. Th., Vrij, A. and Overbeek, J. Th. G., J. Phys. Chem. **75**, 2094 (1971).
[25]   Garvey, M. J., Tadros, Th. F. and Vincent, B., J. Colloid Interface Sci., **55**, 440 (1976).
[26]   Asakura, S. and Oosawa, F., J. Phys. Chem., **22**, 1255 (1954).
[27]   Asakura, S. and Oosawa, F., J. Polym. Sci., **33**, 183 (1958).
[28]   Fleer, G. J., Scheutjens, J. H. H. H. and Vincent, B., ACS Symposium Ser., **246**, 245 (1984).
[29]   Heath, D. Knott, R. D. Knowles D. A. and Tadros, Th. F. ACS Symposium Ser., **254**, 11 (1984).
[30]   Tadros, Th. F., "Applied Surfactants", Wiley-VCH, Germany (2005).

9 Rheology of suspensions

9.1 Introduction

Rheological measurements are useful tools for probing the microstructure of suspensions [1]. This is particularly the case if measurements are carried out at low stresses or strains as will be discussed below. In this case the spacial arrangement of particles is only slightly perturbed by the measurement. In other words, the convective motion due to the applied deformation is less than the Brownian diffusion. The ratio of the stress applied, σ, to the "thermal stress" (that is equal to $kT/6\pi a^3$, where k is the Boltzmann constant, T is the absolute temperature and a is the particle radius) is defined in terms of a dimensionless Peclet number Pe,

$$Pe = \frac{6\pi a^3 \sigma}{kT} . \tag{9.1}$$

For a colloidal particle with radius of 100 nm, σ should be less than 0.2 Pa to ensure that the microstructure is relatively undisturbed. In this case, Pe < 1.

In order to remain in the linear viscoelastic region, the structural relaxation by diffusion must occur on a time scale comparable to the experimental time [1]. The ratio of the structural relaxation time, τ, to the experimental measurement time, t_e, is given by the dimensionless Deborah number De,

$$De = \frac{\tau}{t_e} . \tag{9.2}$$

De $\approx$ 1 and the suspension appears viscoelastic (see below).

The rheology of suspensions depends on the balance between three main forces [1]: Brownian diffusion; hydrodynamic interaction; and interparticle forces. These forces are determined by three main parameters: (i) the volume fraction ϕ (total volume of the particles divided by the volume of the dispersion); (ii) the particle size and shape distribution; and (iii) the net energy of interaction, G_T, i.e. the balance between repulsive and attractive forces.

In this chapter I start with a section describing the various techniques that can be applied for studying the rheology of suspensions. This is followed by description of the rheology of suspensions, starting with very dilute suspensions (with volume fraction $\phi \leq 0.01$), moderately concentrated suspensions ($0.2 < \phi \leq 0.1$) and concentrated suspensions ($\phi > 0.2$). The role of interparticle forces in determining the rheology of a suspension is also described. Four main types of systems are considered [1]: (i) hard-sphere suspensions, where both repulsion and attraction are screened (neutral stability); (ii) systems with "soft" electrostatic repulsion, where the rheology is determined by the double layer repulsion; (iii) systems containing adsorbed nonionic surfactants or polymers resulting in steric repulsion; (iv) flocculated systems where

DOI 10.1515/9783110486872-010

the net interaction is attractive. One can distinguish between weakly flocculated systems where the attraction is a few kT units and strongly flocculated suspensions with hundreds kT attractive energy.

9.2 Rheological techniques

Evaluation of the stability/instability of suspensions without any dilution (which can cause significant changes in the structure of the system) requires carefully designed techniques that should cause as little disturbance to the structure as possible. The most powerful techniques that can be applied in any industrial laboratory are rheological measurements [1–7]. These measurements provide accurate information on the state of the system, such as stability and flocculation. These measurements are also applied for the prediction of the long-term physical stability of the suspension. The various rheological techniques that can be applied and the measurement procedures are listed below.

–   *Steady state shear stress σ–shear rate γ measurements*: This requires the use of a shear rate controlled instrument. The results obtained can be fitted to models to obtain the yield value σ_β and the viscosity η as a function of shear rate. Time effects (thixotropy) can also be investigated.

–   *Constant stress (creep) measurements*: A constant is stress is applied on the system and the strain γ or compliance J (γ/σ) is followed as a function of time. By measuring creep curves at increasing stress values one can obtain the residual (zero shear) viscosity $\eta(0)$ and the critical stress σ_{cr}, that is the stress above which the structure starts to break down. σ_{cr} is sometimes referred to as the "true" yield value.

–   *Dynamic (oscillatory) measurements*: A sinusoidal stress or strain with amplitudes σ_0 and γ_0 is applied at a frequency ω (rad s^{-1}) and the stress and strain are simultaneously measured. For a viscoelastic system, as is the case with most formulations, the stress and strain amplitudes oscillate with the same frequency, but out of phase. The phase angle shift δ is measured from the time shift of the strain and stress sine waves. From σ_0, γ_0 and d one can obtain the complex modulus $|G^*|$, the storage modulus G' (the elastic component) and the loss modulus G'' (the viscous component). The results are obtained as a function of strain amplitude and frequency.

9.2.1 Steady state measurements

Most suspensions, particularly those with high volume fraction and/or containing rheology modifiers, do not obey Newton's law. This can be clearly shown from plots of shear stress σ versus shear rate as illustrated in Fig. 9.1. Five different flow curves can be identified: (a) Newtonian; (b) Bingham plastic; (c) pseudoplastic (shear thinning); (d) dilatant (shear thickening); (e) yield stress and shear thinning. The variation of viscosity with shear rate for the above five systems is shown in Fig. 9.2. Apart from the Newtonian flow (a), all other systems show a change of viscosity with applied shear rate.

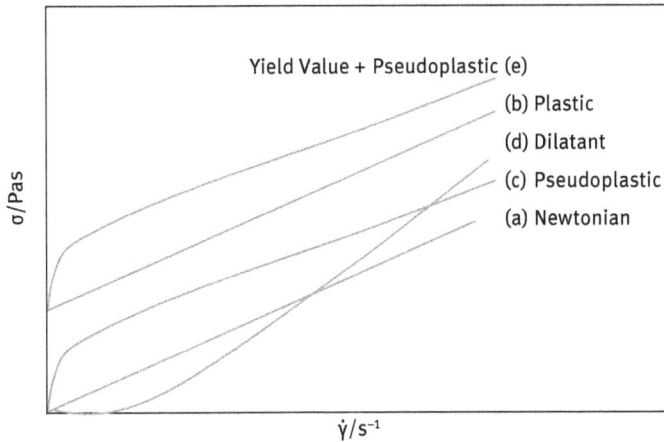

Fig. 9.1: Flow curves for various systems.

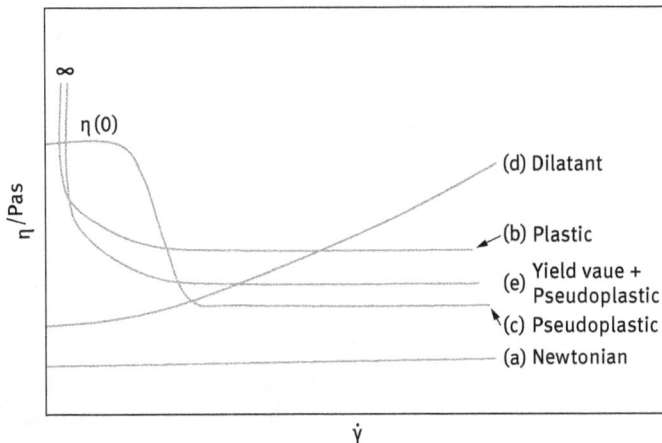

Fig. 9.2: Viscosity-shear rate relationship.

9.2.2 Rheological models for analysis of flow curves

9.2.2.1 Newtonian systems

$$\sigma = \eta \dot{\gamma} . \tag{9.3}$$

η is independent of the applied shear rate, e.g. simple liquids and very dilute dispersions.

9.2.2.2 Bingham plastic systems [8]

$$\sigma = \sigma_\beta + \eta_{pl} \dot{\gamma} . \tag{9.4}$$

The system shows a (dynamic) yield stress, σ_β, that can be obtained by extrapolation to zero shear rate. Clearly, at and below σ_β the viscosity $\eta \to \infty$. The slope of the linear curve gives the plastic viscosity η_{pl}. Some systems like clay suspensions may show a yield stress above a certain clay concentration.

The Bingham equation describes the shear stress/shear rate behaviour of many shear thinning materials at low shear rates. Unfortunately, the value of σ_β obtained depends on the shear rate ranges used for the extrapolation procedure.

9.2.2.3 Pseudoplastic (shear thinning) system

In this case the system does not show a yield value. It shows a limiting viscosity $\eta(0)$ at low shear rates (that is referred to as residual or zero shear viscosity). The flow curve can be fitted to a power law fluid model (Ostwald de Waele)

$$\sigma = k \dot{\gamma}^n , \tag{9.5}$$

where k is the consistency index and n is the shear thinning index $-n < 1$.

By fitting the experimental data to equation (9.5) one can obtain k and n. The viscosity at a given shear rate can be calculated,

$$\eta = \frac{\sigma}{\dot{\gamma}} = \frac{k \dot{\gamma}^n}{\dot{\gamma}} = k \dot{\gamma}^{n-1} . \tag{9.6}$$

The power law model (equation (9.5)) fits the experimental results for many non-Newtonian systems over two or three decades of shear rate. Thus, this model is more versatile than the Bingham model, although one should be careful in applying this model outside the range of data used to define it. In addition, the power law fluid model fails at high shear rates, where the viscosity must ultimately reach a constant value, i.e. the value of n should approach unity.

9.2.2.4 Dilatant (shear thickening) system

In some cases the very act of deforming a material can cause rearrangement of its microstructure such that the resistance to flow increases with increasing shear rate. In other words, the viscosity increases with applied shear rate and the flow curve can be fitted with the power law, equation (9.5), but in this case n > 1. The shear thickening regime extends over only about a decade of shear rate. In almost all cases of shear thickening, there is a region of shear thinning at low shear rates.

Several systems can show shear thickening such as wet sand, corn starch dispersed in milk and some polyvinyl chloride sols. Shear thickening can be illustrated when one walks on wet sand: some water is "squeezed out" and the sand appears dry. The deformation applied by one's foot causes rearrangement of the close-packed structure produced by the water motion. This process is accompanied by volume increase (hence the term dilatancy) as a result of "sucking in" of the water. The process amounts to a rapid increase in the viscosity.

9.2.2.5 Herschel–Bulkley general model [9]

Many systems show a dynamic yield value followed by a shear thinning behaviour. The flow curve can be analysed using the Herschel–Bulkley equation:

$$\sigma = \sigma_\beta + k\dot{\gamma}^n .\tag{9.7}$$

When $\sigma_\beta = 0$, equation (9.7) reduces to the power fluid model. When $n = 1$, equation (9.7) reduces to the Bingham model. When $\sigma_\beta = 0$ and $n = 1$, equation (9.7) becomes the Newtonian equation. The Herschel–Bulkley equation fits most flow curves with a good correlation coefficient and hence it is the most widely used model.

Several other models have been suggested of which the following are worth mentioning.

9.2.2.6 The Casson model [10]

This is a semi-empirical linear parameter model that has been applied to fit the flow curves of many paints and printing ink formulations,

$$\sigma^{1/2} = \sigma_C^{1/2} + \eta_C^{1/2}\dot{\gamma}^{1/2} .\tag{9.8}$$

Thus, a plot of $\sigma^{1/2}$ versus $\dot{\gamma}^{1/2}$ should give a straight line from which σ_C and η_C can be calculated from the intercept and slope of the line. One should be careful in using the Casson equation since straight lines are only obtained from the results above a certain shear rate.

9.2.2.7 The Cross equation [11]

This can be used to analyse the flow curve of shear thinning systems that show a limiting viscosity $\eta(0)$ in the low shear rate regime and another limiting viscosity $\eta(\infty)$ in the high shear rate regime. These two regimes are separated by a shear thinning behaviour as schematically shown in Fig. 9.3.

$$\frac{\eta - \eta(\infty)}{\eta(0) - \eta(\infty)} = \frac{1}{1 + K\dot{\gamma}^m} , \tag{9.9}$$

where K is a constant parameter with dimension of time and m is a dimensionless constant.

An equivalent equation to (9.9) is,

$$\frac{\eta_0 - \eta}{\eta - \eta_\infty} = (K\dot{\gamma}^m) . \tag{9.10}$$

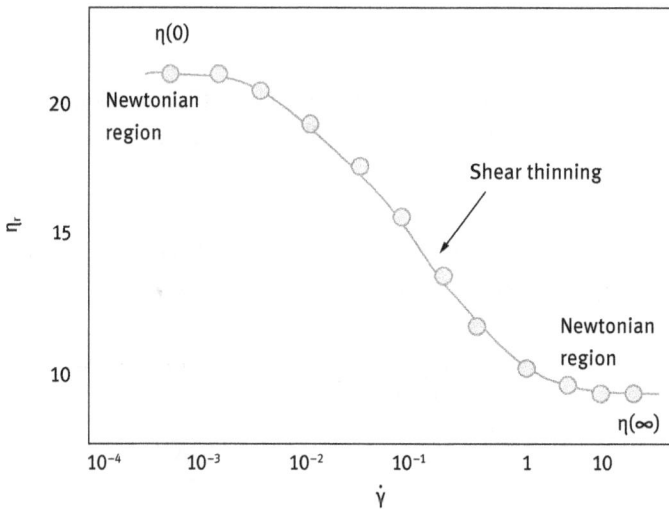

Fig. 9.3: Viscosity versus shear rate for shear thinning system.

9.2.3 Time effects during flow – thixotropy and negative (or anti-) thixotropy

When a shear rate is applied to a non-Newtonian system, the resulting stress may not be achieved simultaneously [1]. (i) The molecules or particles will undergo spatial rearrangement to follow the applied flow field. (ii) The structure of the system may change: breaking of weak bonds; aligning of irregularly shaped particles; collision of particles to form aggregates.

The above changes are accompanied by decreasing or increasing viscosity with time at any given shear rate. These changes are referred to as thixotropy (if viscosity decreases with time) or negative thixotropy or anti-thixotropy (if viscosity increases with time).

Thixotropy refers to the reversible time-dependent decease in viscosity. When the system is sheared for some time, viscosity decreases, but when the shear is stopped (the system is left to rest) the viscosity of the system is restored. Practical examples for systems that show thixotropy are: paint formulations (sometimes referred to as thixotropic paints); tomato ketchup; some hand creams and lotions.

Negative thixotropy or anti-thixotropy: when the system is sheared for some time, viscosity increases, but when the shear is stopped (the system is left to rest) viscosity decreases. A practical example of the above phenomenon is corn starch suspended in milk.

Generally speaking, two methods can be applied to study thixotropy in a suspension. The first and the most commonly used procedure is the loop test in which the shear rate is increased continuously and linearly in time from zero to some maximum value and then decreased to zero in the same way. This is illustrated in Fig. 9.4.

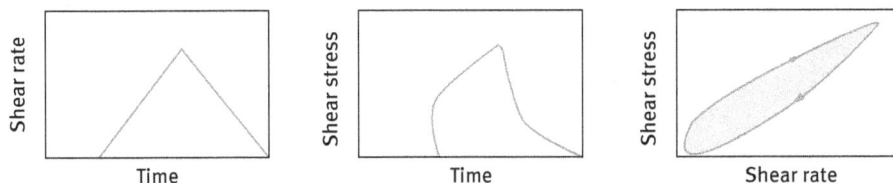

Fig. 9.4: Loop test for studying thixotropy.

The main problem with the loop test is the difficulty of interpreting the results. The nonlinear approach used is not ideal for developing loops, because by decoupling the relaxation process from the strain one does not allow the recovery of the material. However, the loop test gives a qualitative behaviour of the suspension thixotropy.

An alternative method for studying thixotropy is to apply a step change test, in which the suspension is suddenly subjected to a constant high shear rate and the stress is followed as a function of time whereby the structure breaks down and an equilibrium value is reached. The stress is further followed as a function of time to evaluate the rebuilding of the structure. A schematic representation of this procedure is shown in Fig. 9.5.

9.2.4 Constant stress (creep) measurements

A constant stress σ is applied on the system (that may be placed in the gap between two concentric cylinders or a cone and plate geometry) and the strain (relative deformation) γ or compliance J (= γ/σ, Pa^{-1}) is followed as a function of time for a period

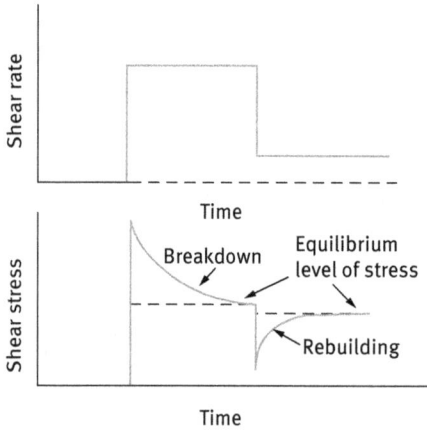

Fig. 9.5: Step change for studying thixotropy.

of t. At t = t, the stress is removed and the strain γ or compliance J is followed for another period t [1].

The above procedure is referred to as "creep measurement". From the variation of J with t when the stress is applied and the change of J with t when the stress is removed (in this case J changes sign) one can distinguish between viscous, elastic and viscoelastic responses, as illustrated in Fig. 9.6.

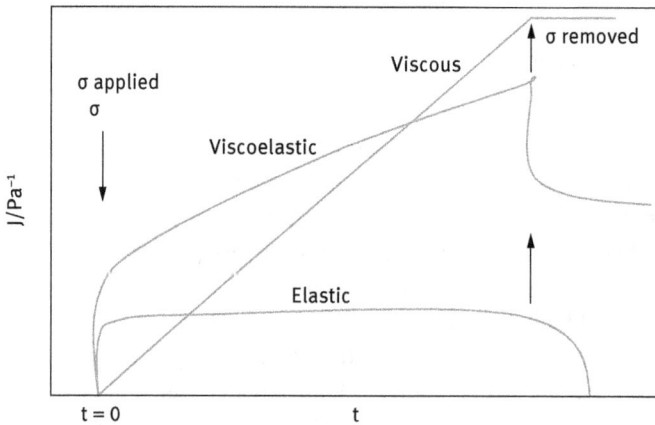

Fig. 9.6: Creep curves for viscous, elastic and viscoelastic responses.

Viscous response: In this case, the compliance J shows a linear increase with increasing time, reaching a certain value after time t. When the stress is removed after time t, J remains the same, i.e. in this case no creep recovery occurs.

Elastic response: In this case, the compliance J shows a small increase at t = 0 and it remains almost constant for the whole period t. When the stress is removed,

J changes sign and it reaches 0 after some time t, i.e. complete creep recovery occurs.

Viscoelastic response: At t = 0, J shows a sudden increase and this is followed by a slower increase for the time applied. When the stress is removed, J changes sign and shows an exponential decrease with increasing time (creep recovery), but it does not reach 0 as with the case of an elastic response.

9.2.4.1 Analysis of creep curves
9.2.4.1.1 Viscous fluid
The linear curve of J versus t gives a slope that is equal to the reciprocal viscosity

$$J(t) = \frac{\gamma}{\sigma} = \frac{\dot{\gamma}t}{\sigma} = \frac{t}{\eta(0)} \, . \tag{9.11}$$

9.2.4.1.2 Elastic solid
The increase of compliance at t = 0 (rapid elastic response) J(t) is equal to the reciprocal of the instantaneous modulus G(0)

$$J(t) = \frac{1}{G(0)} \, . \tag{9.12}$$

9.2.4.1.3 Viscoelastic response
Viscoelastic liquid. Figure 9.7 shows the case for a viscoelastic liquid whereby the compliance J(t) is given by two components: an elastic component, J_e, that is given by the reciprocal of the instantaneous modulus and a viscous component, J_v, that is given by $t/\eta(0)$

$$J(t) = \frac{1}{G(0)} + \frac{t}{\eta(0)} \, . \tag{9.13}$$

Figure 9.7 also shows the recovery curve, which gives $\sigma_0 J_e^0$, and when this is subtracted from the total compliance gives $\sigma_0 t/\eta(0)$.

The driving force for relaxation is spring and the viscosity controls the rate. The Maxwell relaxation time τ_M is given by,

$$\tau_M = \frac{\eta(0)}{G(0)} \, . \tag{9.14}$$

Viscoelastic solid. In this case complete recovery occurs as illustrated in Fig. 9.8 The system is characterized by a Kelvin retardation time τ_k that is also given by the ratio of $\eta(0)/G(0)$.

9.2.4.2 Creep procedure
In creep experiments one starts with a low applied stress (below the critical stress σ_{cr}, see below) at which the system behaves as a viscoelastic solid with complete recovery

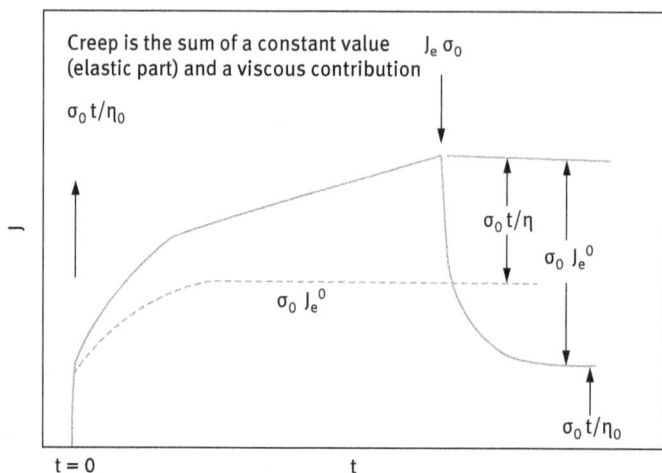

Fig. 9.7: Creep curve for a viscoelastic liquid.

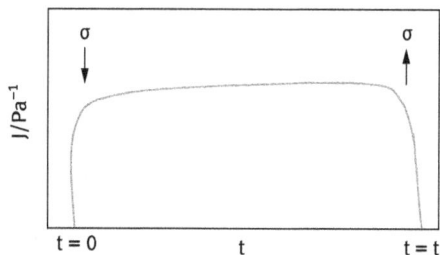

Fig. 9.8: Creep curve for a viscoelastic solid.

as is illustrated in Fig. 9.8. The stress is gradually increased and several creep curves are obtained. Above σ_{cr}, the system behaves as a viscoelastic liquid showing only partial recovery as illustrated in Fig. 9.7. Figure 9.9 shows a schematic representation of the variation of compliance J with time t at increasing σ (above σ_{cr}).

From the slopes of the lines one can obtain the viscosity η_s at each applied stress. A plot of η_s versus σ is shown in Fig. 9.10. This shows a limiting viscosity $\eta(0)$ below σ_{cr}, and above σ_{cr} the viscosity shows a sharp decrease with a further increase in σ. $\eta(0)$ is referred to as the residual or zero shear viscosity, which is an important parameter for predicting sedimentation. σ_{cr} is the critical stress above which the structure "breaks down". It is sometimes referred to as the "true" yield stress.

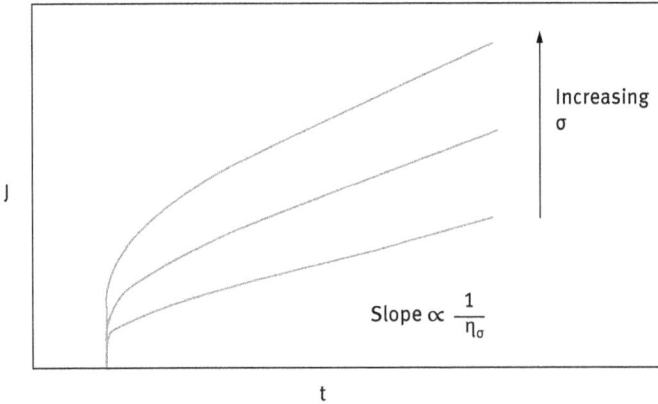

Fig. 9.9: Creep curves at increasing applied stress.

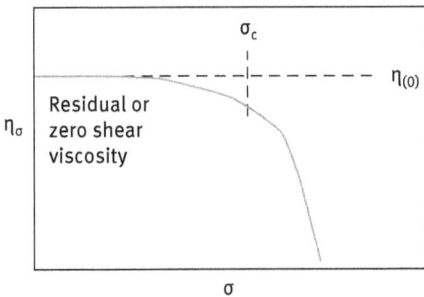

Fig. 9.10: Variation of viscosity with applied stress.

9.2.5 Dynamic (oscillatory) measurements [1]

This is the response of the material to an oscillating stress or strain. When a sample is constrained in, say, a cone and plate or concentric cylinder assembly, an oscillating strain at a given frequency ω (rad s^{-1}) ($\omega = 2v\pi$, where v is the frequency in cycles s^{-1} or Hz) can be applied to the sample. After an initial start-up period, a stress develops in response to the applied strain, i.e. it oscillates with the same frequency. The change of the sine waves of the stress and strain with time can be analysed to distinguish between elastic, viscous and viscoelastic response. Analysis of the resulting sine waves can be used to obtain the various viscoelastic parameters as discussed below.

Three cases can be considered:

Elastic response: where the maximum of the stress amplitude is at the same position as the maximum of the strain amplitude (no energy dissipation). In this case there no time shift between stress and strain sine waves.

Viscous response: where the maximum of the stress is at the point of maximum shear rate (i.e. the inflection point) where there is maximum energy dissipation. In this case the strain and stress sine waves are shifted by $\omega t = \pi/2$ (referred to as the phase angle shift δ which in this case is 90°).

Viscoelastic response: in this case the phase angle shift δ is greater than 0 but less than 90°.

9.2.5.1 Analysis of oscillatory response for a viscoelastic system

Let us consider the case of a viscoelastic system. The sine waves of strain and stress are shown in Fig. 9.11. The frequency ω is in rad s^{-1} and the time shift between strain and stress sine waves is Δt. The phase angle shift δ is given by (in dimensionless units of radians)

$$\delta = \omega \Delta t. \qquad (9.15)$$

As discussed before:

Perfectly elastic solid	$\delta = 0$
Perfectly viscous liquid	$\delta = 90°$
Viscoelastic system	$0 < \delta < 90°$

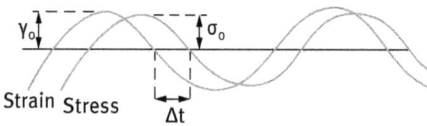

Δt = time shift for sine waves of stress and strain
$\Delta t\, \omega = \delta$, phase angle shift
ω = frequency in radian s^{-1}
$\omega = 2\,\pi\,\nu$
Perfectly elasic solid $\delta = 0$
Perfectly viscous liquid $\delta = 90°$
Viscoelastic system $0 < \delta < 90°$

Fig. 9.11: Strain and stress sine waves for a viscoelastic system.

The ratio of the maximum stress σ_0 to the maximum strain γ_0 gives the complex modulus $|G^*|$

$$|G^*| = \frac{\sigma_0}{\gamma_0}. \qquad (9.16)$$

The complex modulus can be resolved into G' (the storage or elastic modulus) and G'' (the loss or viscous modulus) using vector analysis and the phase angle shift δ as shown below.

9.2.5.2 Vector analysis of the complex modulus

$$G' = |G^*|\cos\delta\,, \tag{9.17}$$

$$G'' = |G^*|\sin\delta\,, \tag{9.18}$$

$$\tan\delta = \frac{G''}{G'}\,. \tag{9.19}$$

Dynamic viscosity η'

$$\eta' = \frac{G''}{\omega}\,. \tag{9.20}$$

Note that $\eta \to \eta(0)$ as $\omega \to 0$.

Both G' and G'' can be expressed in terms of frequency ω and Maxwell relaxation time τ_m by,

$$G'(\omega) = G\frac{(\omega\tau_m)^2}{1+(\omega\tau_m)^2}\,, \tag{9.21}$$

$$G''(\omega) = G\frac{\omega\tau_m}{1+(\omega\tau_m)^2}\,, \tag{9.22}$$

where G is the plateau modulus.

In oscillatory techniques one has to carry two types of experiments:

Strain sweep: the frequency ω is kept constant and G^*, G' and G'' are measured as a function of strain amplitude.

Frequency sweep: the strain is kept constant (in the linear viscoelastic region) and G^*, G' and G'' are measured as a function of frequency.

9.2.5.3 Strain sweep

The frequency is fixed, say at $1\,\mathrm{Hz}$ (or $6.28\,\mathrm{rad\,s^{-1}}$), and G^*, G' and G'' are measured as a function of strain amplitude γ_0. This is illustrated in Fig. 9.12. G^*, G' and G'' remain constant up to a critical strain γ_{cr}. This is the linear viscoelastic region where the moduli are independent of the applied strain. Above γ_{cr}, G^* and G' start to decrease whereas G'' starts to increase with a further increase in γ_0. This is the nonlinear region.

γ_{cr} may be identified with the critical strain above which the structure starts to "break down". It can also be shown that above another critical strain, G'' becomes higher than G'. This is sometimes referred to as the "melting strain" at which the system becomes more viscous than elastic.

9.2.5.4 Oscillatory sweep

The strain γ_0 is fixed in the linear region (taking a mid-point, i.e. not a too low strain where the results may show some "noise" and far from γ_{cr}). G^*, G' and G'' are then

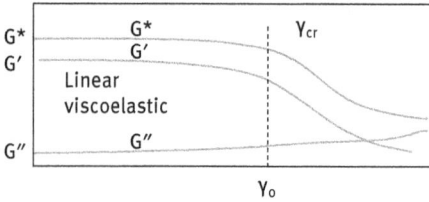

Linear viscoelastic region
G^*, G' and G'' are independent of strain amplitude
γ_{cr} is the critical strain above which systems shows
non-linear response (break down of structure)

Fig. 9.12: Schematic representation
of strain sweep.

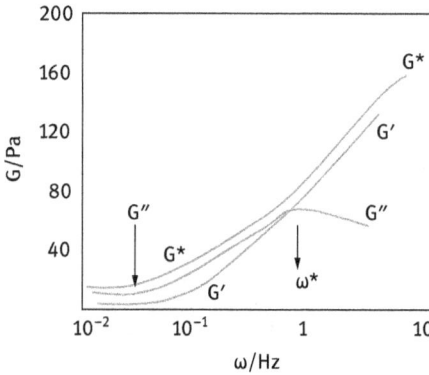

Fig. 9.13: Schematic representation of oscillatory measurements for a viscoelastic liquid.

measured as a function of frequency (a range of $10^{-3}-10^2$ rad s^{-1} may be chosen depending on the instrument and operator patience). Figure 9.13 shows a schematic
representation of the variation of G^*, G' and G'' with frequency ω (rad s^{-1}) for a
viscoelastic system that can be represented by a Maxwell model. One can identify a
characteristic frequency, ω^*, at which $G' = G''$ (the crossover point) and which can
be used to obtain the Maxwell relaxation time τ_m

$$\tau_m = \frac{1}{\omega^*}. \tag{9.23}$$

In the low frequency regime, i.e. $\omega < \omega^*$, $G'' > G'$. This corresponds to a long time experiment (time is reciprocal of frequency) and hence the system can dissipate energy
as viscous flow. In the high frequency regime, i.e. $\omega > \omega^*$, $G' > G''$. This corresponds
to a short time experiment where energy dissipation is reduced. At sufficiently high
frequency, $G' \gg G''$. At such high frequency $G'' \to 0$ and $G' \approx G^*$. The high frequency
modulus $G'(\infty)$ is sometimes referred to as the "rigidity modulus" where the response
is mainly elastic.

For a viscoelastic solid, G' does not become zero at low frequency. G'' still shows
a maximum at intermediate frequency. This is illustrated in Fig. 9.14.

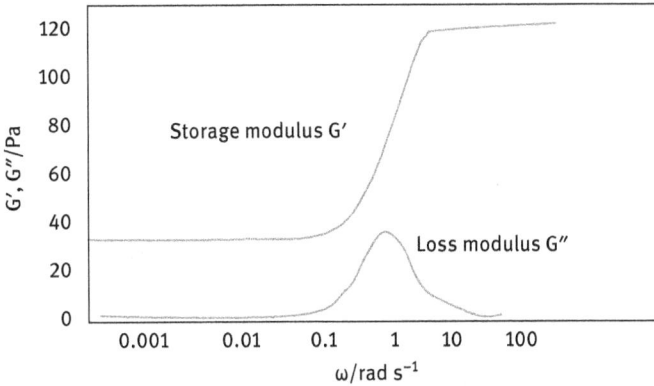

Fig. 9.14: Schematic representation for oscillatory measurements for a viscoelastic solid.

9.2.5.5 The cohesive energy density E_c

The cohesive energy density, which is an important parameter for identification of the "strength" of the structure in a dispersion, can be obtained from the change of G' with γ_0 (see Fig. 9.12).

$$E_c = \int_0^{\gamma_{cr}} \sigma \, d\gamma, \tag{9.24}$$

where σ is the stress in the sample that is given by,

$$\sigma = G'\gamma. \tag{9.25}$$

$$E_c = \int_0^{\gamma_{cr}} G'\gamma_{cr} \, d\gamma = \frac{1}{2}\gamma_{cr}^2 G'. \tag{9.26}$$

Note that E_c is given in $J\,m^{-3}$.

9.3 Rheology of suspensions

9.3.1 Dilute suspensions ($\phi \leq 0.01$) – the Einstein equation

The earliest theory for prediction of the relationship between the relative viscosity η_r and f was described by Einstein and is applicable to $\phi \leq 0.01$. Einstein [12] assumed that the particles behave as hard spheres (with no net interaction). The flow field has to dilate because the liquid has to move around the flowing particles. At $\phi \leq 0.01$ the disturbance around one particle does not interact with the disturbance around another. η_r is related to ϕ by the following expression [12],

$$\eta_r = 1 + [\eta]\phi = 1 + 2.5\phi , \tag{9.27}$$

where $[\eta]$ is referred to as the intrinsic viscosity and has the value 2.5.

For the above hard-sphere very dilute dispersions, the flow is Newtonian, i.e. the viscosity is independent of shear rate. At higher ϕ values ($0.2 > \phi > 0.1$) one has to consider the hydrodynamic interaction suggested by Batchelor [2] that is still valid for hard spheres.

9.3.2 Moderately concentrated suspensions ($0.2 > \phi > 0.01$) – the Bachelor equation [13]

When $\phi > 0.01$, hydrodynamic interaction between the particles become important. When the particles come close to each other the nearby streamlines and the disturbance of the fluid around one particle interact with that around a moving particle.

Using the above picture Batchelor [13] derived the following expression for the relative viscosity,

$$\eta_r = 1 + 2.5\phi + 6.2\phi^2 + 9\phi^3 . \tag{9.28}$$

The third term in equation (9.28), i.e. $6.2\phi^2$, is the hydrodynamic term whereas the fourth term is due to higher order interactions.

9.3.3 Rheology of concentrated suspensions

When $\phi > 0.2$, η_r becomes a complex function of ϕ. At such high volume fractions the system mostly shows non-Newtonian flow ranging from viscous to viscoelastic to elastic response. Three responses can be considered: (i) viscous response; (ii) elastic response; (iii) viscoelastic response. These responses for any suspension depend on the time or frequency of the applied stress or strain [1].

Four different types of systems (with increasing complexity) can be considered as described below.

(1) *Hard-sphere suspensions*: these are systems where both repulsive and attractive forces are screened.

(2) *Systems with "soft" interaction*: these are systems containing electrical double layers with long-range repulsion. The rheology of the suspension is determined mainly by the double layer repulsion.

(3) *Sterically stabilized suspensions*: the rheology is determined by the steric repulsion produced by adsorbed nonionic surfactant or polymer layers – the interaction can be "hard" or "soft" depending on the ratio of adsorbed layer thickness to particle radius (δ/R).

(4) *Flocculated systems*: these are systems where the net interaction is attractive. One can distinguish between weak (reversible) and strong (irreversible) flocculation depending on the magnitude of the attraction.

9.3.3.1 Rheology of hard-sphere suspensions

Hard-sphere suspensions (neutral stability) were developed by Krieger and co-workers [14, 15] using polystyrene latex suspensions whereby the double layer repulsion was screened by using NaCl or KCl at a concentration of 10^{-3} mol dm^{-3} or replacing water by a less polar medium such as benzyl alcohol.

The relative viscosity η_r $(= \eta/\eta_0)$ is plotted as a function of reduced shear rate (shear rate x time for a Brownian diffusion t_r),

$$\dot{\gamma}_{red} = \dot{\gamma} t_r = \frac{6\pi\dot{\gamma}a^3}{kT}, \tag{9.29}$$

where a is the particle radius, η_0 is the viscosity of the medium, k is the Boltzmann constant and T is the absolute temperature.

A plot of (η/η_0) versus $(\eta_0 a^3/(kT))$ is shown in Fig. 9.15 at $\phi = 0.4$ for particles with different sizes. At a constant ϕ, all points fall on the same curve. The curves are shifted to higher values for larger ϕ and to lower values for smaller ϕ.

The curve in Fig. 9.15 shows two limiting (Newtonian) viscosities at low and high shear rates that are separated by a shear thinning region. In the low shear rate regime, Brownian diffusion predominates over hydrodynamic interaction and the system shows a "disordered" three-dimensional structure with high relative viscosity. As the shear rate is increased, these disordered structure starts to form layers coincident

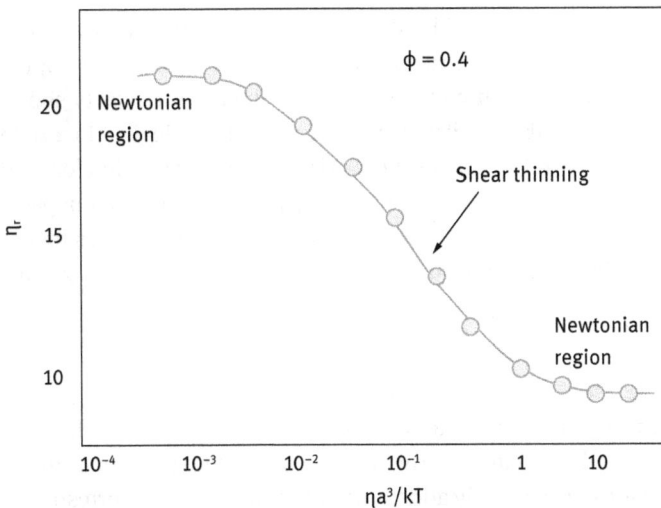

Fig. 9.15: Reduced viscosity versus reduced shear rate for hard-sphere suspensions.

with the plane of shear and this results in the shear thinning region. In the high shear rate regime, the layers can "slide" freely and hence a Newtonian region (with much lower viscosity) is obtained. In this region, hydrodynamic interaction predominates over Brownian diffusion.

If the relative viscosity in the first or second Newtonian region is plotted versus the volume fraction one obtains the curve shown in Fig. 9.16.

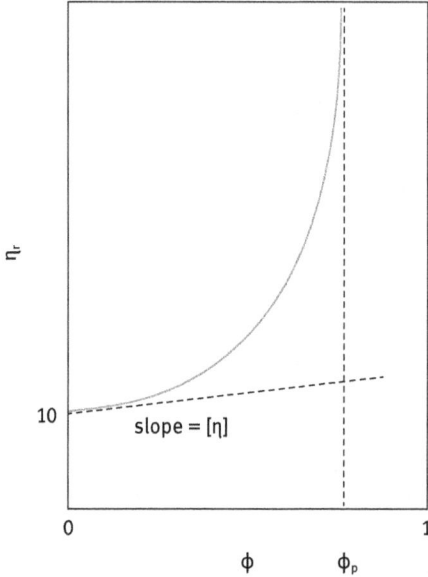

Fig. 9.16: Relative viscosity-volume fraction curve for hard-sphere dispersions.

The curve in Fig. 9.16 has two asymptotes: The slope of the linear portion at low ϕ values (the Einstein region) that gives the intrinsic viscosity $[\eta]$ that is equal to 2.5; the asymptote that occurs at a critical volume fraction ϕ_p at which the viscosity shows a sharp increase with increasing ϕ (the dispersed particles become locked in a rigid structure and flow ceases). ϕ_p is referred to as the maximum packing fraction for hard spheres: for hexagonal packing of equal sized spheres, $\phi_p = 0.74$. For random packing of equal sized spheres, $\phi_p = 0.64$. For polydisperse systems, ϕ_p reaches higher values as is illustrated in Fig. 9.17 for one-size, two-size, three-size and four-size suspensions.

9.3.3.1.1 Analysis of the viscosity-volume fraction curve
The best analysis of the $\eta_r - \phi$ curve is due to Dougherty and Krieger [14, 15] who used a mean field approximation by calculating the increase in viscosity as small increments of the suspension are consecutively added. Each added increment corresponds to replacement of the medium by more particles. They arrived at the following sim-

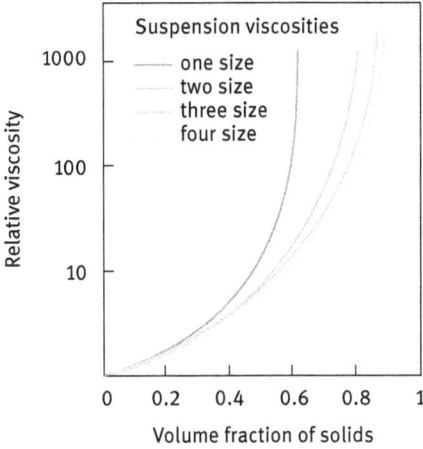

Fig. 9.17: Viscosity-volume fraction curves for polydisperse suspensions.

ple semi-empirical equation that could fit the viscosity data over the whole volume fraction range:

$$\eta_r = \left(1 - \frac{\phi}{\phi_p} \right)^{-[\eta]\phi_p} . \tag{9.30}$$

Equation (9.30) is referred to as the Dougherty–Krieger equation [14, 15] and is commonly used for analysis of the viscosity data.

9.3.3.2 Rheology of systems with "soft" or electrostatic interaction

In this case the rheology is determined by the double layer repulsion, particularly with small particles and extended double layers [16]. In the low shear rate regime the viscosity is determined by Brownian diffusion and the particles approach each other to a distance of the order of $\approx 4.5\, \kappa^{-1}$ (where κ^{-1} is the "double layer thickness" that is determined by electrolyte concentration and valency). This means that the effective radius of the particles, R_{eff}, is much higher than the core radius R. For example, for 100 nm particles with a zeta potential ζ of 50 mV dispersed in a medium of 10^{-5} mol dm^{-3} NaCl (κ^{-1} = 100 nm), $R_{eff} \approx 325$ nm. The effective volume fraction, ϕ_{eff}, is also much higher than the core volume fraction. This results in a rapid increase in the viscosity at low core volume fraction [16]. This is illustrated in Fig. 9.18, which shows the variation of η_r with ϕ at 5×10^{-4} and 10^{-3} mol dm^{-3} NaCl (R = 85 nm and ζ = 78 mV). The low shear viscosity $\eta_r(0)$ shows a rapid increase at $\phi \approx 0.2$ (the increase occurs at higher volume fraction at the higher electrolyte concentration). At $\phi > 0.2$, the system shows "solid-like" behaviour, with $\eta_r(0)$ reaching very high values (> 10^7). At such high ϕ values, the system shows near plastic flow.

In the high shear rate regime, the increase in η_r occurs at much higher ϕ values. This is illustrated by the plot of the high shear relative viscosity $\eta_r(\infty)$ versus ϕ. At such high shear rates, hydrodynamic interaction predominates over Brownian dif-

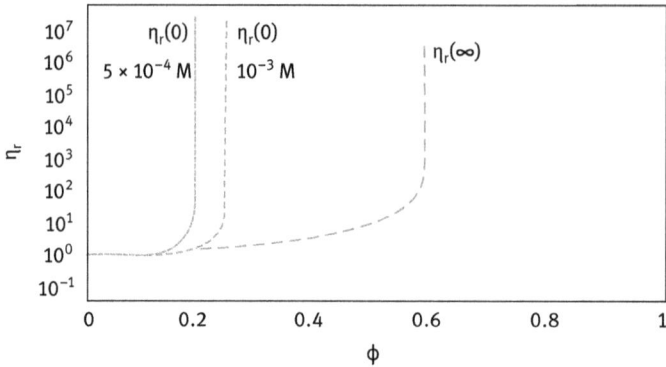

Fig. 9.18: Variation of η_r with ϕ for polystyrene latex dispersions at two NaCl concentrations.

fusion and the system shows a low viscosity denoted by $\eta_r(\infty)$. However, when ϕ reaches a critical value, pseudoplastic flow is observed.

9.3.3.2.1 Viscoelastic behaviour of electrostatically stabilized suspensions

One of the powerful techniques to study the interaction in electrostatically stabilized suspensions is to use dynamic (oscillatory) measurements [17, 18]. This is illustrated in Fig. 9.19, which shows the variation of the complex modulus G^*, storage (elastic) modulus G' and loss (viscous) modulus G'' versus core latex (with radius of 700 nm) volume fraction ϕ. The moduli were measured at low strain (in the linear region) and high frequency (1 Hz).

The trends obtained depend on the NaCl concentration. In 10^{-5} mol dm^{-3} the moduli values increase very rapidly at $\phi > 0.46$ and $G' \gg G''$. In 10^{-3} mol dm^{-3} the moduli values show the rapid increase at $\phi > 0.57$ and G'' is either lower or equal to G'.

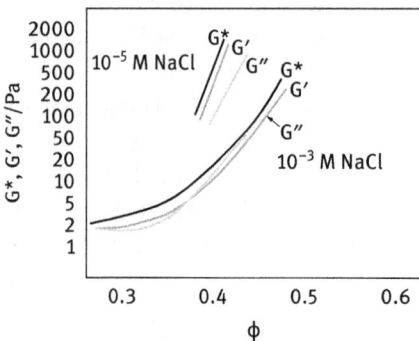

Fig. 9.19: Variation of G^*, G' and G'' with ϕ for electrostatically stabilized latex dispersions.

The above trend reflects the larger ϕ_{eff} values at low NaCl (with a more extended double layer) when compared to the higher NaCl concentration (with a smaller extension).

The above trends can be explained if one considers the presence of the double layer around the particles. To a first approximation, the double layer thickness $(1/\kappa)$ should be added to the particle radius, a, to obtain the effective radius a_{eff}. At any given particle size, a_{eff} depends on the electrolyte concentration C (and valency Z), since the double layer thickness is determined by C and Z. In 10^{-5} mol dm^{-3} NaCl, $(1/\kappa) = 100$ nm and $a_{eff} = 700 + 100 = 800$ nm; in 10^{-3} mol dm^{-3}, $(1/\kappa) = 10$ nm and $a_{eff} = 710$ nm.

The effective volume fraction of the dispersion, ϕ_{eff}, is related to ϕ by the expression,

$$\phi_{eff} = \phi \left[1 + \frac{(1/\kappa)}{a} \right]^3 . \tag{9.31}$$

In 10^{-5} mol dm^{-3} NaCl, $\phi_{eff} = 1.5 \phi$. At the lowest ϕ value studied, namely 0.46, $\phi_{eff} = 0.7$, which is above the maximum packing fraction (0.64 for random packing). In this case double layer interaction is strong and some overlap of the double layers may occur. The response in this case is predominantly elastic. At the highest volume fraction studied, namely 0.524, $\phi_{eff} = 0.79$ and the dispersion behaves as a near elastic solid; in this case $G' \approx G^*$. In 10^{-3} mol dm^{-3} NaCl, $\phi_{eff} = 1.05 \phi$ and up to the maximum volume fraction studied, namely 0.566, ϕ_{eff} is well below the maximum packing fraction. In this case there is little overlap of the double layers and the dispersions show more viscous than elastic response. To achieve predominantly elastic response, the volume fraction has to be increased above 0.6.

9.3.3.2.2 Elastic modulus (G')-distance (h) relation

The above results may be presented in another form by plotting G' versus h, the surface-to-surface separation between the particles in the dispersion [19, 20],

$$h = 2a \left[\left(\frac{\phi_p}{\phi} \right)^{1/3} - 1 \right] . \tag{9.32}$$

A value of $\phi_p = 0.68$ was used in the above calculation of the surface-to-surface distance of separation. Figure 9.20 shows plots of G' versus h for 10^{-5} and 10^3 mol dm^{-3} NaCl. In 10^{-5} mol dm^{-3} NaCl, G' increases very rapidly with decreasing h smaller than 200 nm (twice the double layer thickness). When h is below 100 nm, G' reaches very high values (> 1000 Pa). In 10^{-3} mol dm^{-3} NaCl, high values of G' are only reached when h < 50 nm.

It is clear from Fig. 9.20 that at any given h, G' in 10^{-5} mol dm^{-3} NaCl is several orders of magnitude higher than the corresponding value in 10^{-3} mol dm^{-3} NaCl. This trend is a direct consequence of the double layer repulsion, which at any given distance of separation is much stronger at the lower electrolyte concentration.

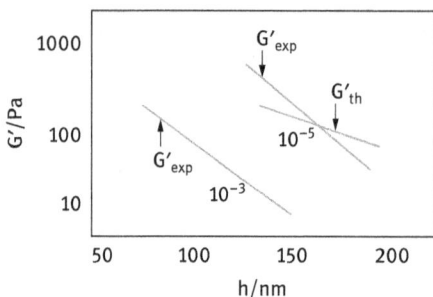

Fig. 9.20: Plots of G'_{th} and G'_{exp} at 10^{-5} and 10^{-3} mol dm^{-3} NaCl.

It is possible to relate the high frequency modulus, G_0, to the total energy of interaction between the particles, G_T,

$$G_0 = \left(\frac{\alpha}{d}\right)\left(\frac{\partial^2 G_T}{\partial d^2}\right), \tag{9.33}$$

$$\alpha = \left(\frac{3}{32}\right)\phi_p n, \tag{9.34}$$

where n is the coordination number and d is the distance of separation between the centres of the particles ($d = 2a + h$).

The total energy of interaction (mainly the double layer repulsion) is given by the expression,

$$G_T = \frac{4\pi\varepsilon\varepsilon_0\psi_d^2}{d}\exp[-\kappa(d - 2a)], \tag{9.35}$$

where ε is the permittivity of the medium, ε_0 is the permittivity of free space and ψ_d is the Stern potential.

By differentiating equation (9.35) twice, one can obtain the theoretical shear modulus, G'_{th},

$$G'_{th} = 4\pi\varepsilon\varepsilon_0 a^2\psi_d^2\left(\frac{\kappa^2 d^2 + 2\kappa d + 2}{d^4}\right)\exp[-\kappa(d - 2a)]. \tag{9.36}$$

Values of G'_{th} were calculated for dispersions in 10^{-5} mol dm^{-3} NaCl ($\kappa a < 10$) since these were highly elastic and the modulus showed little dependence on frequency; a value of $\alpha = 0.833$ was used in these calculations. The results of the calculations are shown in Fig. 9.20. The theoretical G' values increase less rapidly with decreasing h when compared with the experimental results. Calculation of G'_{th} using equation (9.36) is based on a number of assumptions (which may not be completely valid) and the latex dispersions used were relatively large (700 nm). The strains used may not be sufficiently small to ensure that the measurements were made in the linear viscoelastic region.

9.3.3.2.3 Scaling laws for dependence of G' on φ

Another useful way of describing the interaction in concentrated dispersions is to apply scaling laws for the dependence of G' on ϕ,

$$G' = k\phi^n . \tag{9.37}$$

In other words, the storage modulus scales with ϕ with an exponent n which depends on the interparticle interaction. The power n can be obtained from log-log plots of G' versus ϕ; in 10^{-5} mol dm^{-3} NaCl, n = 20, whereas in 10^{-3} mol dm^{-3} NaCl, n = 30. The lower power at the lower electrolyte concentration reflects the softness of the interaction as a result of the extended double layers. At the higher electrolyte concentration, the double layer is significantly compressed and the dispersions behave as near hard spheres.

9.3.3.2.4 Control of rheology of electrostatically stabilized suspensions

Three main parameters can control the rheology: (i) the volume fraction of the dispersion ϕ; (ii) the particle size (a) and shape distribution; (iii) the electrolyte concentration (C) and valency.

To produce dispersions with high ϕ values without much elasticity in the system, one needs to increase a and add small amounts of electrolyte (well below the flocculation value). In this case the double layer thickness is small compared to the particle radius and $\phi_{eff} \approx \phi$. To produce dispersions with high elasticity (and order) at low volume fractions, the particle radius, a, should be kept as small as possible and the electrolyte concentration, C, as low as possible.

The above principles are applied in practice to many industrial dispersions, e.g. charged lattices. These principles can also be applied in the formulation of many chemical products, e.g. paints, personal care products, agrochemicals and pharmaceuticals. In all these cases it is essential to know the various physicochemical parameters in the system, in order to control the rheology.

9.3.3.3 Rheology of sterically stabilized dispersions

These are dispersions where the particle repulsion results from the interaction between adsorbed or grafted layers of nonionic surfactants or polymers [21]. The flow is determined by the balance of viscous and steric forces. Steric interaction is repulsive as long as the Flory–Huggins interaction parameter $\chi < \frac{1}{2}$. With short chains, the interaction may be represented by a hard-sphere type with $R_{eff} = R + \delta$. This is particularly the case with nonaqueous dispersions with an adsorbed layer of thickness smaller than the particle radius (any electrostatic repulsion is negligible in this case). With most sterically stabilized dispersions, the adsorbed or grafted layer has an appreciable thickness (compared to particle radius) and hence the interaction is "soft" in nature as a result of the longer range of interaction. Results for aqueous ster-

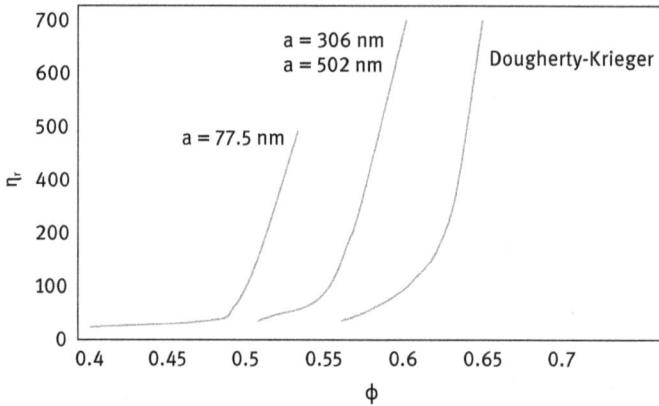

Fig. 9.21: $\eta_r - \phi$ curves for PS latex dispersions containing grafted PEO chains.

ically stabilized dispersions were produced using polystyrene (PS) latex with grafted poly(ethylene oxide) (PEO) layers [22, 23]. As an illustration, Fig. 9.21 shows the variation of η_r with ϕ for latex dispersions with three particle radii (77.5, 306 and 502 nm). For comparison, the $\eta_r - \phi$ curve calculated using the Dougherty–Krieger equation is shown on the same figure. The $\eta_r - \phi$ curves are shifted to the left as a result of the presence of the grafted PEO layers. The experimental relative viscosity data may be used to obtain the grafted polymer layer thickness at various volume fractions of the dispersions. Using the Dougherty–Krieger equation, one can obtain the effective volume fraction of the dispersion. From a knowledge of the core volume fraction, one can calculate the grafted layer thickness at each dispersion volume fraction.

To apply the Dougherty–Krieger equation, one needs to know the maximum packing fraction, ϕ_p. This can be obtained from a plot of $1/(\eta_r)^{1/2}$ versus ϕ and extrapolation to $1/(\eta_r)^{1/2}$, using the following empirical equation,

$$\frac{K}{\eta_r^{1/2}} = \phi_p - \phi. \tag{9.38}$$

The value of ϕ_p using equation (9.38) was found to be in the range 0.6–0.64. The intrinsic viscosity [η] was assigned a value of 2.5.

Using the above calculations, the grafted PEO layer thickness δ was calculated as a function of ϕ for the three latex dispersions. For the dispersions with R = 77.5 nm, δ was found to be 8.1 nm at $\phi = 0.42$, decreasing to 5.0 nm when ϕ was increased to 0.543. For the dispersions with R = 306 nm, $\delta = 12.0$ nm at $\phi = 0.51$, decreasing to 10.1 when ϕ was increased to 0.60. For the dispersions with R = 502 nm, $\delta = 21.0$ nm at $\phi = 0.54$ decreasing to 14.7 as ϕ was increased to 0.61.

9.3.3.3.1 Viscoelastic properties of sterically stabilized suspensions

The rheology of sterically stabilized dispersions is determined by the steric repulsion particularly for small particles with "thick" adsorbed layers. This is illustrated in Fig. 9.22, which shows the variation of G^*, G' and G'' with frequency (Hz) for polystyrene latex dispersions of 175 nm radius containing grafted poly(ethylene oxide) (PEO) with molecular weight of 2000 (giving a hydrodynamic thickness $\delta \approx 20$ nm) [23]. The results clearly show the transition from predominantly viscous response when $\phi \le 0.465$, to predominantly elastic response when $\phi \ge 0.5$. This behaviour reflects the steric interaction between the PEO layers. When the surface-to-surface distance h between the particles becomes $< 2\delta$, elastic interaction occurs and $G' > G''$.

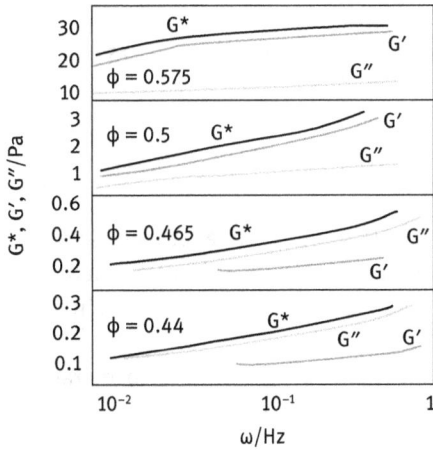

Fig. 9.22: Variation of G^*, G' and G'' with frequency for sterically stabilized dispersions.

The exact volume fraction at which a dispersion changes from predominantly viscous to predominantly elastic response may be obtained from plots of G^*, G' and G'' (at fixed strain in the linear viscoelastic region and fixed frequency) versus the volume fraction of the dispersion. This is illustrated in Fig. 9.23, which shows the results for the above latex dispersions. At $\phi = 0.482$, $G' = G''$ (sometimes referred to as the crossover point), which corresponds to $\phi_{eff} = 0.62$ (close to maximum random packing). At $\phi > 0.482$, G' becomes progressively larger than G'' and ultimately the value of G' approaches G^*, and G'' becomes relatively much smaller than G'. At $\phi = 0.585$, $G' \approx G^* = 4.8 \times 10^3$ and at $\phi = 0.62$, $G' \approx G^* = 1.6 \times 10^5$ Pa. Such high elastic moduli values indicate that the dispersions behave as near elastic solids ("gels") as a result of interpenetration and/or compression of the grafted PEO chains.

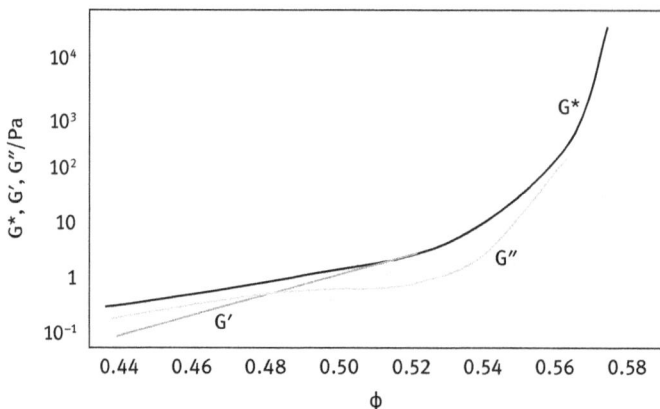

Fig. 9.23: Variation of G*, G′ and G″ (at ω = 1 Hz) with φ for latex dispersions (a = 175 nm) containing grafted PEO chains.

9.3.3.3.2 Correlation of the viscoelastic properties of sterically stabilized suspensions with their interparticle interactions

For this purpose one measures the energy E(D)-distance D curve for two mica cross cylinders containing adsorbed graft copolymer consisting of poly(methylmethacrylate) backbone with PEO chains (with similar molecular weight as that used in latex) [24–27].

The force between mica surfaces bearing the copolymer is converted to interaction energy between flat surfaces using the Deryaguin approximation,

$$E(D) = \frac{F(D)}{2\pi a},\qquad(9.39)$$

where D is the surface separation and a is the cylinder radius.

Figure 9.24 shows the energy-distance curve for mica sheets covered with the graft copolymer. The PEO side chain has molecular weight of 750 Daltons. This figure shows a monotonic exponential decrease of E(D) with increasing surface separation. Steric interaction starts at D ≈ 25 nm, which correspond to an adsorbed layer thickness of ≈ 12.5 nm. The same copolymer was used as a steric stabilizer for latex dispersions (with particle radius of 165 nm) for subsequent rheological measurements.

Using de Gennes scaling theory [28], it is possible to calculate E(D) as a function of D,

$$E(D) = \frac{\beta kT}{s^3}\left[\frac{(2L)^{2.25}}{1.25(D)^{1.25}} + \frac{D^{1.75}}{1.75(2L)^{0.75}}\right] - \left[\frac{2L}{1.25} + \frac{2L}{1.75}\right],\qquad(9.40)$$

where L is the stabilizer thickness on each layer, s is the distance between side chain attachment points, k is the Boltzmann constant, T is the absolute temperature and β is a numerical prefactor. L was taken to be 12.5 nm and this gave good agreement between theory and experiment.

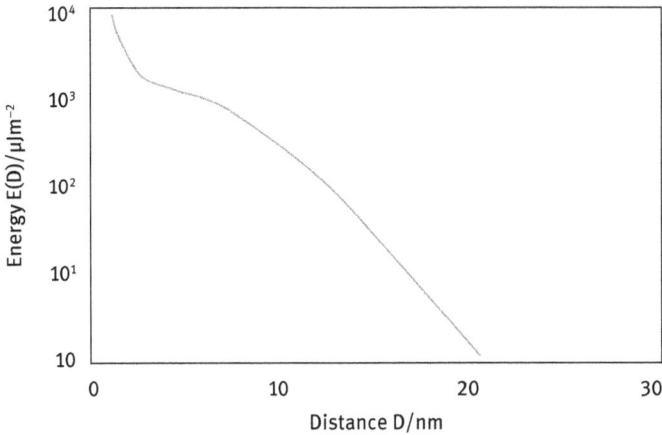

Fig. 9.24: Interaction energy E(D) versus separation distance D.

The high frequency modulus G_∞ was calculated using the following equation,

$$G_\infty = NkT + \frac{\phi_m n}{5\pi R^2}\left[4\frac{dV(R)}{dR} + R\frac{d^2V(R)}{dR^2}\right],\qquad(9.41)$$

where $V(R)$ is the potential of mean force, n is the coordination number and ϕ_m is the maximum packing fraction.

9.3.3.3.3 The high frequency modulus-volume fraction results

The results of Fig. 9.24 are given as interaction energy between flat plates. They can be converted to the force between spheres using the Deryaguin approximation,

$$G_\infty = NkT - \frac{\phi na}{5R^2}\left[4E(D) + R\frac{dE(D)}{dD}\right].\qquad(9.42)$$

Using equation (9.40) for $E(D)$, one can calculate G'_∞ as a function of core volume fraction ϕ. The results of these calculations are shown in Fig. 9.25 together with the experimental data. Figure 9.25 shows the expected trend of variation of G'_∞ with ϕ. However, the calculated values are about two orders of magnitude higher than the experimental results. By adjusting the numerical prefactor one can obtain good agreement between theory and experiment as indicated by the solid line in Fig. 9.25.

9.3.3.4 Rheology of flocculated suspensions

In a flocculated suspension, the flow curve is pseudoplastic, as illustrated in Fig. 9.26. The flow curve is characterized by three main parameters: (i) The shear rate $\dot\gamma_{cr}$ above which the flow curve shows linear behaviour. Above this shear rate collisions occur between the flocs and this may cause interchange between the flocculi (the smaller

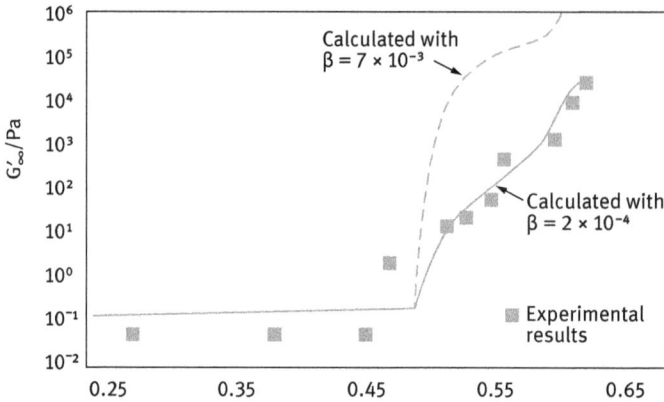

Fig. 9.25: G'_{∞} versus ϕ.

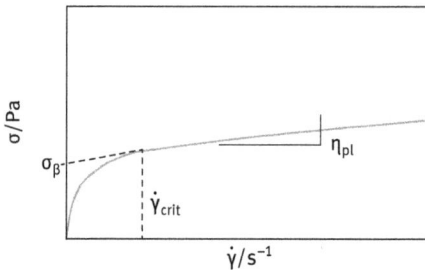

Fig. 9.26: Pseudoplastic flow curve for a flocculated suspension.

floc units that aggregate to form a floc). In this linear region, the ratio of the floc volume to the particle volume (ϕ_F/ϕ_p), i.e. the floc density, remains constant. (ii) σ_{β} the residual stress (yield stress) that arises from the residual effect of inter-particle potential. (iii) η_{pl}: the slope of the linear portion of the flow curve that arises from purely hydrodynamic effects.

Various theories have been put forward to explain the flow behaviour of flocculated suspensions, as will be discussed below. To obtain a satisfactory model of the flow process it is necessary to identify, at a microscopic level, the energy dissipation process; it should then be possible to calculate the forces necessary to produce any postulated structure or deformation from the characteristics of the fluid motion.

In a flocculated suspension it is assumed that a dynamic equilibrium is established between aggregate growth and destruction at any shear rate. High shear rates shift the equilibrium in the direction of better dispersion, whereas at low shear rates aggregation is favoured. Above $\dot{\gamma}_{cr}$ it is generally assumed that the aggregates have been broken down to single flocs or individual particles. Albers and Overbeek [29] postulated that the maximum hydrodynamic force, F_H, exerted on a pair of flow units at $\dot{\gamma}_{cr}$ must be equal to the maximum interaction force, F_{max}, between the flow units.

It is clear from the above discussion that development of a theoretical model for flocculated suspensions is not easy, since calculation of the various interactions for the different floc units is not straightforward. On the practical side, control of the rheology of flocculated and coagulated suspensions is difficult, since the rheology depends not only on the magnitude of the attractive energies but also on how one arrives at the flocculated or coagulated structures in question.

Various structures can be formed, e.g. compact flocs, weak and metastable structures, chain aggregates, etc. At high volume fraction of the suspension, a "three-dimensional" network of particles is formed throughout the sample. Under shear this network is broken down into smaller units of flocculated spheres which can withstand the shear forces [30]. The size of the units that survive is determined by the balance of shear forces which tend to break the units down and the energy of attraction that holds the spheres together [31–33]. The appropriate dimensionless group characterizing this process (balance of viscous and van der Waals force) is $\eta_0 a^4 \dot{\gamma}/A$ (where η_0 is the viscosity of the medium, a is the particle radius, $\dot{\gamma}$ is the shear rate and A is the effective Hamaker constant). Each flocculated unit is expected to rotate in the shear field, and it is likely that these units will tend to form layers as individual spheres do. As the shear stress increases, each rotating unit will ultimately behave as an individual sphere and, therefore, a flocculated suspension will show pseudoplastic flow with the relative viscosity approaching a constant value at high shear rates. The viscosity-shear rate curve will also show a pseudo-Newtonian region at low and high shear rates (similar to the case with stable systems described above). However, the values of the low and high shear rate viscosities (η_0 and η_∞) will of course depend on the extent of flocculation and the volume fraction of the suspension. It is also clear that such systems will show an apparent yield stress (Bingham yield value, σ_β) normally obtained by extrapolation of the linear portion of the $\sigma-\dot{\gamma}$ curve to $\dot{\gamma} = 0$. Moreover, since the structural units of a weakly flocculated system change with changing shear rate, most flocculated suspensions show thixotropy as discussed above. Once shear is initiated, some finite time is required to break the network of agglomerated flocs into smaller units which persist under the shear forces applied. As smaller units are formed, some of the liquid entrapped in the flocs is liberated, thereby reducing the effective volume fraction, ϕ_{eff}, of the suspension. This reduction in ϕ_{eff} is accompanied by a reduction in η_{eff} and this plays a major role in generating thixotropy.

It is convenient to distinguish between two types of unstable systems depending on the magnitude of the net attractive energy. (i) Weakly flocculated suspensions: the attraction in this case is weak (energy of few kT units) and reversible, e.g. in the secondary minimum of the DLVO curve or the shallow minimum obtained with sterically stabilized systems. A particular case of weak flocculation is that obtained on the addition of "free" (nonadsorbing) polymer referred to as depletion flocculation. (ii) Strongly flocculated (coagulated) suspensions: attraction in this case is strong (involving energies of several 100 kT units) and irreversible. This is the case of flocculation in the primary minimum or suspensions flocculated by reduction of solvency of the medium (for sterically stabilized suspensions) to worse than a θ-solvent. Study of the rheology of flocculated suspensions is difficult since the structure of the flocs is at nonequilibrium. Theories for flocculated suspensions are also qualitative and based on a number of assumptions.

9.3.3.4.1 Weakly flocculated suspensions

As mentioned in Chapter 6, weak flocculation may be obtained by the addition of "free" (nonadsorbing) polymer to a sterically stabilized dispersion [35]. Several rheological investigations of such systems have been carried out by Tadros and his collaborators [35–38]. This is exemplified by a latex dispersion containing grafted PEO chains of M = 2000 to which "free" PEO is added at various concentrations. The grafted PEO chains were made sufficiently dense to ensure absence of adsorption of the added free polymer. Three molecular weights of PEO were used: 20 000, 35 000 and 90 000. As an illustration, Fig. 9.27–9.29 show the variation of the Bingham yield value σ_β with volume fraction of "free" polymer ϕ_p at the three PEO molecular weights studied and at various latex volume fractions ϕ_s. The latex radius R in this case was 73.5 nm. The results of Fig. 9.27–9.29 show a rapid and linear increase in σ_β with increasing ϕ_p when the latter exceeds a critical value, ϕ_p^+. The latter is the critical free polymer volume fraction for depletion flocculation. ϕ_p^+ decreases with increasing molecular weight, M, of the free polymer, as expected. There does not seem to be any dependency of ϕ_p^+ on the volume fraction of the latex, ϕ_s.

Similar trends were obtained using larger latex particles (with radii 217.5 and 457.5 nm). However, there was a definite trend of the effect of particle size; the larger the particle size, the smaller the value of ϕ_p^+. A summary of ϕ_p^+ for the various molecular weights and particle sizes is given in Tab. 9.1.

The results in Tab. 9.1 show a significant reduction in ϕ_p^+ when the molecular weight of PEO is increased from 20 000 to 35 000, whereas when M is increased from 35 000 to 100 000, the reduction in ϕ_p^+ is relatively smaller. Similarly, there is a significant reduction in ϕ_p^+ when the particle radius is increased from 73.5 to 217.5 nm, with a relatively smaller decrease on further increase of a to 457.5 nm.

The straight line relationship between the extrapolated yield value and the volume fraction of free polymer can be described by the following scaling law,

$$\sigma_\beta = K\phi_s^m(\phi_p - \phi_p^+),\qquad(9.43)$$

where K is a constant and m is the power exponent in ϕ_s which may be related to the flocculation process. The values of m used to fit the data of σ_β versus ϕ_s are summarized in Tab. 9.2.

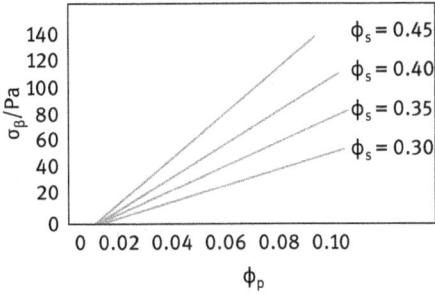

Fig. 9.27: Variation of yield value σ_β with volume fraction ϕ_p of "free polymer" (PEO; M = 20 000) at various latex volume fractions ϕ_s.

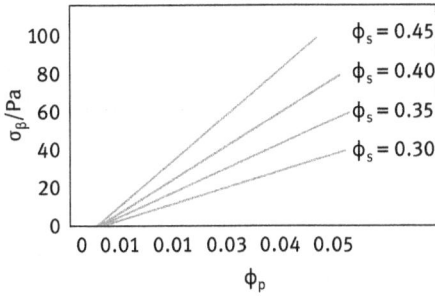

Fig. 9.28: Variation of yield value σ_β with volume fraction ϕ_p of "free polymer" (PEO; M = 35 000) at various latex volume fractions ϕ_s.

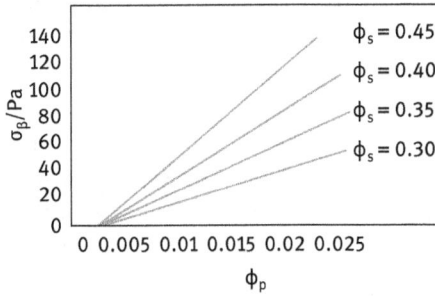

Fig. 9.29: Variation of yield value σ_β with volume fraction ϕ_p of "free polymer" (PEO; M = 100 000) at various latex volume fractions ϕ_s.

Tab. 9.1: Volume fraction of free polymer at which flocculation starts, ϕ_p^+.

Particle Radius / nm	M (PEO)	ϕ_p^+
73.5	20 000	0.0150
73.5	35 000	0.0060
73.5	10 000	0.0055
73.5	20 000	0.0150
217.5	20 000	0.0055
457.5	20 000	0.0050

Tab. 9.2: Power law plot for σ_β versus ϕ_s for various PEO molecular weights and latex radii.

| Latex R = 73.5 nm | | | | | | Latex R = 217.5 | | Latex R = 457.5 | |
| PEO 20 000 | | PEO 35 000 | | PEO 100 000 | | | | | |
ϕ_p	m	ϕ_p	m	ϕ_p	m	ϕ_p	m	ϕ_p	m
0.040	3.0	0.022	2.9	0.015	2.7	0.020	3.0	0.020	2.7
0.060	2.7	0.030	3.0	0.020	2.7	0.040	2.9	0.030	2.7
0.080	2.8	0.040	2.8	0.025	2.8	0.060	2.8	0.040	2.8
0.100	2.8	0.050	2.9	—	—	0.080	2.8	0.050	2.7

It can be seen from Tab. 9.2 that m is nearly constant, being independent of particle size and free polymer concentration. An average value for m of 2.8 may be assigned for such a weakly flocculated system. This value is close to the exponent predicted for diffusion controlled aggregation (3.5 ± 0.2) predicted by Ball and Brown [39, 40] who developed a computer simulation method treating the flocs as fractals that are closely packed throughout the sample.

The near independence of ϕ_p^+ from ϕ_s can be explained on the basis of the statistical mechanical approach of Gast et al. [41] which showed such independence when the osmotic pressure of the free polymer solution is relatively low and/or the ratio of the particle diameter to the polymer coil diameter is relatively large (> 8–9). The latter situation is certainly the case with the latex suspensions with diameters of 435 and 915 nm at all PEO molecular weights. The only situation where this condition is not satisfied is with the smallest latex and the highest molecular weight.

The dependency of ϕ_p^+ on particle size can be explained from a consideration of the dependency of free energy of depletion and van der Waals attraction on particle radius as will be discussed below. Both attractions increase with increasing particle radius. Thus the larger particles would require lower free polymer concentration at the onset of flocculation.

It is possible, in principle, to relates the extrapolated Bingham yield value, σ_β, to the energy required to separate the flocs into single units, E_{sep} [37, 38],

$$\sigma_\beta = \frac{3\phi_s n E_{sep}}{8\pi R^3}, \tag{9.44}$$

where n is the average number of contacts per particle (the coordination number). The maximum value of n is 12, which corresponds to hexagonal packing of the particles in a floc. For random packing of particles in the floc, n = 8. However, it is highly unlikely that such values of 12 or 8 are reached in a weakly flocculated system and a more realistic value for n is probably 4 (a relatively open structure in the floc).

In order to calculate E_{sep} from σ_β, one has to assume that all particle-particle contacts are broken by shear. This is highly likely since the high shear viscosity of the weakly flocculated latex was close to that of the latex before addition of the free polymer. Values of E_{sep} obtained using equation (9.44) with n = 4 are given in Tab. 9.3

Tab. 9.3: Results of E_{sep}, G_{dep} calculated on the basis of AO and FSV models.

ϕ_p	ϕ_s	σ_β / Ps	E_{sep} / kT	G_{dep} / kT	
				AO Model	FSV Model
(a) M(PEO) = 20 000					
0.04	0.30	12.5	8.4	18.2	78.4
	0.35	21.0	12.1	18.2	78.4
	0.40	30.5	15.4	18.2	78.4
	0.45	40.0	18.0	18.2	78.4
(b) M(PEO) = 35 000					
0.03	0.30	17.5	11.8	15.7	78.6
	0.35	25.7	14.8	15.7	78.6
	0.40	37.3	18.9	15.7	78.6
	0.45	56.8	25.5	15.7	78.6
(c) M(PEO) = 100 000					
0.02	0.30	10.0	6.7	9.4	70.8
	0.35	15.0	8.7	9.4	70.8
	0.40	22.0	11.1	9.4	70.8
	0.45	32.5	14.6	9.4	70.8

at the three PEO molecular weights for the latex with the radius of 73.5 nm. It can be seen that E_{sep} at any given ϕ_p increases with increasing volume fraction, ϕ_s, of the latex.

A comparison between E_{sep} and the free energy of depletion flocculation, G_{dep}, can be made using the theories of Asakura and Oosawa (AO) [42, 43] and Fleer, Vincent and Scheutjens (FVS) [43]. Asakura and Oosawa [42, 43] derived the following:

$$\frac{G_{dep}}{kT} = -\frac{3}{2}\phi_2\beta x^2; \quad 0 < x < 1, \tag{9.45}$$

where k is the Boltzmann constant, T is the absolute temperature, ϕ_2 is the volume concentration of free polymer that is given by,

$$\phi_2 = \frac{4\pi\Delta^3 N_2}{3v}. \tag{9.46}$$

Δ is the depletion layer thickness that is equal to the radius of gyration of free polymer, R_g, and N_2 is the total number of polymer molecules in a volume v of solution.

$$\beta = \frac{R}{\Delta}, \tag{9.47}$$

$$x = \frac{[\Delta - (h/2)]}{\Delta}, \tag{9.48}$$

where h is the distance of separation between the outer surfaces of the particles. Clearly when h = 0, i.e. at the point where the polymer coils are "squeezed out" from the region between the particles, x = 1.

Fleer, Scheutjens and Vincent (FSV model) [44] developed a general approach of the interaction of hard spheres in the presence of a free polymer. This model takes into account the dependence of the range of interaction on free polymer concentration and any contribution from the nonideal mixing of polymer solutions. This theory gives the following expression for G_{dep},

$$G_{dep} = 2\pi R \left(\frac{\mu_1 - \mu_1^0}{v_1^0} \right) \Delta^2 \left(1 + \frac{2\Delta}{3R} \right), \tag{9.49}$$

where μ_1 is the chemical potential at bulk polymer concentration ϕ_p, μ_1^0 is the corresponding value in the absence of free polymer and v_1^0 is the molecular volume of the solvent.

The difference in chemical potential $(\mu_1 - \mu_1^0)$ can be calculated from the volume fraction of free polymer ϕ_p and the polymer-solvent (Flory–Huggins) interaction parameter χ,

$$\frac{\mu_1 - \mu_1^0}{kT} = -\left[\frac{\phi_p}{n_2} + \left(\frac{1}{2} - \chi \right) \phi_p^2 \right], \tag{9.50}$$

where n_2 is the number of polymer segments per chain.

A summary of the values of E_{sep}, G_{dep} calculated on the basis of AO and FSV models is given in Tab. 9.3 at three molecular weights for PEO and for a latex with a = 77.5 nm. It can be seen from Tab. 9.3 that at any given ϕ_p, E_{sep} increases with increasing volume fraction ϕ_s of the latex. In contrast, the value of G_{dep} does not depend on ϕ_s, which is valid for the case where the particle radius is much larger than the polymer coil radius, on the value of ϕ_s. The theories on depletion flocculation only show a dependency of G_{dep} on ϕ_p and a. Thus, one cannot make a direct comparison between E_{sep} and G_{dep}. The close agreement between E_{sep} and G_{dep} using Asakura and Oosawa's theory [42, 43] and assuming a value of n = 4 should only be considered fortuitous.

Using Equations (9.43) and (9.44), a general scaling law may be used to show the variation of E_{sep} with the various parameters of the system,

$$E_{sep} = \frac{8\pi R^3}{3\phi_s n} K_1 \phi_s^{2.8} (\phi_p - \phi_p^+) = \frac{8\pi K_1}{3n} R^3 \phi_s^{1.8} (\phi_p - \phi_p^+). \tag{9.51}$$

Equation (9.51) shows the four parameters that determine E_{sep}: the particle radius a, the volume fraction of the suspension ϕ_s, the concentration of free polymer ϕ_p and the molecular weight of the free polymer, which together with a determines ϕ_p^+.

More insight on the structure of flocculated latex dispersions was obtained using viscoelastic measurements [37, 38]. As an illustration, Fig. 9.30 shows the variation of the storage modulus G' with ϕ_p (M = 20 000) at various latex (a = 77.5) volume fractions ϕ_s . Similar trends were obtained for the other PEO molecular weight. All results show the same trend, namely an increase in G' with increasing ϕ_p, reaching a plateau value at high ϕ_p values. These results are different from those obtained using steady state measurements, which show a rapid and linear increase of yield value σ_β. This difference reflects the behaviour when using oscillatory (low deformation) measurements, which cause little perturbation of the structure when using low amplitude and high frequency measurements. Above ϕ_p^+ flocculation occurs and G' increases in magnitude with a further increase in ϕ_p until a three-dimensional network structure is reached and G' reaches a limiting value. Any further increase in free polymer concentration may cause a change in the floc structure but this may not cause a significant increase in the number of bonds between the units formed (which determine the magnitude of G').

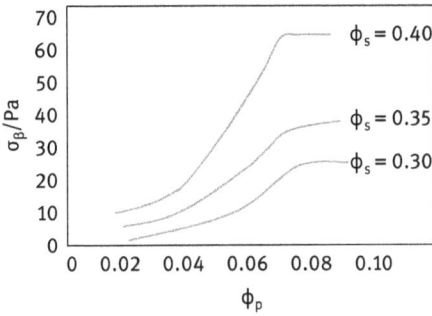

Fig. 9.30: Variation of storage modulus G' with volume fraction of polymer (PEO; M = 20 000).

9.3.3.4.2 Strongly flocculated (coagulated) suspensions

Steady state shear stress-shear rate curves show pseudoplastic flow curve as was illustrated in Fig. 9.26. Several theories were proposed to analyse the flow curve as illustrated below.

Impulse theory: Goodeve and Gillespie [45, 46]. The interparticle interaction effects (given by σ_β) and hydrodynamic effects (given by η_{pl}) are assumed to be additive,

$$\sigma = \sigma_\beta + \eta_{pl} + \dot{\gamma}. \tag{9.52}$$

To calculate σ_β Goodeve proposed that when shearing occurs, links between particles in a flocculated structure are stretched, broken and reformed. An impulse is transferred from a fast-moving layer to a slow-moving layer. Non-Newtonian effects are due to the effect of shear on the number of links, the lifetime of a link and any change in the size of the floc.

According to Goodeve's theory [45], the yield value is given by,

$$\sigma_\beta = \left(\frac{3\phi^2}{2\pi R^3} \right) E_A, \tag{9.53}$$

where ϕ is the volume fraction of the dispersed phase, a is the particle radius and E_A is the total binding energy.

$$E_A = n_L \varepsilon_L,$$ (9.54)

where n_L is the number of links with a binding energy ε_L per link.

According to equation (9.53): $\sigma_\beta \propto \phi^2$; $\propto (1/a^3)$; $\propto E_A$ (the energy of attraction). The relationship between σ_β and ϕ^2 has been confirmed experimentally by Hunter and co-workers [47, 48], by Michaels and Bolger [49] and by Higuchi and Stehle [50]. Moreover, the relationship between σ_β and the energy of interaction between pairs of particles has also been established.

Elastic floc model: Hunter and co-workers [47, 48]. The floc is assumed to consist of an open network of "girders", as schematically shown in Fig. 9.31. The floc undergoes extension and compression during rotation in a shear flow. The bonds are stretched by a small amount Δ (that can be as small as 1 % of the particle radius).

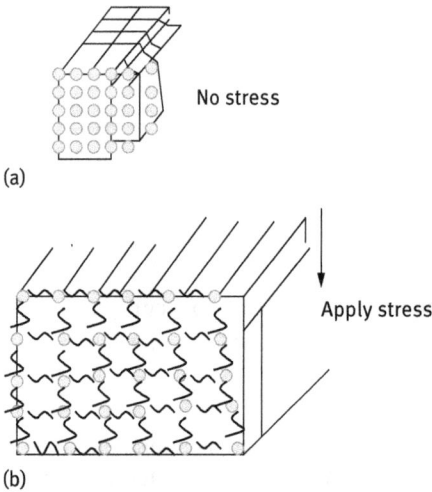

(a) No stress

(b) Apply stress

Fig. 9.31: Schematic picture of the elastic floc.

To calculate σ_β, Hunter and co-workers [47, 48] considered the energy dissipation during rupture of the flocs (assumed to consist of doublets). The most important contributions to the energy dissipation are: (i) a viscous energy dissipation due to the presence of spheres (i.e. spherical flocs); and (ii) the energy dissipation during rupture of the floc doublets. Firth and Hunter [48] considered various floc structures having different floc densities. The floc density, or the amount of branching within a floc, can be described by a quantity C_{FP}, which is the ratio of the volume fraction, ϕ_F, of the floc to ϕ_s, the volume fraction of suspended particles,

$$C_{FP} = \frac{\phi_F}{\phi_s}.$$ (9.55)

Thus, the more open the structure of a floc, the larger its C_{FP} value.

The various floc structures can be thought of as consisting of cubical unit cells, the sides of which contain 2, 3, 4, etc. particles. For simply counting the number of spheres and the number of sphere-sphere bonds in a unit cell, the ratio of the number of spheres, n_s, to the number of bonds, n_b, can be calculated. This ratio can be reasonably related to C_{FP} by the empirical relationship,

$$\frac{n_s}{n_b} = \frac{C_{FP} - 1}{C_{FP} + 0.7} .$$
(9.56)

The number of spheres per unit floc can be found using equation (9.55),

$$n_s = \frac{R_{floc}^3}{C_{FP}R^3} ,$$
(9.57)

where R_{floc} is the floc radius and R is the particle radius. Substituting (9.57) into (9.56) gives the number of bonds per floc,

$$n_b = \frac{R_{floc}^3}{R^3} \left[\frac{C_{FP} + 0.7}{C_{FP}(C_{FP} - 1)} \right] .$$
(9.58)

It has been observed [48] that the floc density C_{FP} is independent of the shear rate, but depends on the history of the system. Vadas et al. [51] showed that increasing the shear rate favours the formation of more spherical aggregates. This process is irreversible and hence one would expect the density of the floc to be determined by the highest shear rate $\dot{\gamma}_{CFP}$ to which the system is subjected. Experimental observations of C_{FP} at $\dot{\gamma} < \dot{\gamma}_{CFP}$ showed no measurable dependency on $\dot{\gamma}$, and hence the flocs formed at $\dot{\gamma}_{CFP}$ must, at $\dot{\gamma} < \dot{\gamma}_{CFP}$, be grouped in a rather close packed structure to maintain their C_{FP} value. The flocs formed at $\dot{\gamma}_{CFP}$ are called "flocculi" [47, 48], whereas at a given shear rate less than $\dot{\gamma}_{CFP}$, several flocculi form a close packed structure, called a "floc", and having the same C_{FP} value as the "flocculi". At a still lower shear rate, $\dot{\gamma} < \dot{\gamma}_{cr}$ the "flocs" form larger aggregates (floc doublets).

In order to calculate the Bingham yield value σ_β, one must consider the energy dissipation during rupture of the floc doublets. In the elastic floc model [47, 48], it is assumed that two elastic flocs when colliding can form a floc doublet. To break such a doublet, a certain amount of energy, which is supplied by the shear field, is required. This energy is obtained in two ways, namely the energy required to break the bonds between two flocs forming a doublet and the energy needed to stretch the bonds within the flocs (as the tension must be transmitted from the shear field to the floc-floc interface). Since the number of bonds between flocs is relatively small, the energy required to break these bonds is relatively small when compared with the energy due to bond stretching, as this involves many bonds. Clearly, the energy required to stretch one bond is smaller than that required to break it. Thus to calculate σ_β, one needs to calculate the energy required to stretch the bonds in a floc during collision within another floc. This energy consists of three parts: (i) the elastic energy required to overcome the interparticle forces which keeps the particles say in a primary energy minimum; (ii) the energy required to overcome the viscous drag as a result of displacement of particles within a floc due to stretching; (iii) the energy involved in floc shape and/or volume change during stretching, which is accompanied by some internal movement of the liquid within the floc.

The elastic energy E_e of bond stretching is given by [48],

$$E_e = \frac{R}{2} \left(\frac{n_F}{n_c} \right)^2 (d_i - d_o) Q_r , \tag{9.59}$$

where n_F is the number of particle links between two colliding flocs per unit area and n_c is the number of particle chains. d_i is the distance of closest approach between two particles and d_o is the distance at which the attractive force between them is a maximum. Q_r is the maximum attraction force between the particles with Q given by the sum of the attractive and the repulsive,

$$Q = \frac{A}{12 d_i^2} + \beta(d_i) \zeta^2 , \tag{9.60}$$

where A is the Hamaker constant, $\beta(d_i)$ is a function of distance d_i, the permittivity and ionic strength of the solution and ζ is the zeta potential.

The viscous energy E_s due to displacement of a particle within a floc can be calculated using Stokes' law. The force F_i needed to overcome the Stokes resistance during displacement of particles within a floc (as a result of stretching) is given by,

$$F_i = 6 \pi \lambda \eta V_i R , \tag{9.61}$$

where V_i is the velocity of stretching, η is the viscosity of the medium and λ is a correction factor, of the order of unity, due to the fact that the shear is shielded by other spheres [52]. The velocity V_i is equal to $i\delta$ (where δ is the increase in distance between the spheres as a result of stretching) divided by the time needed for stretching (i.e. about a quarter of floc rotation, i.e. $\pi/\dot{\gamma}$ in s). Hence,

$$V_i = \frac{i \delta \dot{\gamma}}{\pi} . \tag{9.62}$$

The energy E_i required to overcome the Stokes resistance is then $F_i i\delta$,

$$E_i = 6 \lambda \eta \dot{\gamma} R \delta^2 i^2 . \tag{9.63}$$

Because $2R_i = h$, the energy E_h for a sphere at a distance h from the floc centre equals $3\lambda\dot{\gamma}\delta^2 h^2 / 2R$. Averaging over all spheres in a floc yields, for the average energy dissipation E_s per sphere,

$$E_s = \frac{3}{4 \pi \rho R_{floc}^3} \int_{\infty}^{2} E_h 4 \pi \rho h^2 \, dh , \tag{9.64}$$

where ρ is the number of spheres per unit volume of floc. Assuming λ to be approximately constant, equation (9.64) reduces to,

$$E_s = \frac{0.9 \lambda \eta R_{floc}^2 \delta^2}{R} . \tag{9.65}$$

The viscous energy due to fluid movement inside the flocs during collision can be similarly calculated to be,

$$E = 0.9\pi^2 \lambda \eta \dot{\gamma} R R_{floc}^2 , \tag{9.66}$$

assuming that all the liquid inside the floc is freely mobile. If only a fraction, ΔV, of the total volume V of the fluid inside the floc is mobile, the energy dissipation, E_f, due to fluid motion inside a floc is given by,

$$E_f = \left(\frac{\Delta V}{V} \right) 0.9\pi^2 \lambda \eta \dot{\gamma} R R_{floc}^2 . \tag{9.67}$$

$\Delta V/V$ is approximately equal to δ/R and hence,

$$E_f = 0.9\pi^2 \lambda \eta \dot{\gamma} R_{floc}^2 \delta . \tag{9.68}$$

In the absence of floc-floc interaction, the energy dissipation is given by,

$$E_v = \dot{\gamma}^2 \eta_{pl} + \frac{\dot{\gamma}^* \lambda \eta \dot{\gamma}^2 R_{floc}^2 \delta \phi_p}{R^3} . \tag{9.69}$$

The first term on the right-hand side of equation (9.69) gives the energy dissipation for the flocs, whilst the second term is the energy dissipation due to the mobile liquid inside the floc. The asterisk in $\dot{\gamma}^*$ is meant to show that for sufficiently high ϕ_p, $\dot{\gamma}^* \ll \dot{\gamma}$.

The total energy dissipation during a collision is the sum of E_e, E_s and E_f. In order to know the energy dissipation per unit time, one has to know the collision frequency. This, for example, is given approximately by Smoluchowski's theory [53] as $3\phi_p^2 \dot{\gamma}/\pi^2 R_{floc}^3$. In this approximate expression both interparticle forces and hydrodynamic effects are neglected. Van de Ven and Mason [54] have improved Smoluchowski's theory. The number of doublets per unit area can be written as $3\alpha_0 \phi_p^2 \dot{\gamma}/\pi^2 R_{floc}^3$, where the orthokinetic capture efficiency, α_0, can be calculated numerically and is, in general, a function of the nature of the particles and the suspending fluid and the shear rate.

The energy dissipation, E_{DR}, dissipated per unit volume per unit time due to the rupture of the floc doublets is given using equations (9.59), (9.65) and (9.68),

$$E_{DR} = \frac{3\alpha_0 \phi_F^2 \dot{\gamma}}{\pi^2 R_{floc}^3} \left[\frac{2R_{floc}^3}{R^3} \frac{C_{FP} + 0.7}{C_{FP}(C_{FP} - 1)} E_e + \frac{2R_{floc}^3}{C_{FP} R^3}(E_s + E_f) \right] . \tag{9.70}$$

Substituting (9.55), (9.59), (9.65) and (9.68) into (9.70) gives,

$$E_{DR} = \frac{6\alpha_0 C_{FP} \phi_p^2 \dot{\gamma}}{\pi R_{floc}^3} \left[\frac{C_{FP} + 0.7}{C_{FP} - 1} \frac{R}{2} \left(\frac{n_f}{n_c} \right)^2 (d_1 - d_0)Q + 0.9\lambda \eta D \delta R_{floc}^2 \left(\frac{\delta}{R} + \pi^2 \right) \right] \tag{9.71}$$

Even for flocs consisting of very small spheres (of the order of few nm), $\delta/R \ll \pi$ so that $E_s \ll E_f$. Also inserting realistic values for A, d_0, d and R in equation (9.59) and noting that $n_f < n_c$ one finds $E_e \ll E_s$ in all cases. Thus, in most cases $E_e \ll E_s \ll E_f$

and hence the most dominant contribution to the energy dissipation is the internal liquid movement in the flocs and hence equation (9.71) simplifies to,

$$E_{DR} = \frac{\alpha_0 \beta \lambda \eta C_{FP} \dot{\gamma}^2 R_{floc}^2 \delta \phi_p^2}{R^3},$$ (9.72)

where $\beta = 27/5$.

The total energy dissipation, E_{tot}, is given by,

$$E_{tot} = E_v + E_{DR}$$ (9.73)

The shear stress $\sigma = E_{tot}/\dot{\gamma}$ and hence using equations (9.69), (9.72) and (9.73) this leads to,

$$\sigma = \lambda \eta \dot{\gamma} \frac{R_{floc}^2 \delta}{R^3} \phi_p (\alpha_0 \beta C_{FP} \phi_p + \dot{\gamma}^*) + \eta_{pl} \dot{\gamma}.$$ (9.74)

At high $\dot{\gamma}$, σ is approximately proportional to $\dot{\gamma}$, hence,

$$\sigma = \sigma_\beta + \eta_{pl} \dot{\gamma}.$$ (9.75)

From equations (9.74) and (9.75), the yield value σ_β is given by the expression,

$$\sigma_\beta = \lambda \eta \dot{\gamma} \left(\frac{R_{floc}^2 \delta}{R^3} \right) \phi_p (\alpha_0 \beta C_{FP} + \dot{\gamma}^*).$$ (9.76)

Thus, from equation (9.76), the Bingham yield value, σ_β, depends on the floc size, R_{floc}, the distance of stretching the bond in a floc, δ, and the volume fraction of the floc, ϕ_{floc}. In order to know whether experimental results for σ_β can be explained by equation (9.76), it is essential to know how the floc size, R_{floc}, varies with factors such as ϕ_{floc}, interaction forces between the spheres in a floc, shear rate $\dot{\gamma}$ and particle radius, R. Reich and Vold [55] measured the size of ferric oxide flocs, as a function of volume fraction, using turbidity measurements. They found that the increase of floc size with increasing ϕ_p is very substantial at low ϕ_p (when the concentration is increased from 0.002 to 0.2 %, particle size increases from 3–5 to 90 µm). Furthermore, as ϕ_p approaches 1 % the average distance between the flocs becomes of the same order of magnitude as R_{floc}. The increase in R_{floc} with ϕ_p should depend on the mechanism of breakup of the flocs. From a single statistical argument, one would expect R_{floc} to increase with ϕ_p, if larger flocs are formed by fusion of two flocs. However, if floc growth is determined by capture of single spheres, then R_{floc} should be independent of ϕ_p. The latter case has been observed by Kao and Mason [56], who found that the attractive forces are comparable to the hydrodynamic force on a single sphere. The same picture can be visualized for a system where the flocs are considered to be formed from aggregates of smaller flocculi, as suggested by van de Ven and Hunter [57]. In this case, the hydrodynamic force is comparable to the attractive force between these flocculi. The average number of flocculi in a floc will thus depend on the shear rate. At a given shear rate, dynamic equilibrium is established between the growth and breakup of flocs, particularly above $\dot\gamma_{cr}$. Thus, the whole process is identical to the case of floc growth governed by capture of single spheres, except the latter are replaced by flocculi. In this case R_{floc} becomes independent of ϕ_p [57].

The dependency of R_{floc} on shear rate is rather complex. For example, Fair and Gammel [58] showed that $R_{floc} \propto \dot\gamma^{-1.0}$, whereas Bagster and Tomi [59] reported that $R_{floc} \propto \dot\gamma^{-0.3}$. Van de Ven and Hunter [60] believed that the dependency of R_{floc} on $\dot\gamma$ should depend on whether or not flocs can be broken apart in the shear field. If the shear forces are not strong enough to break up the flocs, they will grow until a critical size is reached, beyond which no capture occurs [60]. The capture of particles in a shear flow is determined by the ratio of interparticle to hydrodynamic forces [60]. Since the interparticle force is proportional to R_{floc}, whereas the hydrodynamic force is proportional to ηDR_{floc}^2 [60], it follows that $R_{floc} \propto \dot\gamma^{-1.0}$, as has been observed for some systems [60, 61]. On the other hand, if R_{floc} is determined by a dynamic equilibrium between growth and breakup of flocs, a smaller dependency of R_{floc} on $\dot\gamma$ is expected and this will depend on the connections between flocculi forming a floc. Using the experimental data of Michaels and Bolger [49] for flocculated kaolin suspensions, $R_{floc} \propto \dot\gamma^{-0.3}$ [59].

The dependency of R_{floc} on interaction forces between the particles, namely the repulsive and attractive forces, depends on the magnitude of these forces. When the repulsive forces between the spheres are below a critical value (i.e. at low ζ-potential and/or high ionic strength), they play no part in the doublet formation frequency. Under these conditions, R_{floc} is independent of G_{elec}, but obviously depend on G_A, i.e. the Hamaker constant of the system.

Experimental measurements by Hunter et al. [47, 48] revealed that σ_β is proportional to ϕ_p^2, varies linearly with ζ^2 and G_A and is independent of $\dot\gamma$. Using equation (9.76), the average floc size has been calculated at $\dot\gamma = 10^3$ s^{-1} from the measured σ_β, assuming $\delta = 0.5$ nm, $\lambda = 1/3$, $\alpha_0 = 1$ and $\dot\gamma^* = 0$. The results are summarized in Tab. 9.4.

It can be seen from Tab. 9.4 that the floc sizes of all systems studied are of the order of micrometres. Also the order of magnitude of $(R_{floc}^2\delta/R^3)$ is, for all systems, rather similar, despite the large range of particle sizes involved. Microscopic observation of

Tab. 9.4: Average floc sizes calculated from the Bingham yield value [60].

Particle radius R / µm	$(R_{floc}^2\delta/R^3)$	R_{floc} / µm
0.11	36	10
0.22	27	24
0.07	390	16
0.28–1.25	0.6–1.2	6–48
0.0045–0.0115	165–550	0.3–0.6
1.3	18	276

the third system ($R = 0.07$) showed that the flocs range from 8 to 22 μm. This shows that the calculated value is within that range and, hence, the calculated floc sizes using equation (9.76) are of the correct order of magnitude. Thus, it seems that the elastic floc model described by Hunter et al. [47, 48] describes the flow behaviour of sheared flocculated suspensions in an adequate manner. It follows from this model that the energy dissipation due to liquid motion in flocs together with the viscous energy dissipation (due to the presence of flocs) are the most important factors in determining the flow behaviour of flocculated suspensions.

9.3.3.4.3 Fractal concept for flocculation

The floc structure can be treated as fractals whereby an isolated floc with radius a_F can be assumed to have uniform packing throughout that floc [62, 63].

In the above case the number of particles in a floc is given by

$$n_f = \phi_{mf} \left(\frac{R_F}{R} \right)^3 , \tag{9.77}$$

where ϕ_{mf} is the packing fraction of the floc.

If the floc does not have constant packing throughout its structure, but is dendritic in form, the packing density of the floc begins to reduce as one goes from the centre to the edge. If this reduction is with a constant power law D,

$$n_F = \left(\frac{R_F}{R} \right)^D , \tag{9.78}$$

where $0 < D \leq 3$.

D is called the packing index and it represents the packing change with distance from the centre. Two cases may be considered:

(i) *Rapid aggregation* (Diffusion Limited Aggregation, DLA). When particles touch, they stick. Particle-particle aggregation gives $D = 2.5$ while aggregate-aggregate aggregation gives $D = 1.8$. The lower the value of D, the more open the floc structure is.

(ii) *Slow aggregation* (Rate Limited Aggregation, RLA). The particles have a lower sticking probability; some are able to rearrange and densify the floc such that $D \approx 2.0$–2.2.

The lower the value of D, the more open the floc structure is. Thus by determining D, one can obtain information on the flocculation behaviour. If flocculation of a suspension occurs by changing the conditions (e.g. increasing temperature) one can visualize sites for nucleation of flocs occurring randomly throughout the whole volume of the suspension.

The total number of primary particles does not change and the volume fraction of the floc is given by,

$$\phi_F = \phi \left(\frac{R_F}{F} \right)^{3-D}. \tag{9.79}$$

Since the yield stress σ_β and elastic modulus G' depend on the volume fraction one can use a power law in the form,

$$\sigma_\beta = K\phi^m, \tag{9.80}$$

$$G' = K\phi^m, \tag{9.81}$$

where the exponent m reflects the fractal dimension.

Thus by plotting $\log \sigma_\beta$ or $\log G'$ versus $\log\phi$, one can obtain m from the slope, which can be used to characterize the floc nature and structure, $m = 2/(3 - D)$.

9.3.3.5 Examples of strongly flocculated (coagulated) suspension
9.3.3.5.1 Coagulation of electrostatically stabilized suspensions by addition of electrolyte

As mentioned in Chapter 4, electrostatically stabilized suspensions become coagulated when the electrolyte concentration is increased above the critical coagulation concentration (ccc). This is illustrated by using a latex dispersion (prepared using surfactant-free emulsion polymerization) to which $0.2\,\mathrm{mol\,dm^{-3}}$ NaCl is added (well above the ccc, which is $0.1\,\mathrm{mol\,dm^{-3}}$ NaCl).

Figure 9.32 shows the strain sweep results for latex dispersions at various volume fractions ϕ and in the presence of $0.2\,\mathrm{mol\,dm^{-3}}$ NaCl.

It can be seen from Fig. 9.32 that G^* and G' (which are very close to each other) remain independent of the applied strain (the linear viscoelastic region), but above a critical strain, γ_{cr}, G^* and G' show a rapid reduction with a further increase in strain (the nonlinear region). In contrast, G'' (which is much lower than G') remains constant showing an ill-defined maximum at intermediate strains. Above γ_{cr} the flocculated structure becomes broken down with applied shear.

Figure 9.33 shows the variation of G' (measured at strains in the linear viscoelastic region) with frequency ν (in Hz) at various latex volume fractions. As mentioned above, G' is almost equal to G^* since G'' is very low.

In all cases $G' \gg G''$ and it shows little dependency on frequency. This behaviour is typical of highly elastic (coagulated) structures, whereby a "continuous gel" network structure is produced at such high volume fractions. Scaling laws can be applied for the variation of G' with the volume fraction of the latex ϕ. A log-log plot of G' versus ϕ is shown in Fig. 9.34. This plot is linear and can be represented by the following scaling equation,

$$G' = 1.98 \times 10^7 \phi^{6.0}. \tag{9.82}$$

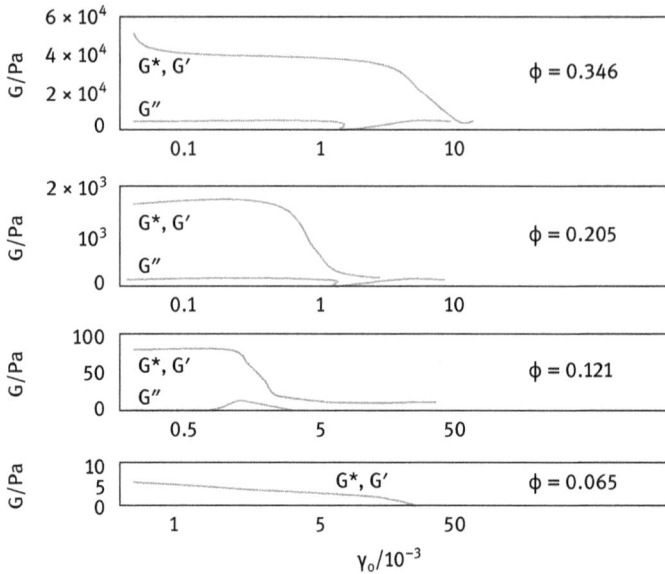

Fig. 9.32: Strain sweep results for latex dispersions at various volume fractions φ in the presence of 0.2 mol dm^{-3} NaCl.

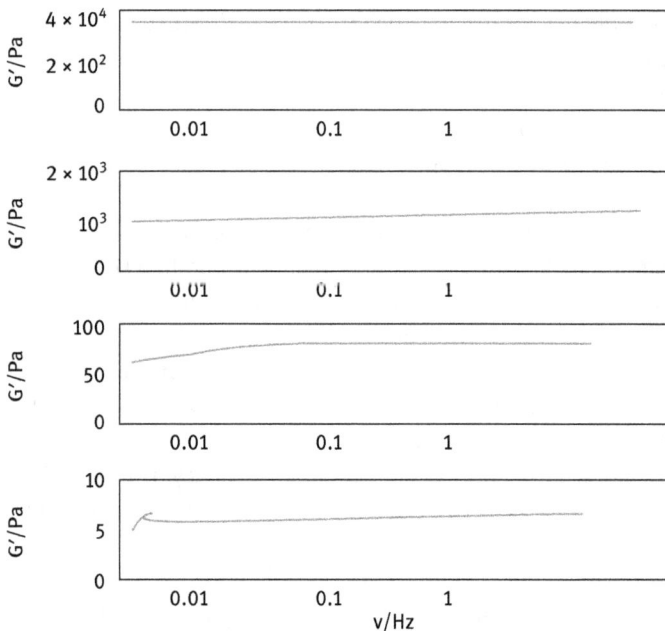

Fig. 9.33: Variation of G′ with frequency at various latex volume fractions.

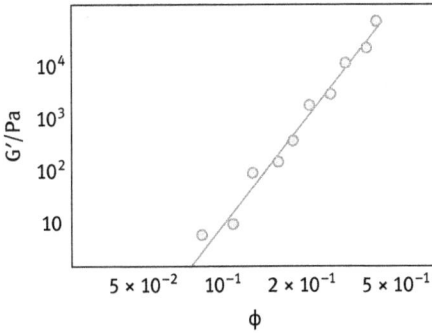

Fig. 9.34: Log-log plots of G' versus ϕ for coagulated polystyrene latex suspensions.

The high power in ϕ is indicative of a relatively compact coagulated structure. This power give a fractal dimension of 2.67 that confirms the compact structure.

It is also possible to obtain the cohesive energy of the flocculated structure, E_c, from a knowledge of G' (in the linear viscoelastic region) and γ_{cr}. E_c is related to the stress σ in the coagulated structure by the following equation,

$$E_c = \int_0^{\gamma_{cr}} \sigma \, d\gamma . \tag{9.83}$$

Since $\sigma = G'\gamma_0$, then,

$$E_c = \int_0^{\gamma_{cr}} \gamma_0 G' \, d\gamma = \frac{1}{2}\gamma_{cr}^2 G' . \tag{9.84}$$

A log-log plot of E_c versus ϕ is shown in Fig. 9.35 for such coagulated latex suspensions.

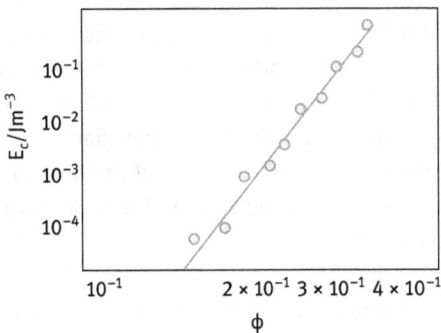

Fig. 9.35: Log-log plots of E_c versus ϕ for coagulated polystyrene latex suspensions.

The E_c versus ϕ curve can be represented by the following scaling relationship,

$$E_c = 1.02 \times 10^3 \phi^{9.1} . \tag{9.85}$$

The high power in ϕ is indicative of the compact structure of these coagulated suspensions.

9.3.3.6 Strongly flocculated sterically stabilized systems

9.3.3.6.1 Influence of addition of electrolyte

As mentioned in Chapter 8, sterically stabilized suspensions show strong flocculation (sometimes referred to as incipient flocculation) when the medium for the stabilizing chain becomes worse than a θ-solvent (the Flory–Huggins interaction parameter $\chi > 0.5$). Reduction of solvency for a PEO stabilizing chain can be achieved by addition of a nonsolvent for the chains or by addition of electrolyte such as Na_2SO_4. Above a critical Na_2SO_4 concentration (to be referred to as the critical flocculation concentration, CFC), the χ parameter exceeds 0.5 and this results in incipient flocculation. This process of flocculation can be investigated using rheological measurements without diluting the latex. This dilution may result in a change in the floc structure and hence investigations without dilution ensure absence of change of the floc structure, in particular when using low deformation (oscillatory) techniques [1].

Figure 9.36 shows the variation of extrapolated yield value, σ_β, as a function of Na_2SO_4 concentration at various latex volume fractions ϕ_s at 25 °C. The latex had a z-average particle diameter of 435 nm and it contained grafted PEO with M = 2000. It is clear that when $\phi_s < 0.52$, σ_β is virtually equal to zero up to 0.3 mol dm^{-3} Na_2SO_4, above which it shows a rapid increase in σ_β with a further increase in Na_2SO_4 concentration. When $\phi_s > 0.52$, a small yield value is obtained below 0.3 mol dm^{-3} Na_2SO_4, which may be attributed to the possible elastic interaction between the grafted PEO chains when the particle-particle separation is less than 2δ (where δ is the grafted PEO layer thickness). Above 0.3 mol dm^{-3} Na_2SO_4, there is a rapid increase in σ_β. Thus, the CFC of all concentrated latex dispersions is around 0.3 mol dm^{-3} Na_2SO_4. It should be mentioned that at Na_2SO_4 below the CFC, σ_β shows a measurable decrease with increasing Na_2SO_4 concentration. This is due to the reduction in the effective radius of the latex particles as a result of the reduction in the solvency of the medium for the chains. This accounts for a reduction in the effective volume fraction of the dispersion which is accompanied by a reduction in σ_β.

Figure 9.37 shows the results for the variation of the storage modulus G′ with Na_2SO_4 concentration. These results show the same trend as those shown in Fig. 9.36, i.e. an initial reduction in G′ due to the reduction in the effective volume fraction, follows by a sharp increase above the CFC (which is 0.3 mol dm^{-3} Na_2SO_4). Log-log plots of σ_β and G′ versus ϕ_s at various Na_2SO_4 concentrations are shown in Fig. 9.38

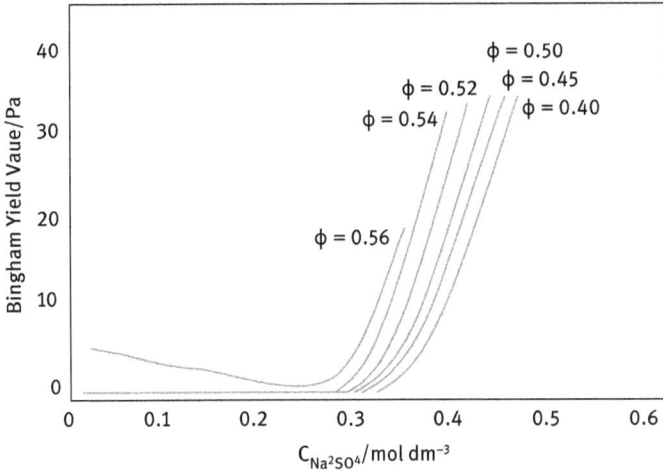

Fig. 9.36: Variation of Bingham yield value with Na_2SO_4 concentration at various volume fractions ϕ of latex.

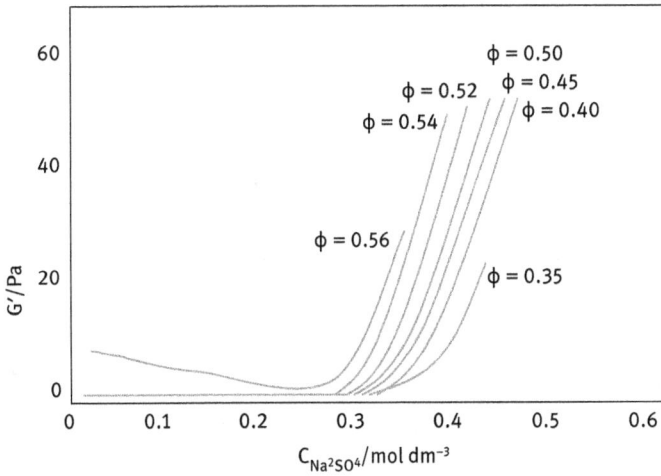

Fig. 9.37: Variation of storage modulus G' with Na_2SO_4 concentration at various volume fractions ϕ of latex.

and 9.39. All the data are described by the following scaling equations,

$$\sigma_\beta = k\phi_s^m \,, \tag{9.86}$$

$$G' = k'\phi_s^n \,, \tag{9.87}$$

with $0.35 < \phi_s < 0.53$.

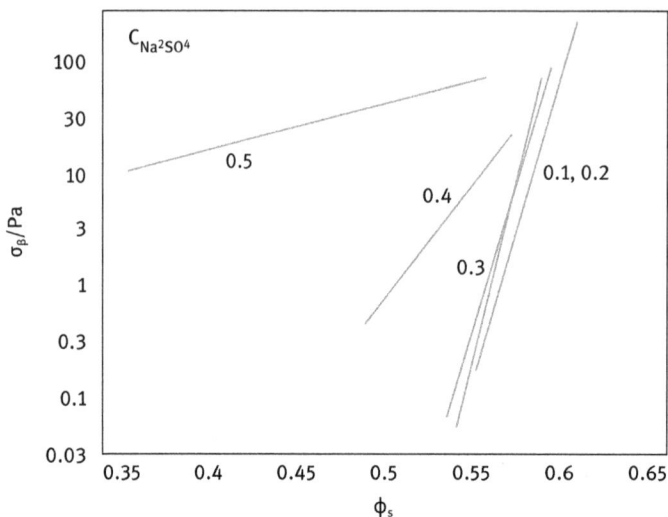

Fig. 9.38: Log-log plots of σ_β versus ϕ_s.

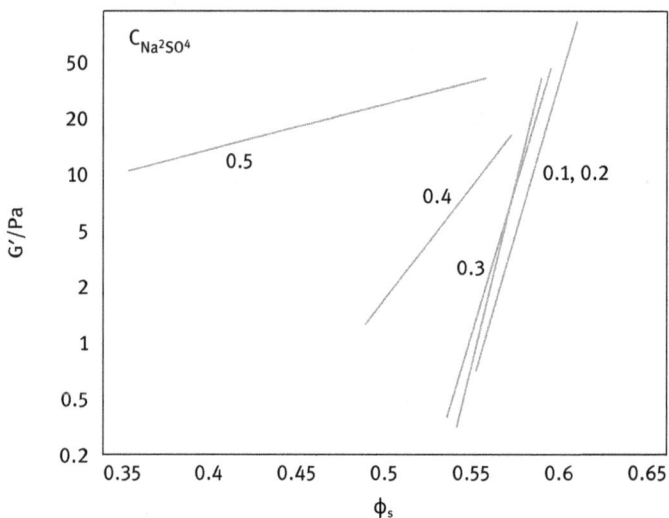

Fig. 9.39: Log-log plots of G' versus ϕ_s.

The values of m and n are very high at Na_2SO_4 concentrations below the CFC, reaching values in the range of 30–50, which indicate the strong steric repulsion between the latex particles. When the Na_2SO_4 concentration exceeds the CFC, m and n decrease very sharply reaching a value of m = 9.4 and n = 12 when the Na_2SO_4 concentration increases to 0.4 mol dm^{-3} and a value of m = 2.8 and n = 2.2 at 0.5 mol dm^{-3} Na_2SO_4.

These slopes can be used to calculate the fractal dimensions (see above) giving a value of D = 2.70–2.75 at 0.4 mol dm^{-3} and D = 1.6–1.9 at 0.5 mol dm^{-3} Na$_2$SO$_4$.

The above results of fractal dimensions indicate a different floc structure when compared with the results obtained using electrolyte to induce coagulation. In the latter case, D = 2.67, indicating a compact structure similar to that obtained at 0.4 mol dm^{-3} Na$_2$SO$_4$. However, when the Na$_2$SO$_4$ concentration exceeded the CFC value (0.5 mol dm^{-3} Na$_2$SO$_4$) a much more open floc structure with a fractal dimension less than 2 was obtained.

9.3.3.6.2 Influence of increasing temperature

Sterically stabilized dispersions with PEO chains as stabilizers undergo flocculation on increasing the temperature. At a critical temperature (Critical Flocculation Temperature, CFT) the Flory–Huggins interaction parameter becomes higher than 0.5 resulting in incipient flocculation. This is illustrated in Fig. 9.40, which shows the variation of the storage modulus G' and loss modulus G'' with increasing temperature for a latex dispersion with a volume fraction φ = 0.55 and at Na$_2$SO$_4$ concentration of 0.2 mol dm^{-3}. At this electrolyte concentration the latex is stable in the temperature range 10–40 °C. However, above this temperature (CFT) the latex is strongly flocculated.

The results of Fig. 9.40 show an initial systematic reduction in the moduli values with increasing temperature up to 40 °C. This is the result of reduction in solvency of the chains with increasing temperature. The latter increase causes a breakdown in the hydrogen bonds between the PEO chains and water molecules. This results in

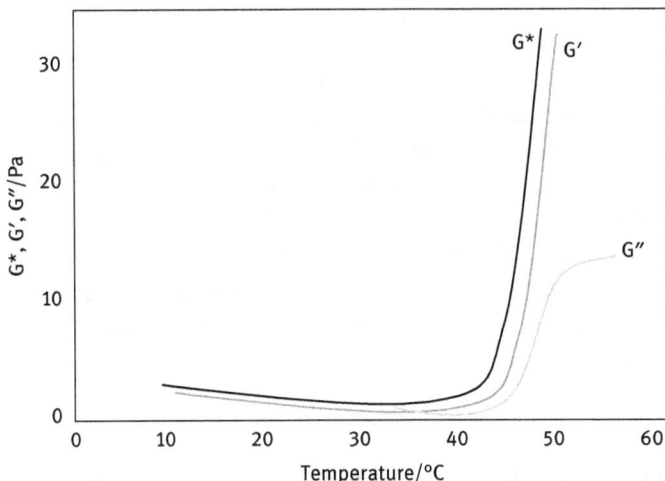

Fig. 9.40: Variation of G*, G' and G'' with temperature for latex dispersions (φ$_s$ = 0.55) in 0.2 mol dm^{-3} Na$_2$SO$_4$.

a reduction in the thickness of the grafted PEO chains and hence a reduction in the effective volume fraction of the dispersion. The latter causes a decrease in the moduli values. However, above 40 °C there is a rapid increase in the moduli values with a further increase in temperature. The latter indicates the onset of flocculation (the CFT). Similar results were obtained at 0.3 and 0.4 mol dm^{-3} Na$_2$SO$_4$, but in these cases the CFT was 35 and 15 °C respectively.

9.3.3.7 Models for interpretation of rheological results
9.3.3.7.1 Doublet floc structure model

Firth, Neville and Hunter [65] introduced a doublet floc model to deal with sterically stabilized dispersions which have undergone flocculation. They assumed that the major contribution to the excess energy dissipation in such pseudoplastic systems comes from the shear field which provides energy to separate contacting particles in a floc. The extrapolated yield value can be expressed as,

$$\sigma_\beta = \frac{3\phi_H^2}{2\pi^2(R+\delta)^2} E_{sep} , \qquad (9.88)$$

where ϕ_H is the hydrodynamic volume fraction of the particles that is equal to the effective volume fraction,

$$\phi_H = \phi_s \left[1 + \frac{\delta}{R}\right]^3 . \qquad (9.89)$$

$(R+\delta)$ is the interaction radius of the particle and E_{sep} is the energy needed to separate a doublet, which is the sum of van der Waals and steric attractions,

$$E_{sep} = \frac{AR}{12H_0} + G_s . \qquad (9.90)$$

At a particle separation of ≈ 12 nm (twice the grafted polymer layer thickness), the van der Waals attraction is very small (1.66 kT, where k is the Boltzmann constant and T is the absolute temperature) and the contribution of G_s to the attraction is significantly larger than the van der Waals attraction. Therefore, E_{sep} may be approximated to G_s.

From equation (9.88) one can estimate E_{sep} from σ_β. The results are shown in Tab. 9.5, which show an increase in E_{sep} with increasing ϕ_s.

The values of E_{sep} given in Tab. 9.5 are unrealistically high and hence the assumptions made for calculating E_{sep} are not fully justified and hence the data of Tab. 9.5 must be only considered as qualitative.

9.3.3.7.2 Elastic floc model

This model [57, 66, 67] has already been described above and it is based on the assumption that the structural units are small flocs of particles (called flocculi) which are characterized by the extent to which the structure is able to entrap the dispersion medium. A floc is made from an aggregate of several flocculi. The latter may range

Tab. 9.5: Results of E_{sep} calculated from σ_β for a flocculated sterically stabilized latex dispersion at various latex volume fractions.

0.4 mol dm^{-3} Na$_2$SO$_4$			0.4 mol dm^{-3} Na$_2$SO$_4$		
ϕ_s	σ_β / Pa	E_{sep} / kT	ϕ_s	σ_β / Pa	E_{sep} / kT
0.43	1.3	97	0.25	3.5	804
0.45	2.4	165	0.29	5.4	910
0.51	3.3	179	0.33	7.4	961
0.54	5.3	262	0.37	11.4	1170
0.55	7.3	336	0.41	14.1	1190
0.57	9.1	397	0.44	17.0	1240
0.58	17.4	736	0.47	21.1	1380
			0.49	23.1	1390
			0.52	28.3	1510

from a loose open structure (if the attractive forces between the particles are strong) to a very close-packed structure with little entrapped liquid (if the attractive forces are weak). In the system of flocculated sterically stabilized dispersions, the structure of the flocculi depends on the volume fraction of the solid and how far the system is from the critical flocculation concentration (CFC). Just above the CFC, the flocculi are probably close packed (with relatively small floc volume), whereas far above the CFC a more open structure is found which entraps a considerable amount of liquid. Both types of flocculi persist at high shear rates, although the flocculi with weak attraction may become more compact by maximizing the number of interactions within the flocculus.

As discussed above, the Bingham yield value is given by equation (9.30), which allows one to obtain the floc radius a_{floc} provided one can calculate the floc volume ratio C_{FP} (ϕ_F/ϕ_s) and assumes a value for Δ (the distance through which bonds are stretched inside the floc by the shearing force).

At high volume fractions, ϕ_F and hence C_{FP} can be calculated using the Krieger equation [14, 15],

$$\eta_{pl} = \eta_0 \left(1 - \frac{\phi_F}{\phi_s^m} \right)^{-[\eta]\phi_s^m} , \tag{9.91}$$

where η_0 is the viscosity of the medium, ϕ_s^m is the maximum packing fraction which may be taken as 0.74 and $[\eta]$ is the intrinsic viscosity taken as 2.5.

Assuming a value of Δ of 0.5 nm, the floc radius R_{floc} was calculated using equation (9.76). Figure 9.41 shows the variation of R_{floc} with latex volume fraction at the two Na$_2$SO$_4$ concentrations studied. At any given electrolyte concentration, the floc radius increases with increaing latex volume fraction, as expected. This can be understood by assuming that the larger flocs are formed by fusion of two flocs and the smaller flocs by "splitting" of the larger ones. From simple statistical arguments, one can predict that a_{floc} will increase with increasing ϕ_s because in this case larger flocs are favoured over smaller ones. In addition, at any given volume fraction of latex, the floc radius increases with increasing electrolyte concentration. This is consistent with the scaling results as discussed above.

The above results show clearly the correlation of viscoelasticity of flocculated dispersions with their interparticle attraction. These measurements allow one to obtain the CFC and CFT of concentrated flocculated dispersions with reasonable accuracy. In addition, the results obtained can be analysed using various models to obtain some characteristics of the flocculated structure, such as the "openness" of the network, the liquid entrapped in the floc structure and the floc radius. Clearly, several assumptions have to be made, but the trends obtained are consistent with expectations from theory.

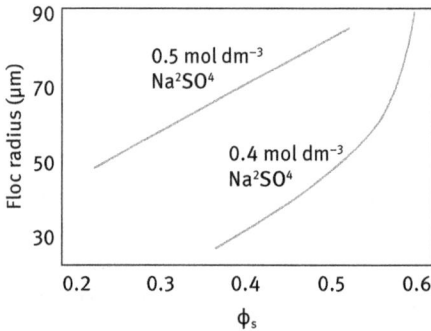

Fig. 9.41: Floc radius (a_{floc}) as a function of latex volume fraction at 0.4 and 0.5 mol dm^{-3} (above the CFC).

References

[1]   Tadros, Th. F., "Rheology of Dispersions", Wiley-VCH, Germany (2010).
[2]   Mackosko, C. W., "Rheology, Principles, Measurement and Applications", Wiley-VCH, New York (1994).
[3]   Goodwin, J. W., in "Surfactants", Th. F. Tadros (ed.), Academic Press, London (1984).
[4]   Goodwin, J. W. and Hughes, R. W., Advances Colloid Interface Sci., **42**, 303 (1992).
[5]   Tadros, Th. F., Advances Colloid Interface Science, **68**, 97 (1996).
[6]   Goodwin, J. W. and Hughes, R. W., "Rheology for Chemists", Royal Society of Chemistry Publication, Cambridge (2000).
[7]   Whorlow, R. W., "Rheological Techniques", John Wiley & Sons, New York (1980).
[8]   Bingham, E. C., "Fluidity and Plasticity", McGraw Hill, N. Y. (1922).
[9]   Herschel, W. H. and Bulkley, R., Proc. Amer. Soc. Test Materials, **26**, 621 (1926); Kolloid Z., **39**, 291 (1926).
[10] Casson, N., "Rheology of Disperse Systems", C. C. Mill (ed.), Pergamon Press, N. Y. (1959), pp. 84–104.
[11] Cross, M. M., J. Colloid Interface Sci., **20**, 417 (1965).
[12] Einstein, A., Ann. Physik., **19**, 289 (1906); **34**, 591 (1911).
[13] Bachelor, G. K., J. Fluid Mech., **83**, 97 (1977).
[14] Krieger, I. M. and Dougherty, T. J., Trans. Soc. Rheol., **3**, 137 (1959)
[15] Krieger, I. M., Advances Colloid and Interface Sci., **3**, 111 (1972).

[16] Goodwin, J. W. and Hughes, R. W., "Rheology for Chemists", Royal Society of Chemistry Publication, Cambridge (2000).

[17] Tadros, Th. F., "Rheology of Concentrated Stable and Flocculated Suspension", in "Flocculation and Dewatering", B. M. Mougdil and B. J. Scheiter (eds.), Engineering Foundation Publishers (1989), pp. 43–87.

[18] Tadros, Th. F., Advances Colloid Interface Sci., **68**, 97 (1996).

[19] Tadros, Th. F., Langmuir, **6**, 28 (1990).

[20] Tadros, Th. F. and Hopkinson, A., Faraday Disc. Chem. Soc., **90**, 41 (1990).

[21] Napper, D. H., "Polymeric Stabilisation of Colloidal Dispersions", Academic Press, London (1983).

[22] Liang, W., Tadros, Th. F. and Luckham, P. F., J. Colloid Interface Sci., **153**, 131 (1992).

[23] Prestidge, C. and Tadros, Th. F., J. Colloid Interface Sci., **124**, 660 (1988).

[24] de L. Costello, B. A., Luckham, P. F. and Tadros, Th. F., Colloids and Surfaces, **34**, 301 (1988/1989).

[25] Luckham, P. F., Ansarifar, M. A., de L. Costello, B. A. and Tadros, Th. F., Powder Technol., **65**, 371 (1991).

[26] Tadros, Th. F., Liang, W., de L. Costello, B. A. and Luckham, P. F., Colloids and Surfaces, **79**, 105 (1993).

[27] Luckham, P. F., Powder Technol., **58**, 75 (1989).

[28] de Gennes, P. G., Advances Colloid Inter. Sci,. **27**, 189 (1987).

[29] Albers, W. and Overbeek, J. Th. G., J. Colloid Sci., **15**, 589 (1960).

[30] Hoffman, R. L. in "Science and Technology of Polymer Colloids", G. W. Poehlein, R. H. Ottewill and J. W. Goodwin (eds.), Martinus Nijhoff Publishers, Boston, The Hague, Vol. II (1983), p. 570.

[31] Firth, B. A. and Hunter, R. J., J. Colloid Interface Sc., **57**, 248 (1976).

[32] van de Ven, T. G. M. and Hunter, R. J., Rheol. Acta, **16**, 534 (1976).

[33] Hunter, R. J. and Frayane, J., J. Colloid Interface Sci., **76**, 107 (1980).

[34] Heath, D. and Tadros, Th. F., Faraday Disc. Chem. Soc., **76**, 203 (1983).

[35] Prestidge, C. and Tadros, Th. F., Colloids and Surfaces, **31**, 325 (1988).

[36] Tadros, Th. F. and Zsednai, A., Colloids and Surfaces, **49**, 103 (1990).

[37] Liang, W., Tadros, Th. F. and Luckham, P. F., J. Colloid Interface Sci., **155** (1993).

[38] Liang, W., Tadros, Th. F. and Luckham, P. F., J. Colloid Interface Sci., **160**, 183 (1993).

[39] Ball, R. and Brown, W. D., Personal Communication.

[40] Buscall, R. and Mill, P. D. A., J. Chem. Soc. Faraday Trans. I, **84**, 4249 (1988).

[41] Gast, A. P., Hall, C. K. and Russel, W. B., J. Colloid Interface Sci., **96**, 251 (1983).

[42] Asakura, S. and Oosawa, F., J. Chem. Phy., **22**, 1255 (1954).

[43] Asakura, S. and Oosawa, F., J. Polym. Sci., **33**, 183 (1958).

[44] Fleer, G. J., Scheutjens, J. H. H. H. and Vincent, B., ACS Symposium Ser., **240**, 245 (1984).

[45] Goodeve, C. V., Trans. Faraday Soc., **35**, 342 (1939).

[46] Gillespie, T., J. Colloid Sci., **15**, 219 (1960).

[47] Hunter, R. J. and Nicol, S. K., J. Colloid Interface Sci., **28**, 200 (1968).

[48] Firth, B. A. and Hunter, R. J., J. Colloid Interface Sci., **57**, 248, 257, 266 (1976).

[49] Michaels, A. S. and Bolger, J. C., Ind. Eng. Chem. Fundamentals, **1**, 153 (1962).

[50] Higuchi, W. I. and Stehle, R. G., J. Pharm. Sci., **54**, 265 (1965).

[51] Vadas, E. B., Goldsmith, H. L. and Mason, S. G., J. Colloid Interface Sci., **43**, 630 (1973).

[52] Gluckman, M. J., Pfeffer, H. and Weinbaum, S., J. Fluid Mech., **50**, 705 (1971).

[53] von Smoluchowski, M., Z. Phys. Chem., **92**, 129 (1917).

[54] van de Ven, T. G. M. and Mason, S. G., Colloid and Polym. Sci., **255**, 468 (1977).

[55] Reich, T. and Vold, R. D., J. Phys. Chem., **63**, 1497 (1959).

[56] Kao, S. V. and Mason, S. G., Nature, **253**, 619 (1975).

[57] van de Ven, T. G. M. and Hunter, R. J., J. Colloid Interface Sci., **68**, 135 (1974).
[58] Fair, G. M. and Gemmel, R. S., J. Colloid Sci., **19**, 360 (1964).
[59] Bagster, D. F. and Tomi, D., Chem. Eng. Sci., **29**, 1773 (1974).
[60] van de Ven, T. G. M. and Hunter, R. J., Rheol. Acta, **16**, 4346 (1977).
[61] van de Ven, T. G. M. and Mason, S. G., J. Colloid Interface Sci., **57**, 505 (1976).
[62] Mills, P. D. A., Goodwin, J. W. and Grover, B., Colloid Polym. Sci., **269**, 949 (1991).
[63] Goodwin, J. W. and Hughes, R. W., Advances Colloid Interface Sci., **42**, 303 (1992).
[64] Liang, W., Tadros, Th. F. and Luckham, P. F., Langmuir, **9**, 2077 (1983).
[65] Firth, B. A., Neville, P. C. and Hunter, R. J., J. Colloid Interface Sci., **49**, 214 (1974).
[66] Hunter, R. J., Advances Colloid Interface Sci., **17**, 197 (1982).
[67] Friend, J. P. and Hunter, R. J., J. Colloid Interface Sci., **37**, 548 (1971).

10 Nonaqueous suspension concentrates

10.1 Introduction

Nonaqueous (oil-based) suspensions are currently used for formulation of many industrial applications, e.g. agrochemicals, in particular those which are chemically unstable in aqueous media, cosmetics, paints, etc. With agrochemical oil-based suspensions one can use oils (such as methyl oleate) that may enhance the biological efficacy of the active ingredient [1]. In addition, one may incorporate water insoluble adjuvants in the formulation. Many oil-based paints contain suspension particles that are used for decorative purposes. Cosmetic oil-based suspensions are frequently used in sunscreen applications, colour cosmetics, etc.

The most important criterion for the oil used is to have minimum solubility of the active ingredient, otherwise Ostwald ripening or crystal growth will occur on storage as discussed in Chapter 7.

With agrochemicals, the oil-based suspension concentrates have to be diluted in water to produce an oil-in-water emulsion. A self-emulsifiable system has to be produced and this requires the presence of the appropriate surfactants for self-emulsification. The surfactants used for self-emulsification should not interfere with the dispersing agent that is used to stabilize the suspension particles in the nonaqueous media. Displacement of the dispersing agent with the emulsifiers can lead to flocculation of the suspension.

To prevent sedimentation of the particles (since the density of the active ingredient is higher than that of the oil in which it is dispersed), an appropriate rheology modifier (antisettling agent) that is effective in the nonaqueous medium must be incorporated in the suspension as discussed in Chapter 8. This rheology modifier should not interfere with the self-emulsification process of the oil-based suspension for agrochemicals.

Two main types of nonaqueous suspensions may be distinguished: (i) Suspensions in polar media such as alcohol, glycols, glycerol and esters. These media have a relative permittivity $\varepsilon_r > 10$. In this case double layer repulsion plays an important role, in particular when using ionic dispersing agents. (ii) Suspensions in nonpolar media, $\varepsilon_r < 10$, such as hydrocarbons (paraffinic or aromatic oils) which can have a relative permittivity as low as 2. In this case charge separation and double layer repulsion are not effective and hence one has to depend on the use of dispersants that produce steric stabilization.

DOI 10.1515/9783110486872-011

10.2 Stability of suspensions in polar media

This follows the same principles as aqueous suspensions as discussed in Chapter 4, but one must take into account the "incomplete" dissociation of the ionic species which is the case when $\varepsilon_r < 40$. This could result in a low surface charge and hence a low zeta potential. However, the latter may be sufficient to produce an effective energy barrier that prevents any flocculation as explained by the Deryaguin–Landau–Verwey–Overbeek (DLVO) theory [2, 3].

Due to the lack of complete dissociation, one cannot use the electrolyte concentration for calculation of the thickness of the double layer. In this case one has to obtain the effective electrolyte concentration that must take into account the incomplete dissociation. The effective ionic concentration C can be determined from conductivity measurements and this allows one to calculate the double layer thickness $(1/\kappa)$.

$$\frac{1}{\kappa} = \left(\frac{\varepsilon_r \varepsilon_0 kT}{2e^2 N_A C 10^3} \right)^{1/2}, \tag{10.1}$$

where ε_0 is the permittivity of free space, k is the Boltzmann constant, T is the absolute temperature and N_A is the Avogadro constant.

Since the effective ionic concentration in a polar medium is less than that in water, the double layer will be more extended [4]. For polar substances in polar media, such as TiO_2 in glycol, the double layer charge may be sufficient for effective repulsion to occur and in this case a stable dispersion may be produced. For hydrophobic particles, such as is the case with agrochemicals stabilized with ionic surfactants or polyelectrolytes, the double layer charge is also effective for stabilization of the suspension.

In the above case, one may describe stability using the classical theory of colloid stability due to Deyaguin, Landau, Verwey and Overbeek (DLVO theory) [2, 3]. For systems where $\kappa a > 1$ (where a is the particle radius) and weak interaction, the electrostatic repulsion, G_{elec}, is given by the following expression,

$$G_{elec} = 4\pi \varepsilon_r \varepsilon_0 a \psi_0^2 \ln[1 + \exp(-\kappa h)]. \tag{10.2}$$

h is the surface-to-surface separation and ψ_0 is the surface potential which can be replaced by the measured zeta potential. G_{elec} decreases exponentially with increasing h and the rate of decay increases with increasing electrolyte concentration.

The van der Waals attraction is simply given by the expression [5],

$$G_A = -\frac{Aa}{12h}, \tag{10.3}$$

where A is the effective Hamaker constant,

$$A = (A_{11}^{1/2} - A_{22}^{1/2})^2. \tag{10.4}$$

A_{11} is the Hamaker constant of the particles and A_{22} is the Hamaker constant of the medium. G_A increases with decreasing h and at very short distances it reaches very high values.

The basis of the DLVO theory is to sum G_{elec} and G_A at all distances,

$$G_T = G_{elec} + G_A. \tag{10.5}$$

The general form of the G_T–h curve is schematically shown in Fig. 10.1, which shows the case at low electrolyte concentrations (see Chapter 4). The manner in which G_T varies with interparticle distance is determined by the way in which G_{el} and G_A vary with h. G_{el} decays exponentially with h and approaches 0 at large separation distances. G_A decays with h as an inverse power law and a residual attraction remains at large values of h. The G_T–h curve shows two minima (at long and short distances of separation) and one maximum (at intermediate distance), as illustrated in Fig. 10.1.

At large h, $G_A > G_{el}$, resulting in a shallow secondary minimum. At small h, $G_A \gg G_{el}$, resulting in a deep primary minimum. At intermediate h, $G_{el} > G_A$, resulting in an energy maximum G_{max}. Three cases can be distinguished: (i) G_{max} large (> 25 kT), resulting in a colloidally stable suspension. (ii) G_{max} small or absent, resulting in an unstable suspension (coagulated). (iii) G_{max} intermediate and G_{sec} deep (1–10 kT), resulting in weak (reversible) flocculation.

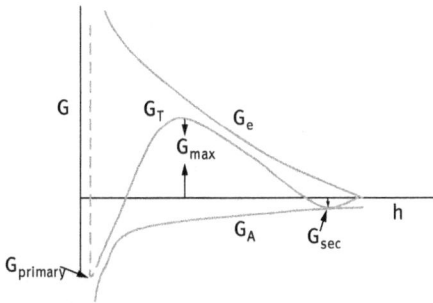

Fig. 10.1: Energy-distance curves according to DLVO theory.

It should be mentioned that for hydrophobic particles in polar solvents, stabilized by ionic surfactants, the surface charge may be very low and hence electrostatic stabilization cannot be achieved. For this case it is essential to use a dispersant that strongly adsorb on the particle surface (with an anchor B chain). This dispersant should contain stabilizing chains (A chains) that are strongly solvated by the medium. The use of block and graft copolymers for stabilization of particles in polar media follows the same principles that are applied for stabilization in nonpolar media. This stabilizing mechanism is referred to as steric stabilization that will be discussed below.

The choice of dispersants for stabilization of suspensions in polar media is not an easy task. The "anchor" chain B needs to be insoluble in the polar medium and has high affinity to the surface. Since most polymer chains have some solubility in nonaqueous polar media, it is very difficult to have strong adsorption to the particle surface. The stabilizing chains A have to be highly soluble in the polar solvent and strongly solvated by its molecules. This is not as difficult as the choice of the B chains.

For these reasons it is difficult to find commercially available dispersant for stabilization of hydrophobic particles in polar media.

10.3 stability of suspensions in nonpolar media

For nonpolar media, such as hydrocarbon oils (e.g. paraffinic oils), or some modified vegetable oils such as methyl oleate, that are commonly used for formulation of oil-based suspensions, the relative permittivity is lower than 10 and can reach values as low as 2. In this case the double layer is very extended, since the effective ionic concentration is very small ($\approx 10^{-10}$ mol dm^{-3}) and $(1/\kappa)$ can be as high as 10 µm. In this case, double layer repulsion plays a very minor role. Effective stabilization requires the presence of an adsorbed layer of an effective dispersant with particular properties (sometimes referred to as "protective colloid").

As mentioned in Chapter 5, the most effective dispersants are polymeric surfactants which may be classified into two main categories [6]: (i) homopolymers and (ii) block and graft copolymers. The homopolymers adsorb as random coils with tail-train-loop configurations. In most cases, there is no specific interaction between the homopolymer and the particle surface and it seldom can provide effective stabilization. The block and graft copolymers can provide effective stabilization providing they satisfy the following criteria: (i) Strong adsorption of the dispersant to the particle surface. This can be provided by a block B that is chosen to be insoluble in the nonaqueous medium and has some affinity to the particle surface. In the case when the affinity to the surface is not strong, one can rely on "rejection anchoring" whereby the insoluble B chain is rejected towards the surface as a result of its insolubility in the nonaqueous medium. (ii) Strongly solvated A blocks that provide effective steric stabilization as a result of their unfavourable mixing and loss of entropy when the particles approach each other in the suspension. (iii) A reasonably thick adsorbed layer to prevent any strong flocculation.

Based on the above principles various dispersants have been designed for suspensions in nonaqueous medium. One of the most effective stabilizing chains in nonaqueous media is poly(12-hydroxystearic acid) (PHS) that has a molar mass in the region of 1000–2000 Daltons. This chain is strongly solvated in most hydrocarbon oils (paraffinic or aromatic oils). It is also strongly solvated in many esters such as methyl oleate that is commonly used for oil-based suspensions. For the B chain, one can choose a polar chain such as polethylene imine or poyvinylpyrrolidone which is insoluble in most oils. The B chain could also be polystyrene or polymethylmethacrylate which is insoluble in aliphatic hydrocarbons and may have some affinity to the hydrophobic agrochemical particle.

10.4 Characterization of the adsorbed polymer layer

For full characterization of the adsorbed polymer layer one needs to determine the following parameters [7]: (i) Amount of polymer adsorbed per unit area, Γ. This requires determination of the adsorption isotherms at various temperatures. The influence of solvency for the stabilizing chain should be studied. (ii) Number of segments in direct contact with the surface, p (in trains). This can be obtained using pulse gradient NMR [6]. (iii) The adsorption energy per segment χ^s. For polymer adsorption to occur, a minimum value for χ^s is required to overcome the entropy loss when the polymer adsorbs at the surface. The adsorption energy can be determined using microcalorimetry. (iv) Extension of the layer from the surface, i.e. segment density distribution $\rho(z)$ or the adsorbed layer thickness δ. The most convenient method for determination of δ is to use dynamic light scattering, usually referred to as Photon Correlation Spectroscopy (PCS) [8]. For this purpose, model small spherical particles such as polymethylmethacrylate can be used.

10.5 Theory of steric stabilization

This theory was described in detail in Chapter 5 and a summary is given here. When two particles with adsorbed polymer layers, each with thickness δ, approach to a distance of separation, h, that is smaller than twice the adsorbed layer thickness 2δ, the layers will either overlap with each other or they become compressed [6]. This is illustrated in Fig. 10.2.

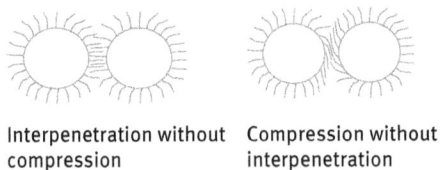

Interpenetration without compression   Compression without interpenetration

Fig. 10.2: Schematic representation of the interaction between particles containing adsorbed polymer layers.

In both cases, there will be an increase in the local segment density of the polymer chains in the interaction region. Provided the dangling chains (the A chains in A-B, A-B-A block or BA_n graft copolymers) are in a good solvent, this local increase in segment density in the interaction zone will result in strong repulsion as a result of two main effects: (i) Increase in the osmotic pressure in the overlap region as a result of the unfavourable mixing of the polymer chains, when these are in good solvent conditions [9]. This is referred to as osmotic repulsion or mixing interaction and it is described by a free energy of interaction G_{mix}. (ii) Reduction of the configurational entropy of the chains in the interaction zone; this entropy reduction results from the decrease in the volume available for the chains when these are either overlapped or compressed. This is referred to as volume restriction interaction, entropic or elastic interaction and it is described by a free energy of interaction G_{el}.

The combination of G_{mix} and G_{el} is usually referred to as the steric interaction free energy, G_s, i.e.,

$$G_s = G_{mix} + G_{el} . \tag{10.6}$$

The sign of G_{mix} depends on the solvency of the medium for the chains. If in a good solvent, i.e. the Flory–Huggins interaction parameter χ is less than 0.5, then G_{mix} is positive and the mixing interaction leads to repulsion. In contrast, if $\chi > 0.5$ (i.e. the chains are in a poor solvent condition), G_{mix} is negative and the mixing interaction becomes attractive. G_{el} is always positive and hence in some cases one can produce stable dispersions in a relatively poor solvent (enhanced steric stabilization).

In the overlap region, the chemical potential of the polymer chains is now higher than in the rest of the layer (with no overlap). This amounts to an increase in the osmotic pressure in the overlap region; as a result, solvent will diffuse from the bulk to the overlap region, thus separating the particles and hence a strong repulsive energy arises from this effect. This repulsive energy can be calculated by considering the free energy of mixing of two polymer solutions, as for example treated by Flory and Krigbaum [10]. The free energy of mixing is given by two terms: (i) an entropy term that depends on the volume fraction of polymer and solvent and (ii) an energy term that is determined by the Flory–Huggins interaction parameter χ.

Using the above theory one can derive an expression for the free energy of mixing of two polymer layers (assuming a uniform segment density distribution in each layer) surrounding two spherical particles as a function of the separation distance, h, between the particles.

The expression for G_{mix} is,

$$\frac{G_{mix}}{kT} = \left(\frac{2V_2^2}{V_1} \right) v_2^2 \left(\frac{1}{2} - \chi \right) \left(\delta - \frac{h}{2} \right)^2 \left(3R + 2\delta + \frac{h}{2} \right), \tag{10.7}$$

where k is the Boltzmann constant, T is the absolute temperature, V_2 is the molar volume of polymer, V_1 is the molar volume of solvent and n_2 is the number of polymer chains per unit area.

The sign of G_{mix} depends on the value of the Flory–Huggins interaction parameter χ: if $\chi < 0.5$, G_{mix} is positive and the interaction is repulsive; if $\chi > 0.5$, G_{mix} is negative and the interaction is attractive. The condition $\chi = 0.5$, i.e. $G_{mix} = 0$, is referred to as the θ-condition.

The elastic interaction arises from the loss in configurational entropy of the chains on the approach of a second particle [11]. As a result of this approach, the volume available for the chains becomes restricted, resulting in loss of the number of configurations.

For two flat plates, G_{el} is given by the following expression [11],

$$\frac{G_{el}}{kT} = -2v_2 \ln\left[\frac{\Omega(h)}{\Omega(\infty)}\right] = 2v_2 R_{el}(h), \qquad (10.8)$$

where $R_{el}(h)$ is a geometric function whose form depends on the segment density distribution.

It should be stressed that G_{el} is always positive and could play a major role in steric stabilization. It becomes very strong when the separation distance between the particles becomes comparable to the adsorbed layer thickness δ.

Combining G_{mix} and G_{el} with G_A gives the total energy of interaction G_T (assuming there is no contribution from any residual electrostatic interaction), i.e.,

$$G_T = G_{mix} + G_{el} + G_A. \qquad (10.9)$$

A schematic representation of the variation of G_{mix}, G_{el}, G_A and G_T with surface-surface separation distance h is shown in Fig. 10.3.

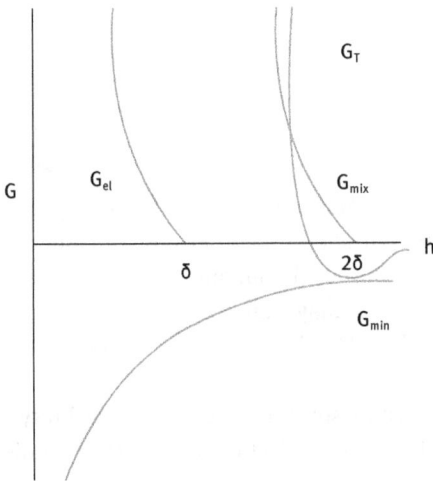

Fig. 10.3: Energy-distance curves for sterically stabilized systems.

G_{mix} increases very sharply with decreasing h; when $h < 2\delta$. G_{el} increases very sharply with decreasing h; when $h < \delta$. G_T versus h shows a minimum, G_{min}, at separation distances comparable to 2δ. When $h < 2\delta$, G_T shows a rapid increase with decreasing h.

Unlike the G_T–h curve predicted by DLVO theory (which shows two minima and one energy maximum), the G_T–h curve for systems that are sterically stabilized shows only one minimum, G_{min}, followed by a sharp increase in G_T with decreasing h (when $h < 2\delta$). The depth of the minimum depends on the Hamaker constant A, the particle radius R and adsorbed layer thickness δ. G_{min} increases with increasing A and R. At a given A and R, G_{min} increases with decreasing δ (i.e. with decreasing molecular

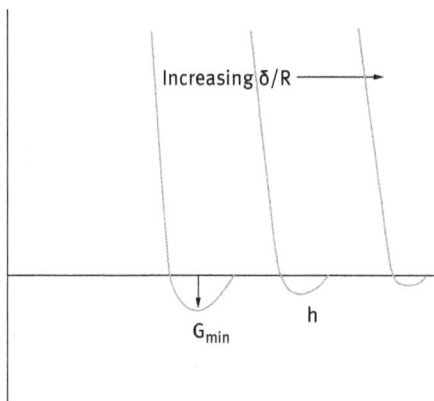

Fig. 10.4: Variation of G_{min} with δ/R.

weight, M_w, of the stabilizer). The larger the value of δ/R, the smaller the value of G_{min}, as illustrated in Fig. 10.4. In this case, the system may approach thermodynamic stability, as is the case with nanodispersions.

10.6 Criteria for effective steric stabilization

(i) Full coverage of the particles by the polymer. The amount of polymer added should correspond to the plateau of the adsorption isotherm. Any bare patches may lead to flocculation as a result of van der Waals attraction and/or bridging.

(ii) Strong adsorption of the polymer to the particle surface. Strong B "anchor" chain – this requires the chain to be insoluble in the medium and have some affinity to the surface. Lack of a strong "anchor" may lead to chain displacement on approach of the particles in a suspension. This is particularly important for concentrated suspensions.

(iii) The stabilizing chain(s) A should be in good solvent conditions. The Flory–Huggins interaction parameter χ should be less than 0.5 under all storage conditions, e.g. temperature changes.

(iv) An optimum layer thickness δ to ensure that G_{min} is not deep. This is particularly the case with concentrated suspensions where weak flocculation may occur at relatively small G_{min}. In most cases, an adsorbed layer thickness in the region 5–10 nm is sufficient. This is particularly the case with graft copolymers whereby the A chains are stretched ("comb-like" or "brush" structure).

10.7 Settling of suspensions and preventing the formation of dilatant sediments

This follows the same principles as described for aqueous suspension concentrates (see Chapter 8) and only a summary is given here.

Settling (or sedimentation) results from gravity forces [12]. When $\rho_{particles} > \rho_{medium}$, the gravity force $(4/3)\pi a^3 \Delta\rho g L$ (where g is the acceleration due to gravity and L is the height of the container) will overcome the thermal (Brownian) motion leading to particle sedimentation. Only when $kT > (4/3)\pi a^3 \Delta\rho g L$ does sedimentation becomes insignificant. This condition is only satisfied with very small particles ($< 0.1\,\mu m$) and $\Delta\rho < 0.1$. Thus, with most colloidally stable suspensions, sedimentation is the rule rather than the exception. The particles may sediment individually to the bottom of the container and the repulsive forces necessary to maintain stability allow them to move past each other with the result of the formation of a very compact sediment. The compact sediment, that is technically referred to as "clay" or "cake", is very difficult to redisperse by shaking. Such clays are dilatant (shear thickening) and must be avoided.

For an infinitely dilute suspension of noninteracting particles, the sedimentation velocity v_0 is given by Stokes' law,

$$v_0 = \frac{2}{9}\frac{a^2 \Delta\rho g}{\eta}. \tag{10.10}$$

For $1\,\mu m$ particles with $\Delta\rho = 0.2$ and $\eta = 10^{-3}$ Pa s (low viscosity oil), $v_0 = 4.4 \times 10^{-7}\,m\,s^{-1}$, whereas for $10\,\mu m$ particles $v_0 = 4.4 \times 10^{-5}\,m\,s^{-1}$. In a 100 cm container the $1\,\mu m$ particles will settle in about 3 days and the $10\,\mu m$ particles will settle in about 40 minutes.

For more concentrated suspensions, the sedimentation velocity, v, is reduced below the Stokes velocity v_0 as a result of interparticle interaction. For suspensions with volume fraction $\phi \approx 0.1$, v is related to v_0 by the equation [13],

$$v = v_0(1 - k\phi), \tag{10.11}$$

where $k = 6.55$, a constant that accounts for hydrodynamic interaction.

For suspensions with $\phi > 0.1$, v becomes a complex function of ϕ and only empirical equations are used to describe the sedimentation velocity [12]. When v is plotted against ϕ, an exponential reduction of v with increasing ϕ is found and eventually v approaches zero at some limiting volume fraction [14], the so-called maximum packing fraction ϕ_p. At this maximum packing fraction, the relative viscosity of the suspension approaches infinity.

In most practical suspensions, a volume fraction in the range 0.1–0.5 is used and hence one should use a suspending agent to prevent sedimentation and formation of dilatant clays. The suspending agent (sometimes referred to as an antisettling agent) should provide a restoring force to overcome gravity.

It is useful to calculate the stress exerted by a particle σ_p in a suspension [15],

$$\sigma_p = \frac{\text{Gravity force}}{\text{Area}} = \frac{(4/3)\pi a^3 \Delta\rho g}{4\pi a^2} = \frac{a\Delta\rho g}{3}. \tag{10.12}$$

For 10 μm particles with $\Delta\rho = 1$, $\sigma_p = 3.3 \times 10^{-2}$ Pa. This clearly illustrates the need to measure the viscosity at such low stresses (using a constant stress rheometer).

10.8 Examples of suspending agents that can be applied for prevention of settling in nonaqueous suspensions

The main criteria for an effective suspending agent are as follows:
(a) It should produce a "three-dimensional" gel network in the continuous phase with optimum rheological characteristics.
(b) It should have a very high viscosity at low shear rates. A residual (zero shear) viscosity in excess of 100 Pa s is desirable. The residual viscosity can be measured using constant stress (creep) measurements.
(c) It should have sufficient "bulk" modulus to prevent separation of the suspension and syneresis. The bulk modulus is related to the shear modulus which can be measured using dynamic (oscillatory) techniques.
(d) The suspending agent should produce a shear thinning system such that on application of the suspension, the high shear viscosity is not too high.

Several suspending systems can be used and these are summarized below [1]:
(i) Hydrophobically modified clays (Bentones). Clays such as sodium montmorillonite can be made hydrophobic by addition of long chain alkyl ammonium surfactants (e.g. dodecyl or cetyl trimethyl ammonium chloride). The alkyl ammonium cation exchanges with the Na^+ ions producing a hydrophobic surface and the bentone particles can be dispersed in the nonaqueous medium. The exchange is carried out in an optimum manner such that some hydrophilic sites remain on the bentone particles. On addition of a polar solvent, such as propylene carbonate or alcohol, a gel is produced. The most likely mechanism of gel formation in this case is hydrogen bonding by the polar molecules between the polar sites on the bentone particles.
(ii) Fumed silica. These are commercially available under the trade name "Aerosil" or "Cabosil". They are produced by reaction of silicon tetrachloride with steam. Fine particles (primary particle size < 0.02 μm) are produced with a surface containing silanol groups that are separated by siloxane bonds. When dispersing the fumed silica powder in nonaqueous media a gel is produced by hydrogen bonding between the silanol groups. Chain aggregates are produced and, by controlling the dispersion procedure, adequate rheological characteristics are produced.

(iii) Trihydroxystearin (Thixin). This product is derived from castor oil and it needs both shear and heat for full activation to produce a gel.

(iv) Aluminium magnesium hydroxide stearate. This has been used to gel a number of cosmetic oils and it can provide suspending properties. It could also be applied for oil-based suspensions of agrochemicals. However, one should be careful in applying this system since in many cases the gel strength is too high for adequate dispersion on application.

(v) Use of high molecular weight polymers. Polyethylenes and copolymers of polyethylene can be used to gel mineral oils and several aliphatic solvents. Some of these materials need high incorporation temperatures and one must keep in mind that the cooling procedures affect the final appearance and rheological characteristics of the gel.

The above systems are the most commonly used rheology modifiers in most nonaqueous suspensions. However, two other methods may be applied for reduction of settling of nonaqueous suspensions and these are discussed below.

(i) Controlled flocculation (self-structured systems). By reducing the adsorbed layer thickness one can increase the depth of the minimum, G_{min}, to such an extent that weak flocculation may produce a "three-dimensional" gel structure that is sufficient to reduce sedimentation and formation of dilatant clays. The adsorbed layer thickness may be reduced by reducing the molecular weight of the stabilizer or by the addition of a small amount of a nonsolvent. This was illustrated in Fig. 10.4 where the energy-distance curve is plotted as a function of the ratio of adsorbed layer thickness to particle radius. This procedure is particularly useful with large and asymmetric particles.

(ii) Depletion flocculation [16]. This is obtained by the addition of "free" (nonadsorbing) polymer to a sterically stabilized suspension. Above a critical volume fraction, ϕ_p^+, of the free polymer, weak flocculation occurs. Above ϕ_p^+, the polymer chains are "squeezed out" from between the particles leaving a polymer-free zone in the interstices. The higher osmotic pressure outside the particle surfaces causes weak flocculation. This is schematically shown in Fig. 10.5.

The critical polymer volume fraction above which flocculation occurs decreases with increasing molecular weight of the free polymer. Thus, high molecular weight oil soluble polymers should be used in order to reduce the amount required for flocculation. The amount of free polymer required for flocculation also decreases with increasing volume fraction of the suspension. Therefore, depletion flocculation is more applicable for concentrated suspensions.

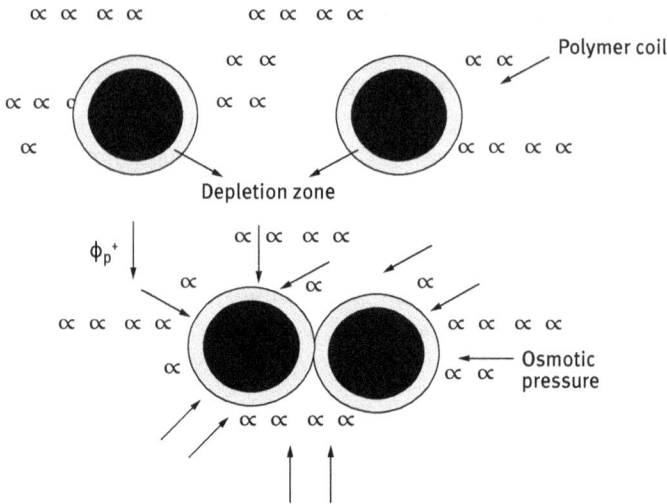

Fig. 10.5: Schematic representation of depletion flocculation.

10.9 Emulsification of oil-based suspensions

The nonaqueous suspension concentrate containing the active ingredient particles and any antisettling system is emulsified into the spray tank before application [1]. Here the same principles applied for self-emulsification of emulsifiable concentrates (ECs) are applied as discussed below. Alternatively, the nonaqueous suspension concentrate is emulsified into an aqueous solution containing another water soluble active ingredient (normally an electrolyte such as glyphosate) to produce a combined mixture for two active ingredients. In the first case, spontaneous emulsification is necessary since only gentle agitation in the spray tank is possible. In the second case, one may apply a normal emulsification procedure if a combined formulation is required. In some cases spontaneous emulsification into an aqueous electrolyte solution may also be required.

The above oil-based suspension concentrates require careful formulation to achieve the required properties: (i) a stable nonaqueous suspension with adequate weak flocculation (to produce a shear thinning system) and no settling; (ii) the antisettling agent used in the formulation should be water dispersible (e.g. fumed silica, Aerosil); (iii) the emulsifying system used should not interfere with the stabilizing polymeric surfactant used for preparation of the nonaqueous suspension; (iv) the viscosity of the formulation at intermediate shear rates should be low enough to ensure spontaneity of emulsification.

10.10 Mechanism of spontaneous emulsification and the role of mixed surfactant film

The first demonstration of spontaneous emulsification was given by Gad [17] who observed that when a solution of lauric acid in oil is carefully placed into an aqueous alkaline solution, an emulsion forms at the interface. It is clear from this experiment that a mixture of lauric acid and sodium laurate is produced and this illustrates the role of the mixed film (which produces an ultra-low interfacial tension). When the right conditions are produced, spontaneous emulsification occurs with minimum agitation. In this experiment, the oil used was Newtonian and it had a relatively low viscosity.

Three mechanisms may be established to explain the process of spontaneous emulsification and these are summarized below.

(i) Interfacial turbulence (Fig. 10.6). Interfacial turbulence occurs as a result of mass transfer from one phase to the other. The interface shows unsteady motion; streams of one phase are ejected and penetrate into the second phase, shedding small droplets. Localized reduction in interfacial tension is caused by nonuniform adsorption of the surfactant at the O/W interface or by mass transfer of surfactant molecules across the interface. When the two phases are not in equilibrium, convection currents may be formed, transferring liquid rich in surfactant towards areas of liquid deficient in surfactant. These convection currents lead to local fluctuations in the interfacial tension causing oscillation of the interface (turbulence).

Such disturbances may amplify themselves leading to violent interfacial perturbations and eventual disintegration of the interface, when liquid droplets of one phase are "thrown" into the other phase. This mechanism requires the presence of two surfactant molecules or a surfactant plus alcohol. This facilitates mass transfer and induces interfacial tension gradients. Several "phases" may be produced at the O/W interface as will be discussed later.

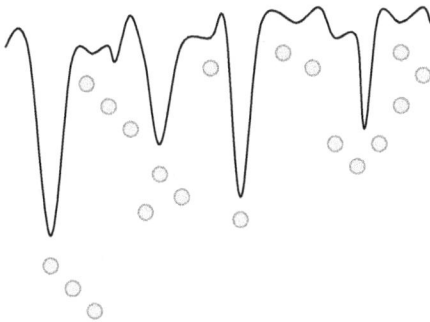

Fig. 10.6: Schematic representation of interfacial turbulence.

(ii) Diffusion and stranding (Fig. 10.7). This is best illustrated by placing an ethanol-toluene mixture (containing say 10 % alcohol) onto water. The aqueous layer eventually becomes turbid as a result of the presence of toluene droplets. In this case, interfacial turbulence does not occur although spontaneous emulsification takes place. The alcohol molecules diffuse into the aqueous phase carrying some toluene molecules in a saturated three-component subphase. At some distance from the interface, the alcohol becomes sufficiently diluted into water and the toluene droplets precipitate as droplets in the aqueous phase.

The above mechanism requires the presence of a third component (sometimes referred to as the cosurfactant) that increases the miscibility of the two previously immiscible phases (toluene and water). Many emulsifiable concentrates contain a high proportion of a polar solvent, such as alcohol or ketone, which facilitates spontaneous emulsification. Its application for emulsification of nonaqueous suspension concentrates needs to be tested (addition of a polar solvent may enhance crystal growth).

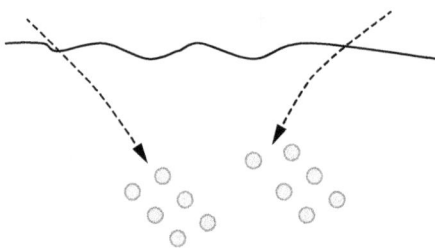

Fig. 10.7: Schematic representation of diffusion and stranding.

(iii) Production of ultra-low (or transiently negative) interfacial tension (Fig. 10.8). This is the same mechanism for producing microemulsions. The ultra-low interfacial tension is produced by a combination of a surfactant and cosurfactant. This is illustrated in Fig. 10.9.

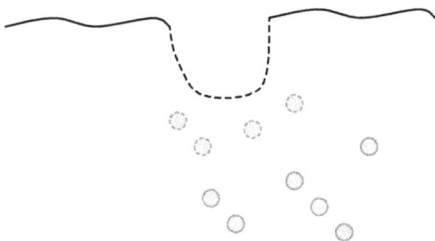

Fig. 10.8: Schematic representation of the production of ultra-low interfacial tension.

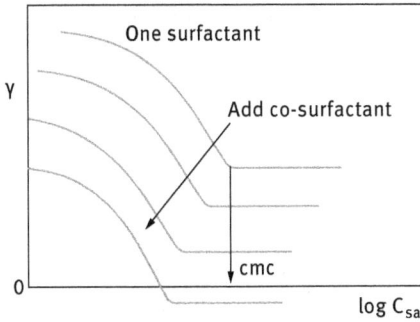

Fig. 10.9: γ–log C curves for surfactant plus cosurfactant.

Addition of surfactant to the aqueous or oil phase causes a gradual lowering of interfacial tension, γ, reaching a limiting value at the critical micelle concentration (cmc). Any further increase in C above the cmc causes little or no further decrease in γ. The limiting value of γ reached with most single surfactants is seldom lower than $0.1\,\mathrm{mN\,m^{-1}}$, which is not sufficiently low for microemulsion formation. If a surfactant mixture with one predominantly water soluble such as sodium dodecyl sulphate, SDS, and one predominantly oil soluble usually referred to as the cosurfactant such as pentanol or hexanol, is used the limiting γ value can reach very low values ($< 10^{-2}\,\mathrm{mN\,m^{-1}}$) or even becomes transiently negative. When a transiently negative interfacial tension is produced, the interface expands spontaneously (producing droplets) adsorbing all surfactant molecules until a small positive γ is reached [18].

The reason for the reduction in interfacial tension when using two molecules can be understood from a consideration of the Gibbs adsorption equation.

The reduction in γ is related to the adsorption of the surfactant molecules, which is referred to as the surface excess Γ ($\mathrm{mol\,m^{-2}}$); for a multicomponent system the reduction in γ, i.e. $d\gamma$, is given by the summation of all surface excesses [18],

$$d\gamma = -\sum \Gamma_i RT\, d\ln C_i. \tag{10.13}$$

Thus, for two surfactants, γ is reduced twice, provided the two molecules are adsorbed simultaneously at the interface. The two surfactant molecules should not interact with each other, otherwise they will lower their respective activities. This explains why the two molecules should vary in nature, one predominantly water soluble and the other predominantly oil soluble.

Several other mechanisms have been proposed to explain the dynamics of spontaneous emulsification. Direct observation using phase contrast and polarizing microscopy showed that in some cases vesicles (closed bilayers) are produced in the oil phase near the interface with the water. These vesicles tend to "explode", thereby pulverizing oil droplets into the aqueous phase. The above structures can be produced for example by using a mixture of nonionic surfactant, such as $C_{12}E_5$, a long chain alcohol such as $C_{12}H_{25}OH$, and an oil such as hexadecane. The oil/surfactant mixture is transparent, but on addition of a small amount of water it becomes turbid and vesicles can be observed under the microscope. The interfacial tension of the oil/surfactant mixture/water is also very low.

10.11 Polymeric surfactants for oil-based suspensions and the choice of emulsifiers

When formulating a nonaqueous suspension concentrate that can be spontaneously emulsified into aqueous solutions one should consider the following criteria: The polymeric emulsifiers should be chosen from the A-B, A-B-A block or BA_n graft copolymer as discussed above. The B chain should be chosen to be insoluble in the oil and should become strongly adsorbed on the particle surface (either by specific interaction or by rejection "anchoring"). As discussed before, for nonpolar oils (hydrocarbons) the B chain could be a polar molecule such as polyethylene imine, which adsorbs on the particle by rejection anchoring. The A chain should be soluble in the oil and highly solvated by its molecules – poly(hydroxystearic acid) or polyisobutylene are ideal.

The emulsifier system should be soluble in the oil phase and it should not cause desorption of the polymeric surfactant. A two component emulsifier system is normally used, one predominantly water soluble (such as an ethoxylated surfactant) and one oil soluble such as a medium or long chain alcohol. In many cases calcium dodecyl benzene sulphonate may be used. The emulsifier system should lower the interfacial tension of O/W to very low values ($< 0.1\,\mathrm{mN\,m^{-1}}$).

The ultra-low interfacial tension may be measured using the spinning drop technique. A drop of oil is injected into the aqueous phase that is placed in a capillary tube that can be rotated at high speed to cause deformation of the spherical drop into a cylinder [18]. From the droplet profile (cylinder radius R) and speed of rotation ω one can calculate γ,

$$\gamma = \frac{\omega^2 \Delta\rho R^4}{4}.$$

(10.14)

$\Delta\rho$ is the density difference between water and oil.

10.12 Emulsification into aqueous electrolyte solutions

For preparation of a combined formulation with one active ingredient suspended in the oil phase and another active ingredient soluble in water (salt), one can emulsify the nonaqueous suspension into the aqueous electrolyte solution using a polymeric surfactant with high HLB number, e.g. Pluronic PF127 (that contains ≈ 55 units polypropylene oxide, PPO, and two polyethylene oxide (PEO) chains with ≈ 100 units each). The PEO-PPO-PEO block copolymer is insoluble in the oil phase and hence it does not interfere with the polymeric surfactant used for preparation of the nonaqueous suspension. Depending on the density of the active ingredient in the oil phase and the amount suspended, the density of the oil drops produced could be smaller or larger than that of the aqueous electrolyte. Thus the combined formulation could undergo creaming or settling on storage. It is, therefore, essential to include an antisettling system in the aqueous electrolyte phase, e.g. xanthan gum (Kelzan or Rhodopol) could be applied in this case.

In the formulation where spontaneous emulsification is required into the aqueous electrolyte solution, one should apply the same principles discussed above, except in this case the emulsifier system should be more hydrophilic (higher HLB number) to prevent "salting-out" of the emulsifier by the electrolyte. The nature and HLB number of the emulsifier system depend on the electrolyte concentration and nature and these could be checked independently using cloud point measurements. The resulting interfacial tension should be low.

10.13 Proper choice of the antisettling system

The antisettling system used for nonaqueous suspensions should be dispersible into water. Fumed silica (Aerosil 200) or microcrystalline cellulose are ideal systems for structuring nonaqueous suspensions. These systems produce a "three-dimensional" gel network in the oil phase by hydrogen bonding between the particles forming chains and cross chains. These gels produce enough "yield stress" to prevent sedimentation of the coarse active ingredient particles. They also produce a very high viscosity at low shear rates, thus preventing sedimentation.

One of the main advantages of these hydrophilic particles is that they can partition into the aqueous phase and this produces two main effects. During partition into the aqueous phase and crossing the interface, they can enhance the interfacial tension gradients and hence promote the turbulence effect described above. This will facilitate self-emulsification. When these particles leave the oil phase, the yield value and viscosity of the suspension is lowered and this helps the self-emulsification process.

Hydrophilic polymers such as hydroxypropyl cellulose may also be used either alone or in combination with silica or microcrystalline cellulose. The combined particulate-polymer system may be advantageous in reducing sedimentation and enhancing the self-emulsification of the oil. The optimum ratio between particles and polymer depends on the nature of the active ingredient suspended in the oil phase.

10.14 Rheological characteristics of the oil-based suspensions

The rheological characteristics of the nonaqueous suspension (its yield value, zero shear viscosity and elastic modulus) need to be carefully adjusted to achieve the following criteria: (i) no settling or separation of the nonaqueous suspension on storage under all storage conditions; (ii) ease of spontaneous emulsification with minimum agitation.

The above two criteria are not compatible and hence one has to control the rheology very carefully. For preventing settling and separation, one needs a high "yield stress", high zero shear viscosity and high modulus. High rheological parameters reduce the ease of spontaneous emulsification. For these reasons, one needs to prepare a highly shear thinning system. The suspension concentrate should have a high viscosity at low shear rates, but once the shear rate exceeds a certain value (say 10–100 s^{-1}) depending on application, the viscosity drops significantly. One can achieve the above effect by proper choice of the antisettling system.

By using an antisettling system that diffuses rapidly into the water phase one can reduce the yield value, zero shear viscosity and modulus during the process of emulsification. Measurement of the rheological parameters is essential when formulating a nonaqueous suspension concentrate and this requires low shear rheometers (constant stress and oscillatory techniques). Details of these techniques are available in several review articles [19, 20].

References

[1] Tadros, Th. F., "Colloids in Agrochemicals", Wiley-VCH, Germany (2009).
[2] Deryaguin, B. V. and Landau, L., Acta Physicochem. USSR, **14**, 633 (1941).
[3] Verwey, E. J. W. and Overbeek, J. Th. G., "Theory of Stability of Lyophobic Colloids", Elsevier, Amsterdam (1948).
[4] Lyklema, J., "Structure of the Solid/Liquid Interface and the Electrical Double Layer", in "Solid/Liquid Dispersions", Th. F. Tadros (ed.), Academic Press, London (1987).
[5] Hamaker, H. C., Physica (Utrecht), **4**, 1058 (1937).
[6] Tadros, Th. F., "Polymer Adsorption and Dispersion Stability", in "The Effect of Polymers on Dispersion Properties", Th. F. Tadros (ed.), Academic Press, London (1981).
[7] Fleer, G. J., Cohen-Stuart, M. A., Scheutjens, J. M. H. M., Cosgrove, T. and Vincent, B., "Polymers at Interfaces", Chapman and Hall, London (1993).
[8] Pusey, P. N., in "Industrial Polymers: Characterisation by Molecular Weights", J. H. S. Green and R. Dietz (eds.), London, Transcripta Books (1973).
[9] Napper, D. H., "Polymeric Stabilisation of Colloidal Dispersions", Academic Press, London (1983).
[10] Flory, P. J. and Krigbaum, W. R., J. Chem. Phys. **18**, 1086 (1950).
[11] Mackor, E. L. and van der Waals, J. H., J. Colloid Sci., **7**, 535 (1951).
[12] Tadros, Th. F., "Settling Suspensions", in "Solid/Liquid Dispersions", Th. F. Tadros (ed.), Academic Press, London (1987).
[13] Bachelor, G. K., J. Fluid Mech., **52**, 245 (1972).
[14] Krieger, I. M., Adv. Colloid Interface Sci., **3**, 45 (1971).
[15] Tadros, Th. F., Advances Colloid and Interface Science, **12**, 141 (1980)
[16] Asakura, A. and Oosawa, F., J. Chem. Phys., **22**, 1235 (1954); J. Polymer Sci., **93**, 183 (1958).
[17] Tadros, Th. F., "Surfactants in Agrochemicals", Marcel Dekker, N. Y. (1994).
[18] Tadros, Th. F., "Applied Surfactants", Wiley-VCH, Germany (2005).
[19] Tadros, Th. F., Advances Colloid and Interface Sci., **46**, 1 (1993).
[20] Tadros, Th. F., Advances in Colloid and Interface Science, **68**, 97 (1996).

11 Characterization, assessment and prediction of stability of suspensions

11.1 Introduction

For full characterization of the properties of suspensions, three main types of investigations are needed: (i) Fundamental investigation of the system at a molecular level. This requires investigation of the structure of the solid/liquid interface, namely the structure of the electrical double layer (for charge stabilized suspensions), adsorption of surfactants, polymers and polyelectrolytes and conformation of the adsorbed layers (e.g. the adsorbed layer thickness). It is important to know how each of these parameters changes with the conditions, such as temperature, solvency of the medium for the adsorbed layers and effect of addition of electrolytes. (ii) Investigation of the state of suspension on standing, namely flocculation rates, flocculation points with sterically stabilized systems, spontaneity of dispersion on dilution and Ostwald ripening or crystal growth. All these phenomena require accurate determination of the particle size distribution as a function of storage time. (iii) Bulk properties of the suspension, which is particularly important for concentrated systems. This requires measurement of the rate of sedimentation and equilibrium sediment height. More quantitative techniques are based on assessment of the rheological properties of the suspension (without disturbing the system, i.e. without its dilution and measurement under conditions of low deformation) and how these are affected by long-term storage.

In this chapter, I start with a summary of the methods that can be applied to assess the structure of the solid/liquid interface, measurement of surfactant and polymeric surfactant adsorption and conformation at the solid/liquid interface. This is followed by more detailed sections on assessment of sedimentation, flocculation and Ostwald ripening. For the latter (flocculation and Oswald ripening), one needs to obtain information on the particle size distribution. Several techniques are available for obtaining this information on diluted systems. It is essential to dilute the concentrated suspension with its own dispersion medium in order not to affect the state of the dispersion during examination. The dispersion medium can be obtained by centrifugation of the suspension whereby the supernatant liquid is produced at the top of the centrifuge tube. Care should be taken on dilution of the concentrated system with its supernatant liquid in order not to break up any flocs that are formed. This dilution process should be carried out with minimum shear. For assessment of the structure of the suspension without dilution (which may affect its structure) rheological techniques (described in Chapter 9) are the most suitable methods. These techniques can also be applied for prediction of the long-term physical stability of the suspension as will be discussed in this chapter.

DOI 10.1515/9783110486872-012

11.2 Assessment of the structure of the solid/liquid interface

11.2.1 Double layer investigation

11.2.1.1 Analytical determination of surface charge

The most common method for investigation of double layers in disperse systems is to use titration techniques that can be applied to obtain the surface charge as a function of surface potential at different electrolyte concentrations and types [1]. The first established example of such a procedure was obtained using silver iodide sols since the charge-determining mechanism is well established. Both stable positive and negative sols can be made and the material is rather inert, insoluble and not particularly sensitive to light. In addition, Ag/AgI electrodes are very stable and the sol can be titrated with Ag^+ and I^- ions while measuring the pAg and pI. Using material balance, one can determine the surface concentration of Ag^+ and I^- ions and this allows one to calculate the surface charge, σ_0.

The titration method was successfully applied for double layer investigations of oxides. σ_0 can be directly determined by titration of an oxide suspension in an aqueous solution of indifferent electrolyte (e.g. KCl) using a cell of the type,

$$E_1 \mid \text{Oxide suspension} \mid E_2$$

where E_1 is an electrode reversible to H^+ and OH^- ions, such as a glass electrode, and E_2 is a reference electrode such as Ag-AgCl. A known mass, m, of solid oxide with a specific surface area A ($m^2\ g^{-1}$) is added to a known volume, V, of electrolyte solution such as KCl or KNO_3 (assuming there is no specific adsorption of ions) of known concentration (e.g. 10^{-2} mol dm^{-3} KCl or KNO_3) [1, 2]. The initial pH (called pH_0) is noted and the sample is titrated with, say, 10^{-2} mol dm^{-3} NaOH and the volume required to achieve each solution pH might be as shown in Fig. 11.1. A suitable time must elapse after each addition to establish equilibrium with the surface. If one superimposes the titration curve for V ml of KCl or KNO_3 alone on the same diagram, the volume v_1 (in cm^3) corresponds to the amount of base taken up by the oxide in order to establish equilibrium with a solution of pH = pH_1. The net increase in negative surface charge per unit area is given by,

$$-(\Gamma_+ - \Gamma_-) = \frac{10^{-5}v_1}{mA}. \qquad (11.1)$$

The quantity obtained from equation (11.1) is the relative amount of OH^- adsorbed at pH_1. If this calculation is repeated for all pH > pH_0 and the comparable data for acid titration are also obtained, a plot of relative charge versus pH can be constructed at various KCl or KNO_3 concentrations as illustrated in Fig. 11.2 (a). The point at which all three isotherms cross one another can be identified as the pzc because only at that point is the surface charge independent of concentration of the supporting electrolyte concentration (assuming there is no specific adsorption). Figure 11.2 (a) can then be

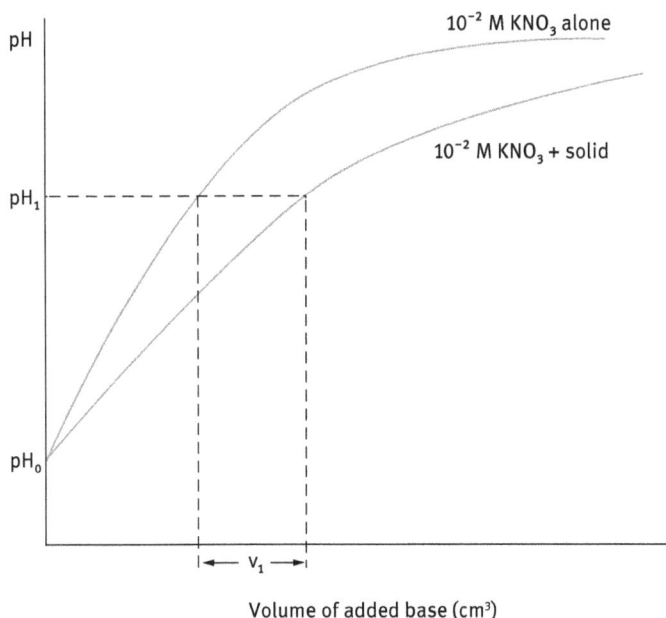

Fig. 11.1: Schematic illustration of potentiometric titration to obtain the surface charge.

redrawn as Fig. 11.2(b) to give the absolute value of σ_0. This procedure amounts to finding the point at which the Esin–Markovic coefficient β [3] is zero [3],

$$\beta = \left(\frac{\partial\sigma_0}{\partial\mu_s}\right)_{\psi_0} = \left(\frac{\partial\sigma_0}{\partial\mu_s}\right)_{a_i}, \tag{11.2}$$

where a_i is the activity of potential determining ion.

The surface charge density σ_0 is given by,

$$\sigma_0 = F(\Gamma_{H^+} - \Gamma_{OH^-}). \tag{11.3}$$

Whereas the surface potential ψ_0 is given by the Nernst equation,

$$\psi_0 = \frac{RT}{F}\ln\frac{a_{H^+}}{(a_{H^+})_{pzc}}, \tag{11.4}$$

where R is the gas constant, T is the absolute temperature, F is the Faraday constant, a_{H^+} is the activity of H^+ ions in bulk solution and $(a_{H^+})_{pzc}$ is the value at the point of zero charge.

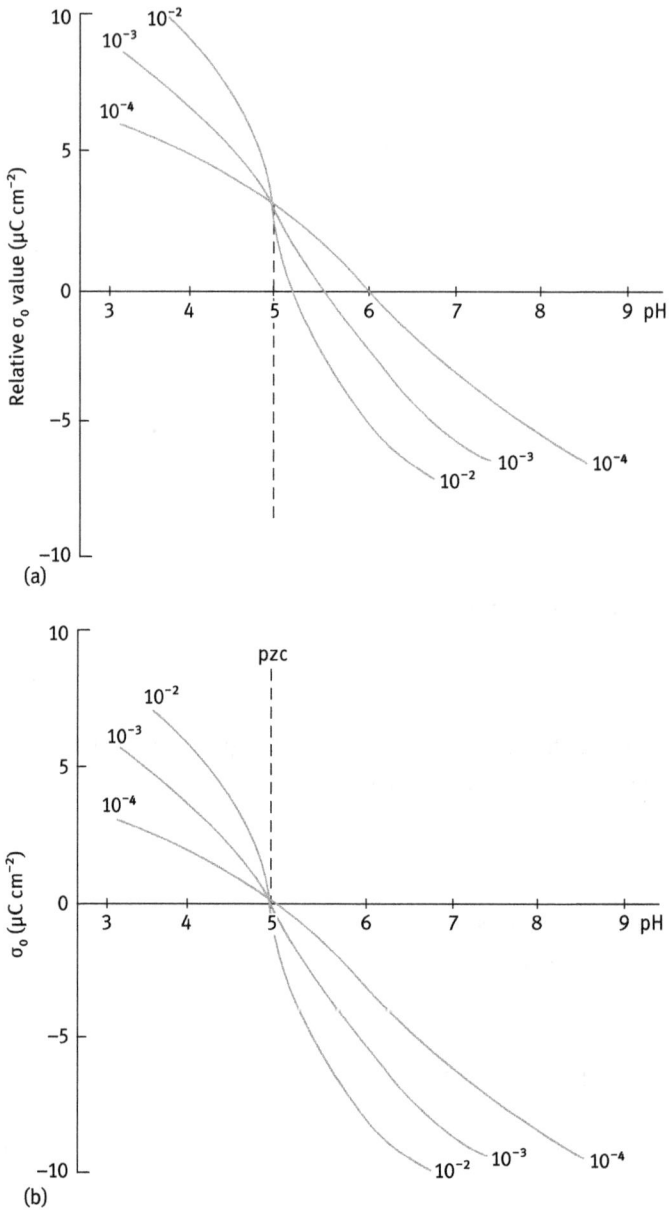

Fig. 11.2: Relative surface charge-pH curves (a) and absolute charge-pH curves (b).

For determining the surface charge density σ_0 (μC cm^{-2}), the specific surface area, A, needs to be accurately determined. The particles are usually irregular in shape and they may undergo Ostwald ripening on standing. This causes problems for determining A from average particle size obtained say by electron microscopy. In addition, most oxide particles are not smooth and this may underestimate the area obtained using gas adsorption and application of the BET equation. Measuring the surface area using dye adsorption requires knowledge of the effective cross-sectional area of the dye molecule and this may vary from one substrate to another. An appropriate method for measuring the surface area is to measure the expulsion of the co-ions from the increase in its concentration in solution, referred to as negative adsorption [4].

As an illustration, Fig. 11.3 shows the results of σ_0–pH curves for homodisperse haematite sols in the three concentrations of KCl (that is not specifically adsorbed) [5]. It can be seen that at low pH, σ_0 becomes more positive with increasing KCl concentration, whereas at high pH it becomes more negative. At one particular pH (pH = 9.1) σ_0 is independent of KCl concentration. At this pH, $\Gamma_{H^+} = \Gamma_{OH^-}$ and this defines the

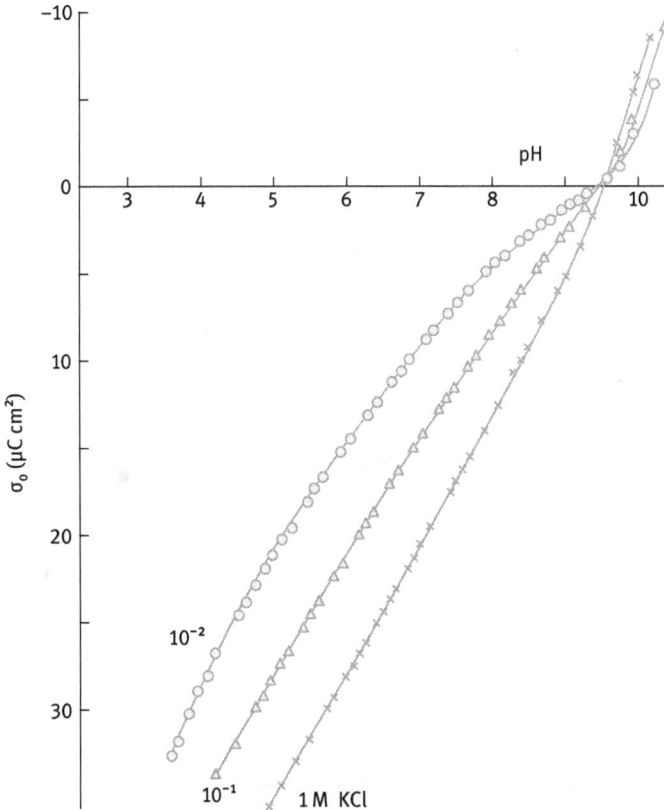

Fig. 11.3: σ_0–pH curves for homodisperse haematite sols in three KCl concentrations.

pzc. At pH < pzc, σ_0 is positive since $\Gamma_{H^+} > \Gamma_{OH^-}$, whereas at pH > pzc, σ_0 is negative since $\Gamma_{H^+} < \Gamma_{OH^-}$.

Most oxides show similar behaviour to that shown in Fig. 11.3. However, this not the case with precipitated silica which shows different trends as illustrated in Fig. 11.4, where σ_0 is plotted versus pH at four different KCl concentrations [6]. There is a common intersection point at pH $\approx$ 3, i.e. the pzc, indicating absence of specific adsorption of K^+ or Cl^- ions. The charge increases progressively with increasing pH reaching very high values at high pH and electrolyte concentrations. This high surface charge was accounted for by assuming the presence of a porous gel layer on the precipitated silica particles [7]. The basic idea is that H^+ and OH^- ions could penetrate the surface layers of the oxide and react there with amphoteric $-OH$ groups. In this way quite large amounts of surface charge could develop whilst retaining a reasonable separation between the charged groups. At the same time, if counterions are permitted to enter this porous layer, the net electrical potential at the outer edge of the porous layer would be reduced considerably in magnitude.

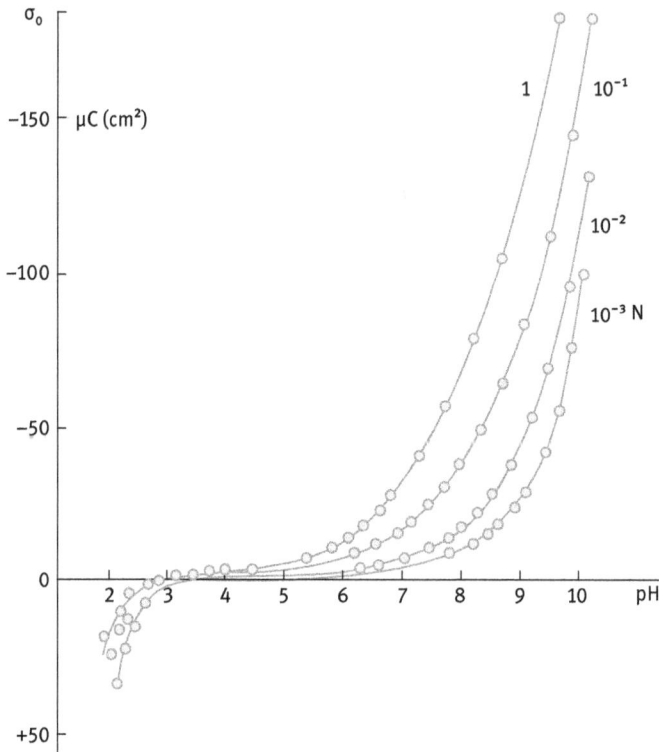

Fig. 11.4: σ_0 versus pH for precipitated silica at four different KCl concentrations.

11.2.1.2 Electrokinetic and zeta potential measurements

The principles of electrokinetic phenomena and measurement of the zeta potential were discussed in detail in Chapter 5. There are essentially three techniques for measurement of the electrophoretic mobility and zeta potential, namely the ultramicroscopic method, laser velocimetry and electroacoustic methods and these are summarized below.

Ultramicroscopic technique (microelectrophoresis)

This is the most commonly used method since it allows direct observation of the particles using an ultramicroscope (suitable for particles that are larger than 100 nm). Microelectrophoresis has many advantages since the particles can be measured in their normal environment [8]. It is preferable to dilute the suspension with the supernatant liquid which can be produced by centrifugation. Basically, a dilute suspension is placed in a cell (that can be circular or rectangular) consisting of a thin walled ($\approx$ 100 μm) glass tube that is attached to two larger bore tubes with sockets for placing the electrodes. The cell is immersed in a thermostat bath (accurate to ± 0.1 °C) that contains attachment for illumination and a microscope objective for observing the particles. It is also possible to use a video camera for directly observing the particles.

Since the glass walls are charged (usually negative at practical pH measurements), the solution in the cell will in general experience electro-osmotic flow. Thus, the observed motion of the particle when the field is applied, v_p, is the sum of its true velocity, v_E, and the total liquid velocity, v_ℓ. The latter varies with the distance r from the axis in accordance with Poiseulle's equation,

$$v_\ell = p\frac{(a^2 - r^2)}{4\eta\ell},$$

(11.5)

where p is the back pressure, a is the tube radius and ℓ is its length.

Thus, only where the electro-osmotic flow is zero, i.e. the so-called stationary level, can the electrophoretic mobility of the particles be measured. To establish the position of the stationary level, let us consider the situation in a microelectrophoresis cell of circular cross section as schematically represented in Fig. 11.5.

The electro-osmotic effects give rise to a solution velocity, v_{eo}, uniform across the cell cross section, towards the electrode of the same sign as the charge on the cell wall. If the cell is closed, a reverse flow will be set up and there will be no net transport of liquid. This is also the case if the cell is not closed, once the necessary hydrostatic pressure is built up. The reverse flow follows Poiseulle's law, where the condition of no net liquid transport is set by the following equation,

$$\int_{r=0}^{r=a} 2\pi vr\,dr = 0$$

(11.6)

$$v = v_{eo} - C(a^2 - r^2),$$

(11.7)

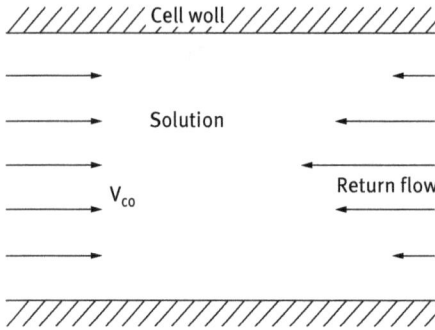

Fig. 11.5: Flow conditions within a closed cylindrical electrophoretic cell with electric field applied.

where C is a constant which from equations (11.15) and (11.16) is given by,

$$C = \frac{2v_{eo}}{a^2} \tag{11.8}$$

$$v = v_{eo}\left[\left(\frac{2r^2}{a^2}\right) - 1\right]. \tag{11.9}$$

At the cell wall r = a and v = v_{eo} as expected. At the centre r = 0 and v = $-v_{eo}$, i.e. the velocity of the liquid flow is equal in magnitude but opposite in direction to that at the wall. The condition for zero liquid velocity is given by the condition,

$$\frac{2r^2}{a^2} = 1 \tag{11.10}$$

or,

$$r = \left(\frac{1}{2}\right)^{1/2} a = 0.707a. \tag{11.11}$$

Thus, the stationary level is located at a distance of 0.707 of the radius from the centre of the tube or 0.146 of the internal diameter from the wall. By focusing the microscope objective at the top and bottom of the walls of the tube, one can easily locate the position of the stationary levels. The average particle velocity is measured at the top and bottom stationary levels by averaging at least 20 measurements in each direction (the eye piece of the microscope is fitted with a graticule).

For large particles (> 1 μm and high density), sedimentation may occur during the measurement. In this case one can use a rectangular cell and observe the particles horizontally from the side of the glass cell. This is illustrated in Fig. 11.6.

The position of the stationary levels within the rectangular cell is more difficult to assign and it depends on the ratio of the two axes a and b in Fig. 11.6. When a/b = ∞, v(x = 0) = 0 when y/b = 0.5774 so that the stationary levels are at 0.211 of the cell thickness 2b from both the front and back walls. The position of the stationary levels for other values of a/b are given in Tab. 11.1.

Several commercial instruments for measuring electrophoretic mobility are available (e.g. Rank Brothers, Bottisham, Cambridge, England and Pen Kem in USA).

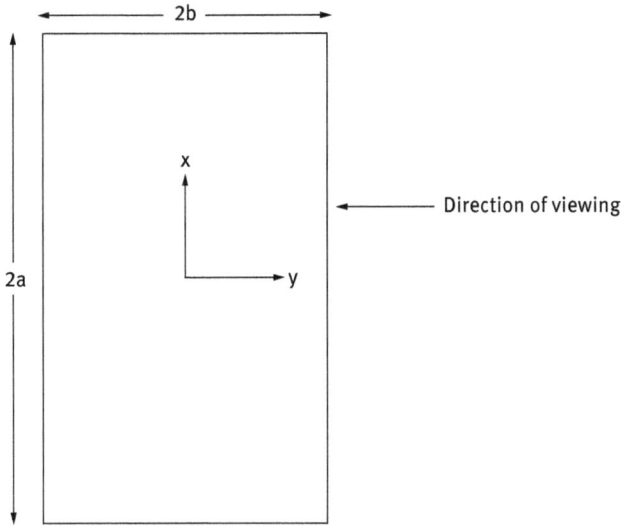

Fig. 11.6: Schematic diagram of the rectangular cell.

Tab. 11.1: Position of stationary levels in rectangular cells.

a/b	(b − y)/2b
∞	0.211
50	0.208
20	0.202
10	0.196

Laser velocimetry technique

This method is suitable for small particles that undergo Brownian motion [2]. The light scattered by small particles will show intensity fluctuation as a result of Brownian diffusion (Doppler shift). When a light beam passes through a colloidal dispersion, an oscillating dipole movement is induced in the particles, thereby radiating the light. Due to the random position of the particles, the intensity of scattered light, at any instant, appears as random diffraction ("speckle" pattern). As the particles undergo Brownian motion, the random configuration of the pattern will fluctuate, such that the time taken for an intensity maximum to become a minimum (the coherence time), corresponds approximately to the time required for a particle to move one wavelength λ. Using a photomultiplier of active area about the diffraction maximum (i.e. one coherent area) this intensity fluctuation can be measured. The analogue output is digitized (using a digital correlator) to measure the photocount (or intensity) correlation function of scattered light. The intensity fluctuation is schematically illustrated in Fig. 11.7.

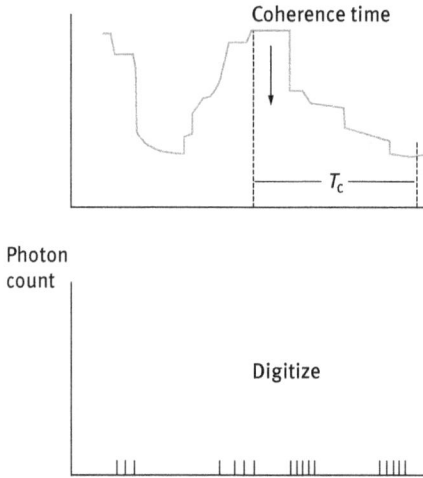

Fig. 11.7: Schematic representation of intensity fluctuation of scattered light.

The photocount correlation function $G^{(2)}(\tau)$ is given by,

$$g^{(2)} = B[1 + \gamma^2 g^{(1)}(\tau)]^2 , \qquad (11.12)$$

where τ is the correlation delay time.

The correlator compares $g^{(2)}(\tau)$ for many values of τ. B is the background value to which $g^{(2)}(\tau)$ decays at long delay times. $g^{(1)}(\tau)$ is the normalized correlation function of the scattered electric field and γ is a constant (≈ 1).

For monodispersed noninteracting particles,

$$g^{(1)}(\tau) = \exp(-\Gamma\gamma) . \qquad (11.13)$$

Γ is the decay rate or inverse coherence time that is related to the translational diffusion coefficient D,

$$\Gamma = DK^2 , \qquad (11.14)$$

where K is the scattering vector,

$$K = \left(\frac{4\pi n}{\lambda_0}\right) \sin\left(\frac{\theta}{2}\right) . \qquad (11.15)$$

The particle radius R can be calculated from D using the Stokes–Einstein equation,

$$D = \frac{kT}{6\pi\eta_0 R} , \qquad (11.16)$$

where η_0 is the viscosity of the medium.

If an electric field is placed at right angles to the incident light and in the plane defined by the incident and observation beam, the line broadening is unaffected but the centre frequency of the scattered light is shifted to an extent determined by electrophoretic mobility. The shift is very small compared to the incident frequency (≈ 100 Hz for and incident frequency of $\approx 6 \times 10^{14}$ Hz), but with a laser source it can be detected by heterodyning (i.e. mixing) the scattered light with the incident beam and detecting the output of the difference frequency. A homodyne method may be applied in which case a modulator to generate an apparent Doppler shift at the modulated frequency is used. To increase the sensitivity of the laser Doppler method, the electric fields are much higher than those used in conventional electrophoresis. Joule heating is minimized by pulsing the electric field in opposite directions. The Brownian motion of the particles also contributes to the Doppler shift and an approximate correction can be made by subtracting the peak width obtained in the absence of an electric field from the electrophoretic spectrum. An He-Ne Laser is used as the light source and the output of the laser is split into two coherent beams which are cross-focused in the cell to illuminate the sample. The light scattered by the particle, together with the reference beam, is detected by a photomultiplier. The output is amplified and analysed to transform the signals to a frequency distribution spectrum. At the intersection of the beams, interferences of known spacing are formed.

The magnitude of the Doppler shift, Δv, is used to calculate the electrophoretic mobility, u, using the following expression,

$$\Delta v = \left(\frac{2n}{\lambda_0}\right) \sin\left(\frac{\theta}{2}\right) uE, \tag{11.17}$$

where n is the refractive index of the medium, λ_0 is the incident wavelength in vacuum, θ is the scattering angle and E is the field strength.

Several commercial instruments are available for measuring electrophoretic light scattering: (i) The Coulter DELSA 440SX (Coulter Corporation, USA) is a multi-angle laser Doppler system employing heterodyning and autocorrelation signal processing. Measurements are made at four scattering angle (8, 17, 25 and 34°) and the temperature of the cell is controlled by a Peltier device. The instrument reports the electrophoretic mobility, zeta potential, conductivity and particle size distribution. (ii) Malvern (Malvern Instruments, UK) has two instruments: the ZetaSizer 3000 and ZetaSizer 5000. The ZetaSizer 3000 is a laser Doppler system using crossed beam optical configuration and homodyne detection with photon correlation signal processing. The zeta potential is measured using laser Doppler velocimetry and the particle size is measured using photon correlation spectroscopy (PCS). The ZetaSizer 5000 uses PCS to measure both movement of the particles in an electric field for zeta potential determination and random diffusion of particles at different measuring angles for size measurement of the same sample. The manufacturer claims that zeta potential for particles in the range 50 nm to 30 μm can be measured. In both instruments, a Peltier device is used for temperature control.

Electroacoustic methods

The mobility of a particle in an alternating field is termed dynamic mobility, to distinguish it from the electrophoretic mobility in a static electric field described above [13]. The principle of the technique is based on the creation of an electric potential by a sound wave transmitted through an electrolyte solution, as described by Debye [9, 10]. The potential, termed the ionic vibration potential (IVP), arises from the difference in the frictional forces and the inertia of hydrated ions subjected to ultrasound waves. The effect of ultrasonic compression is different for ions of different masses and the displacement amplitudes are different for anions and cations. Hence the sound waves create periodically changing electric charge densities. This original theory of Debye was extended to include electrophoretic, relaxation and pressure gradient forces [11, 12].

A much stronger effect can be observed in colloidal dispersions. The sound waves transmitted by the suspension of charged particles generate an electric field because the relative motion of the two phases is different. The displacement of a charged particle from its environment by the ultrasound waves generates an alternating potential, termed colloidal vibration potential (CVP). The IVP and CVP are both called ultrasound vibration potential (UVP).

The converse effect, namely the generation of sound waves by an alternating electric field [13] in a colloidal dispersion, can be measured and is termed the electrokinetic sonic amplitude (ESA). The theory for the ESA effect has been developed by O'Brian and co-workers [14–16]. Dynamic mobility can be determined by measuring either UVP or ESA, although in general the ESA is the preferred method. Several commercial instruments are available for measuring dynamic mobility: (i) the ESA-8000 system from Matec Applied Sciences that can measure both CVP and ESA signals; (ii) the Pen Kem System 7000 Acoustophoretic Titrator that measures the CVP, conductivity, pH, temperature, pressure amplitude and sound velocity.

In the ESA system (from Matec) and the AcoustoSizer (from Colloidal Dynamics) the dispersion is subjected to a high frequency alternating field and the ESA signal is measured. The ESA-8000 operates at constant frequency of $\approx 1\,\text{MHz}$ and the dynamic mobility and zeta potential (but not particle size) are measured. The AcoustoSizer, which operates at various frequencies of the applied electric field, can measure particle mobility, zeta potential and particle size.

The frequency synthesizer feeds a continuous sinusoidal voltage into a grated amplifier that creates a pulse of sinusoidal voltage across the electrodes in the dispersion. The pulse generates sound waves which appear to emanate from the electrodes. The oscillation, the back-and-forth movement of the particle caused by an electric field, is the product of the particle charge times the applied field strength. When the direction of the field is alternating, particles in the suspension between the electrodes are driven away towards the electrodes. The magnitude and phase angle of the ESA signal created is measured with a piezoelectric transducer mounted on a solid nonconductive (glass) rod attached to the electrode as illustrated in Fig. 11.8. The purpose of this non-conductive acoustic delay line is to separate the transducer from the high-frequency electric field in the cell. Three pulses of the voltage signal are recorded, as schematically shown in Fig. 11.9. The first pulse of the signal, shown on the left, is generated when the voltage pulse is applied to the sample and is unrelated to the ESA effect. This first pulse of the signal is received before the sound has sufficient time to pass down the glass rod and is an electronic cross-talk deleted from data processing. The second and third pulses are ESA signals. The second pulse is detected by the nearest electrode. This pulse is used for data processing to determine the particle size and zeta potential. The third pulse originates from the other electrode and is deleted.

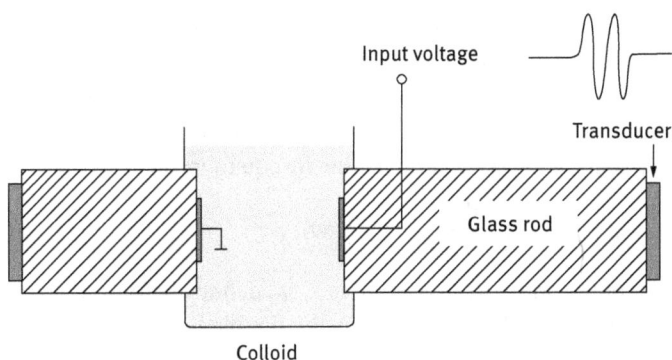

Fig. 11.8: Schematic representation of the AcoustoSizer cell.

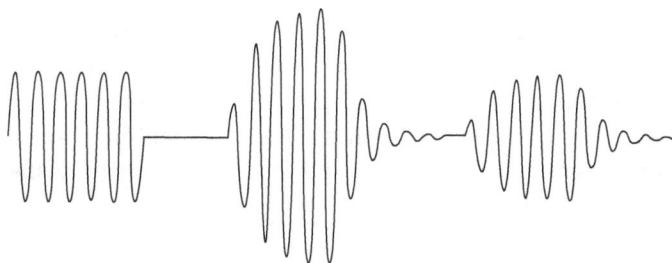

Fig. 11.9: Signals from the right-hand transducer.

In addition to the electrodes, the sample cell of the ESA instruments also houses sensors for pH, conductivity and temperature measurements. It is also equipped with a stirrer and the system is linked to a digital titrator for dynamic mobility and zeta potential measurements as a function of pH.

To convert the ESA signal to dynamic mobility one needs to know the density of the disperse phase and dispersion medium, the volume fraction of the particles and the velocity of sound in the solvent. As shown before, to convert mobility to zeta potential one needs to know the viscosity of the dispersion medium and its relative permittivity. Because of the inertia effects in dynamic mobility measurements, the weight average particle size has to be known.

For dilute suspensions with a volume fraction $\phi = 0.02$, the dynamic mobility u_d can be calculated from the electrokinetic sonic amplitude $A_{ESA}(\omega)$ using the following expression [14–18],

$$A_{ESA}(\omega) = Q(\omega)\phi(\Delta\rho/\rho)(u_d), \tag{11.18}$$

where ω is the angular frequency of the applied field, $\Delta\rho$ is the density difference between the particle (with density ρ) and the medium. $Q(\omega)$ is an instrument-related coefficient independent of the system being measured.

For a dilute dispersion of spherical particles with $\phi < 0.1$, a thin double layer ($\kappa R > 50$) and narrow particle size distribution (with standard deviation < 20 % of the mean size), u_d can be related to the zeta potential ζ by the equation [14–18],

$$u_d = \frac{2\varepsilon\zeta}{3\eta}G\left(\frac{\omega R^2}{v}\right)[1 + f(\lambda, \omega)], \tag{11.19}$$

where ε is the permittivity of the liquid (that is equal to $\varepsilon_r\varepsilon_0$, defined before), R is the particle radius, η is the viscosity of the medium, λ is the double layer conductance and v is the kinematic viscosity ($= \eta/\rho$). G is a factor that represents particle inertia, which reduces the magnitude of u_d and increases the phase lag in a monotonic fashion as the frequency increases. This inertia factor can be used to calculate the particle size from electroacoustic data. The factor $[1 + f(\lambda, \omega)]$ is proportional to the tangential component of the electric field and dependent on the particle permittivity and a surface conductance parameter λ. For most suspensions with large κR, the effect of surface conductance is insignificant and the particle permittivity/liquid permittivity $\varepsilon_p/\varepsilon$ is small. In most cases where the ionic strength is at least 10^{-3} mol dm^{-3} and a zeta potential < 75 mV, the factor $[1 + f(\lambda, \omega)]$ assumes the value 0.5. In this case the dynamic mobility is given by the simple expression,

$$u_d = \frac{\varepsilon\zeta}{\eta}G(\alpha). \tag{11.20}$$

Equation (11.20) is identical to the Smoluchowski equation [8], except for the inertia factor $G(\alpha)$.

The equation for converting the ESA amplitude, A_{ESA}, to dynamic mobility is given by,

$$u_d = \frac{A_{ESA}}{\phi v_s \Delta\rho}G(\alpha)^{-1}. \tag{11.21}$$

The zeta potential ζ is given by,

$$\zeta = \frac{u_d\eta}{\varepsilon}G(\alpha)^{-1} = \frac{A_{ESA}}{\phi v_s \Delta\rho}G(\alpha)^{-1}. \tag{11.22}$$

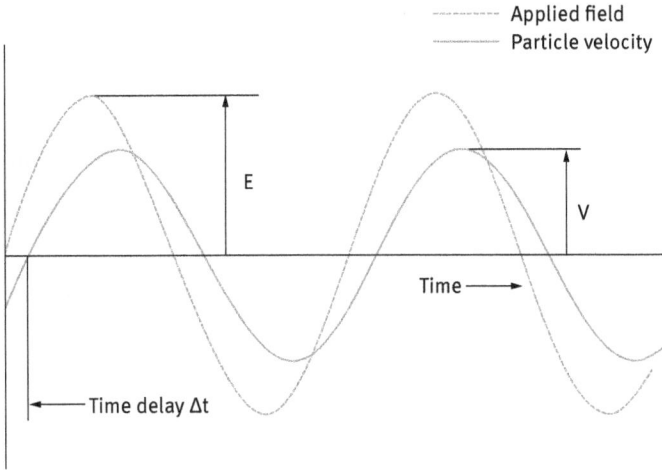

Fig. 11.10: Variation of applied field and particle velocity with time at high frequency.

For a polydisperse system $\langle u_d \rangle$ is given by,

$$\langle u_d(\omega) \rangle = \int_0^\infty u(\omega, R) p(R) \, dR , \qquad (11.23)$$

where $u(\omega, R)$ is the average dynamic mobility of particles with radius R at a frequency ω, and $pR \, dR$ is the mass fraction of particles with radii in the range $R \pm dR/2$.

The ESA measurements can also be applied for determining the particle size in a suspension from particle mobilities. The electric force acting upon a particle is opposed by the hydrodynamic friction and inertia of the particles. At low frequencies of alternating electric field, the inertial force is insignificant and the particle moves in the alternating electric field with the same velocity as it would have moved in a constant field. Particle mobility at low frequencies can be measured to calculate the zeta potential. At high frequencies, the inertia of the particle increases causing the velocity of the particle to decrease and the movement of the particle to lag behind the field. This is illustrated in Fig. 11.10, which shows the variation of applied field and particle velocity with time. Since inertia depends on particle mass, both of these effects depend on the particle mass and consequently on its size. Hence, both zeta potential and particle size can be determined from the ESA signal, if the frequency of the alternating field is sufficiently high. This is the method that is provided by the AcoustoSizer from Colloidal Dynamics.

Several variables affect the ESA measurements and these are listed below.

(i) Particle concentration range: very dilute suspensions generate a weak signal and are not suitable for ESA measurements. The magnitude of the ESA signal is proportional to average particle mobility, the volume fraction of the particles ϕ and the density difference between the particles and the medium $\Delta\rho$. To obtain a signal that is at least one order of magnitude higher than the background electrical noise (≈ 0.002 mPa M/V) the concentration and/or the density difference have to be sufficiently large. If the density difference between the particles and the medium is small, e.g. polystyrene latex with $\Delta\rho \approx 0.05$, then a sufficiently high concentration ($\phi > 0.02$) is needed to obtain a reasonably strong ESA signal. The accuracy of the ESA measurement is also not good at high ϕ values. This is due to the nonlinearity of the ESA amplitude-ϕ relationship at high ϕ values. Such deviation becomes appreciable at $\phi > 0.1$. However, reasonable values of zeta potential can be obtained from ESA measurements up to $\phi = 0.2$. Above this concentration, the measurements are not sufficiently accurate and the results obtained can only be used for qualitative assessment.

(ii) Electrolyte effects: ions in the dispersion generate electroacoustic (IVP) potential and the ESP signal is therefore a composite of the signals created by the particles and ions. However, the ionic contribution is relatively small, unless the particle concentration is low, their zeta potential is low and the ionic concentration is high. The ESA system is therefore not suitable for dynamic mobility and zeta potential measurements in systems with electrolyte concentration higher than 0.3 mol dm^{-3} KCl.

(iii) Temperature: since the viscosity of the dispersion decreases by $\approx 2\%$ per °C and its conductivity increases by about the same amount, it is important that temperature be accurately controlled using a Peltier device. Temperature control should also be maintained during sample preparation, for example when the suspension is sonicated. To avoid overheating the sample should be cooled in an ice bath at regular intervals during sonication.

(iv) Calibration and accuracy: the electroacoustic probe should be calibrated using a standard reference dispersion such as polystyrene latex or colloidal silica (Ludox). The common sources of error are unsuitable particle concentration (too low or too high), irregular particle shape, polydispersity, electrolyte signals, temperature variations, sedimentation, coagulation and entrained air bubbles. The latter in particular can cause erroneous ESA signal fluctuations resulting from weakening of the sound by the air bubbles. In many cases, the zeta potential results obtained using the ESA method do not agree with those obtained using other methods such as microelectrophoresis or laser velocimetry. However, the difference seldom exceeds 20 % and this makes the ESA method more convenient for measurement of many industrial methods. The main advantages are the speed of measurement and the dispersion does not need to be diluted, which may change the state of the suspension.

11.3 Measurement of surfactant and polymer adsorption

Surfactant and polymer adsorption are key to understanding how these molecules affect the stability/flocculation of the suspension. The various techniques that may be applied for obtaining information on surfactant and polymer adsorption have been described before [19]. Surfactant (both ionic and nonionic) adsorption is reversible and the process of adsorption can be described using the Langmuir isotherm [19]. Basically, representative samples of the solid with mass m and surface area A ($m^2\,g^{-1}$) are equilibrated with surfactant solutions covering various concentrations C_1 (a wide concentration range from values below and above the critical micelle concentration, cmc). The particles are dispersed in the solution by stirring and left to equilibrate (preferably overnight while stirred over rollers) after which the particles are removed by centrifugation and/or filtration (using millipore filters). The concentration in the supernatant solution C_2 is determined using a suitable analytical method. The latter must be sensitive enough for determining very low surfactant concentration. The surface area of the solid can be determined using gas adsorption and application of the BET equation. Alternatively, the surface area of the "wet" solid (which may be different from that of the dry solid) can be determined using dye adsorption [19].

From a knowledge of C_1 and C_2, m and A one can calculate the amount of adsorption Γ ($mg\,m^{-2}$ or $mol\,m^{-2}$) as a function of equilibrium concentration C_2 (ppm or $mol\,dm^{-3}$),

$$\Gamma = \frac{C_1 - C_2}{mA}.$$
(11.24)

With most surfactants, a Langmuir-type isotherm is obtained (illustrated in Fig. 11.11). Γ increases gradually with increasing C_2 and eventually reaches a plateau value, Γ_∞, which corresponds to saturation adsorption.

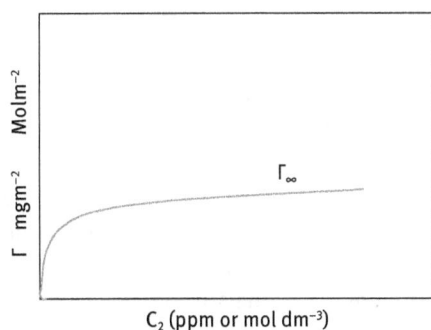

Fig. 11.11: Langmuir-type adsorption isotherm.

The results of Fig. 11.11 can be fitted to the Langmuir equation,

$$\Gamma = \frac{\Gamma_\infty b C_2}{1 + b C_2},$$
(11.25)

where b is a constant that is related to the free energy of adsorption ΔG_{ads},

$$b = \exp(-\Delta G_{ads}/RT).$$
(11.26)

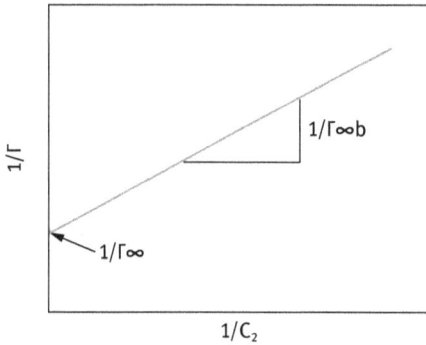

Fig. 11.12: Linearized form of Langmuir equation.

A linearized form of the Langmuir equation may be used to obtain Γ_∞ and b, as illustrated in Fig. 11.12,

$$\frac{1}{\Gamma} = \frac{1}{\Gamma_\infty} + \frac{1}{\Gamma_\infty b C_2} . \tag{11.27}$$

A plot of $1/\Gamma$ versus $1/C_2$ gives a straight line (Fig. 11.12) with intercept $1/\Gamma_\infty$ and slope $1/\Gamma_\infty b$ from which both Γ_∞ and b can be calculated.

From Γ_∞ the area per surfactant ion or molecule can be calculated,

$$\frac{\text{Area}}{\text{Molecule}} = \frac{1}{\Gamma_\infty N_{av}} \ (\text{m}^2) = \frac{10^{18}}{\Gamma_\infty N_{av}} \ (\text{nm}^2) . \tag{11.28}$$

As discussed before [19], the area per surfactant ion or molecule gives information on the orientation of surfactant ions or molecules at the interface. This information is relevant for the stability of the suspension. For example, for vertical orientation of surfactant ions, e.g. dodecyl sulphate anions, which is essential to produce a high surface charge (and hence enhanced electrostatic stability) the area per molecule is determined by the cross-sectional area of the sulphate group which is in the region of $0.4 \, \text{nm}^2$ With nonionic surfactants consisting of an alkyl chain and poly(ethylene oxide) (PEO) head group, adsorption on a hydrophobic surface is determined by the hydrophobic interaction between the alkyl chain and the hydrophobic surface. For vertical orientation of a monolayer of surfactant molecules, the area per molecule depends on the size of the PEO chain. The latter is directly related to the number of EO units in the chain. If the area per molecule is smaller than that predicted from the size of the PEO chain, the surfactant molecules may associate on the surface forming bilayers, hemimicelles, etc. as discussed in detail in Chapter 5. This information can be directly related to the stability of the suspension.

The adsorption of polymers is more complex than surfactant adsorption, since one must consider the various interactions (chain-surface, chain-solvent and surface-solvent) as well as the conformation of the polymer chain on the surface [19]. As discussed before [19], complete information on polymer adsorption may be obtained if one is able to determine the segment density distribution, i.e. the segment concentration, n, in all layers parallel to the surface. However, such information is generally unavailable, and therefore one determines three main parameters: the amount of adsorption Γ per unit area, the fraction p of segments in direct contact with the surface (i.e. in trains) and the adsorbed layer thickness δ.

The amount of adsorption, Γ, can be determined in the same way as for surfactants, although in this case the adsorption process may take a long equilibrium time. Most polymers show a high affinity isotherm as is illustrated in Fig. 11.13.

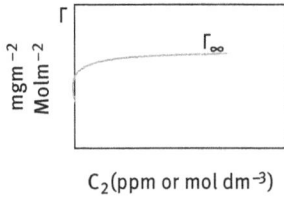

C_2(ppm or mol dm^{-3})   Fig. 11.13: **High affinity isotherm.**

This implies that the first added molecules are completely adsorbed and the isotherm cuts the y-axis at $C_2 = 0$. For desorption to occur, the polymer concentration in the supernatant liquid must approach zero and this implies irreversible adsorption. As discussed in Chapter 6, the magnitude of saturation adsorption depends on the molecular weight of the polymer, the temperature and the solvency of the medium for the chains.

The fraction of segments p in trains can be determined using spectroscopic techniques such IR, ESR and NMR. p depends on surface coverage, polymer molecular weight and solvency of the medium for the chains.

Several techniques may be applied for determining the adsorbed layer thickness δ and these were described in detail before [19].

11.4 Assessment of sedimentation of suspensions

As mentioned in Chapter 8, most suspensions undergo sedimentation on standing due to gravity and the density difference $\Delta\rho$ between the particles and dispersion medium. This is particularly the case when the particle radius exceeds 50 nm and when $\Delta\rho > 0.1$. In this case, Brownian diffusion cannot overcome the gravity force and sedimentation occurs, resulting in an increasing particle concentration from the top to the bottom of the container. As discussed in Chapter 10, to prevent particle sedimentation, "thickeners" (rheology modifiers) are added in the continuous phase. The sedimentation of the suspension is characterized by the sedimentation rate, sediment volume, the change of particle size distribution during settling and the stability of the suspension to sedimentation. Assessment of sedimentation of a suspension depends on the force applied to the particles in the suspension, namely gravitational, centrifugal and electrophoretic. The sedimentation processes are complex and subject to various errors in sedimentation measurements [19]. A suspension is usually agitated before measuring sedimentation, to ensure an initially homogeneous system of particles in random motion. Vigorous agitation or the use of ultrasonic cavitation must be avoided to prevent any breakdown of aggregates and change of the particle size distribution.

Several physical measurements can be applied to assess sedimentation and these methods have been described in detail by Kissa [20]. The simplest method is to measure the density of the settling suspension at a known depth using a hydrometer. Unfortunately, this simple method is highly invasive due to the disturbance of the suspension by the hydrometer. A more accurate method is to use sedimentation balances, whereby the sediment accumulated at the base of the sedimentation column is collected and weighed. Manometric methods that use a capillary side arm for measuring the difference between the densities of the pure sedimentation fluid and that of the suspension can also be applied. Several electrical methods can be applied to assess sedimentation. Most suspensions have complex electrical permittivities and may require measurement of both the capacitance and conductivity to determine the solid volume fraction at depth h and time t. This method has the advantage of being non-invasive, since the sensing electrodes do not have to be in direct contact with the dispersion. A more convenient method is to use ultrasound probes at various heights from the top to the bottom of the sedimentation tube. The ultrasound velocity and attenuation depends on the volume fraction of the suspension, allowing one to obtain the solids content as a function of height in the sedimentation tube. An alternative optical technique is to measure the backscattering of near infrared at various heights from the sedimentation tube. A commercially available apparatus, namely the Turboscan, can be used for this purpose.

Several other techniques have been designed to monitor sedimentation of suspensions of which photosedimentation, X-ray sedimentation and laser anemometry are perhaps worth mentioning. The simplest sedimentation test is based on visual observation of settling. The turbidity of the suspension is estimated visually, or the height of the sediment and sediment volume are recorded as a function of time. This visual estimation of sedimentation is only qualitative but is adequate in many practical situations. However, the characterization of suspensions and the determination of the particle size distribution require quantitative sedimentation methods. Instrumental techniques have been developed for measuring the turbidity of the suspension as a function of time, either by measuring the turbidity of the bulk suspension or by withdrawing a sample at a given height of the settling suspension. The earlier instruments used for measuring the turbidity of suspensions, called nephelometers, have evolved into instruments with a more sophisticated optical system. Photosedimentometers monitor gravitational particle sedimentation by photoelectric measurement of incident light under steady-state conditions. A horizontal beam of parallel light is projected through a suspension in a sedimentation column to a photocell. Double-beam photosedimentometers using matched photocells, one for the sample and the other for the reference beam were later developed. A more sophisticated method was later introduced, using linear charge-coupled photodiode array as the image sensor to convert the light intensity attenuated by the particles into an electric signal. The output of each of the photodetectors is handled by a computer independently. Hence the settling distance between any point in the liquid and the surface of the liquid can be measured accurately without using a mechanical device. As a consequence, particle measurement is rapid, requiring only about 5 minutes to determine a particle size distribution. The use of fibre optics has made it possible to scan the sedimentation column without moving parts or with a fibre optic probe that is moved inside the sedimentation column.

Laser anemometry, also described as laser Doppler velocity measurement (LVD) is a sensitive technique that can extend the range of photosedimentation methods. It has been applied in a sedimentometer to measure particle sizes as low as 0.5 μm.

X-ray sedimentometers measure X-ray absorption to determine concentration gradients in sedimenting suspensions. The use of X-ray and γ-rays has been proposed as transmittance probes that correlate transmitted radiation with the density of the suspension. X-ray transmittance, T, is directly related to the weight of particles by an exponential relationship, analogous to the Lambert–Beer law governing transmittance of visible radiation,

$$\ln T = -A\phi_s , \tag{11.29}$$

where A is a particle-, medium- and equipment constant and ϕ_s is the volume fraction of particles in the suspension.

The concentration of particles remaining in the liquid at various sedimentation depths is determined by using a finely collimated beam of X-rays. The time required for the sedimentation measurement is shortened by continuously changing the effective sedimentation depth. The concentration of particles remaining at various depths is measured as a function of time. X-ray sedimentometers can be used for particles containing elements with atomic numbers above 15 and, therefore, the method cannot be applied to measure sedimentation of organic pigments.

It should be mentioned that gravitational sedimentation is often too slow, particularly if the particles are small and have a density that is not appreciably higher than that of the medium. Application of a centrifugal force accelerates sedimentation, allowing one to obtain results within a reasonable time. However, the data obtained by centrifugation do not always correlate with those resulting from settling under gravity. This is particularly the case with suspensions that are weakly flocculated, where the loose structure may break up on application of a centrifugal force. The interaction between the particles may also change on application of a high gravitational force. This casts doubt on the use of centrifugation as an accelerated test for prediction of sedimentation.

11.5 Assessment of flocculation and Ostwald ripening (crystal growth)

Assessment of flocculation and Ostwald ripening of a suspension requires measurement of the particle size and shape distribution as a function of time. Several techniques may be applied for this purpose and these are summarized below [19].

11.5.1 Optical microscopy

This is by far the most valuable tool for a qualitative or quantitative examination of the suspension. Information on the size, shape, morphology and aggregation of particles can be conveniently obtained with minimum time required for sample preparation. Since individual particles can be directly observed and their shape examined, optical microscopy is considered the only absolute method for particle characterization. However, optical microscopy has some limitations: the minimum size that can be detected. The practical lower limit for accurate measurement of particle size is 1.0 μm, although some detection may be obtained down to 0.3 μm. Image contrast may not be good enough for observation, particularly when using a video camera which is mostly used for convenience. The contrast can be improved by decreasing the aperture of the iris diaphragm but this reduces the resolution. The contrast of the image depends on the refractive index of the particles relative to that of the medium. Hence the contrast can be improved by increasing the difference between the refractive index of the particles and the immersion medium. Unfortunately, changing the medium for the suspension is not practical since this may affect the state of the dispersion. Fortunately, water with a refractive index of 1.33 is a suitable medium for most organic particles with a refractive index usually > 1.4.

The ultramicroscope by virtue of dark field illumination extends the useful range of optical microscopy to small particles not visible in a bright light illumination. Dark field illumination utilizes a hollow cone of light at a large angle of incidence. The image is formed by light scattered from the particles against a dark background. Particles about 10 times smaller than those visible by bright light illumination can be detected. However, the image obtained is abnormal and the particle size cannot be accurately measured. For that reason, the electron microscope (see below) has displaced the ultramicroscope, except for dynamic studies by flow ultramicroscopy.

Three main attachments to the optical microscope are possible:

(1) Phase contrast: this utilizes the difference between the diffracted waves from the main image and the direct light from the light source. The specimen is illuminated with a light cone and this illumination is within the objective aperture. The light illuminates the specimen and generates zero order and higher orders of diffracted light. The zero order light beam passes through the objective and a phase plate which is located at the objective back focal plane. The difference between the op-

tical path of the direct light beam and that of the beam diffracted by a particle causes a phase difference. The constructive and destructive interferences result in brightness changes which enhance the contrast. This produces sharp images allowing one to obtain particle size measurements more accurately. The phase contrast microscope has a plate in the focal plane of the objective back focus. Instead of a conventional iris diaphragm, the condenser is equipped with a ring matched in its dimension to the phase plate.

(2) Differential interference contrast (DIC): this gives a better contrast than the phase contrast method. It utilizes a phase difference to improve contrast, but the separation and recombination of a light beam into two beams is accomplished by prisms. DIC generates interference colours and the contrast effects indicate the refractive index difference between the particle and medium.

(3) Polarized light microscopy: this illuminates the sample with linearly or circularly polarized light, either in a reflection or transmission mode. One polarizing element, located below the stage of the microscope, converts the illumination to polarized light. The second polarizer is located between the objective and the ocular and is used to detect polarized light. Linearly polarized light cannot pass the second polarizer in a crossed position unless the plane of polarization has been rotated by the specimen. Various characteristics of the specimen can be determined, including anisotropy, polarization colours, birefringence, polymorphism, etc.

11.5.2 Sample preparation for optical microscopy

A drop of the suspension is placed on a glass slide and covered with a cover glass. If the suspension has to be diluted, the dispersion medium (that can be obtained by centrifugation and/or filtration of the suspension) should be used as the diluent in order to avoid aggregation. At low magnifications, the distance between the objective and the sample is usually adequate for manipulating the sample, but at high magnification the objective may be too close to the sample. An adequate working distance can be obtained, while maintaining high magnification, by using a more powerful eyepiece with a low power objective. For suspensions encountering Brownian motion (when the particle size is relatively small), microscopic examination of moving particles can become difficult. In this case one can record the image on a photographic film or video tape or disc (using computer software).

11.5.3 Particle size measurements using optical microscopy

The optical microscope can be used to observe dispersed particles and flocs. Particle sizing can be carried out using manual, semiautomatic or automatic image analysis techniques. In the manual method (which is tedious) the microscope is fitted with a minimum of 10× and 43× achromatic or apochromatic objectives equipped with a high numerical apertures (10×, 15× and 20×), a mechanical XY stage, a stage micrometre and a light source. The direct measurement of particle size is aided by a linear scale or globe-and-circle graticules in the ocular. The linear scale is mainly useful for spherical particles, with a relatively narrow particle size distribution. The globe-and-circle graticules are used to compare the projected particle area with a series of circles in the ocular graticule. The size of spherical particles can be expressed by the diameter, but for irregularly shaped particles various statistical diameters are used. One of the difficulties with the evaluation of dispersions by optical microscopy is the quantification of data. The number of particles in at least six different size ranges must be counted to obtain a distribution. This problem can be alleviated by the use of automatic image analysis which can also give an indication of the floc size and its morphology.

11.5.4 Electron microscopy

Electron microscopy utilizes an electron beam to illuminate the sample. The electrons behave as charged particles which can be focused by annular electrostatic or electromagnetic fields surrounding the electron beam. Due to the very short wavelength of electrons, the resolving power of an electron microscope exceeds that of an optical microscope by ≈ 200 times. The resolution depends on the accelerating voltage which determines the wavelength of the electron beam and magnifications as high as 200 000 can be reached with intense beams, but this could damage the sample. Mostly the accelerating voltage is kept below 100–200 kV and the maximum magnification obtained is below 100 000. The main advantage of electron microscopy is the high resolution, sufficient for resolving details separated by only a fraction of a nanometre. The increased depth of field, usually by about 10 μm or about 10 times of that of an optical microscope, is another important advantage of electron microscopy. Nevertheless, electron microscopy also has some disadvantages, such as sample preparation, selection of the area viewed and interpretation of the data. The main drawback of electron microscopy is the potential risk of altering or damaging the sample that may introduce artefacts and possible aggregation of the particles during sample preparation. The suspension has to be dried or frozen and the removal of the dispersion medium may alter the distribution of the particles. If the particles do not conduct electricity, the sample has to be coated with a conducting layer, such as gold, carbon or platinum to avoid negative charging by the electron beam. Two main types of electron microscopes are used: transmission and scanning.

11.5.4.1 Transmission electron microscopy (TEM)

TEM displays an image of the specimen on a fluorescent screen and the image can be recorded on a photographic plate or film. The TEM can be used to examine particles in the range 0.001–5 μm. The sample is deposited on a Formvar (polyvinyl formal) film resting on a grid to prevent charging of the simple. The sample is usually observed as a replica by coating with an electron transparent material (such as gold or graphite). The preparation of the sample for TEM may alter the state of dispersion and cause aggregation. Freeze fracturing techniques have been developed to avoid some of the alterations of the sample during sample preparation. Freeze fracturing allows the dispersions to be examined without dilution and replicas can be made of dispersions containing water. It is necessary to have a high cooling rate to avoid the formation of ice crystals.

11.5.4.2 Scanning electron microscopy (SEM)

SEM can show particle topography by scanning a very narrowly focused beam across the particle surface. The electron beam is directed normally or obliquely at the surface. The backscattered or secondary electrons are detected in a raster pattern and displayed on a monitor screen. The image provided by secondary electrons exhibits good three-dimensional detail. The backscattered electrons, reflected from the incoming electron beam, indicate regions of high electron density. Most SEMs are equipped with both types of detectors. The resolution of the SEM depends on the energy of the electron beam, which does not exceed 30 kV, and hence the resolution is lower than that obtained by the TEM. A very important advantage of SEM is elemental analysis by energy dispersive X-ray analysis (EDX). If the electron beam impinging on the specimen has sufficient energy to excite atoms on the surface, the sample will emit X-rays. The energy required for X-ray emission is characteristic of a given element and since the emission is related to the number of atoms present, quantitative determination is possible.

Scanning transmission electron microscopy (STEM) coupled with EDX has been used for determining of metal particle sizes. Specimens for STEM were prepared by ultrasonically dispersing the sample in methanol and one drop of the suspension was placed onto a Formvar film supported on a copper grid.

11.5.5 Confocal laser scanning microscopy (CLSM)

CLSM is a very useful technique for identifying suspensions. It uses a variable pinhole aperture or variable width slit to illuminate only the focal plane by the apex of a cone of laser light. Out-of-focus items are dark and do not distract from the contrast of the image. As a result of extreme depth discrimination (optical sectioning) the resolution is considerably improved (up to 40 % when compared with optical microscopy). The

CLSM technique acquires images by laser scanning or uses computer software to subtract out-of-focus details from the in-focus image. Images are stored as the sample is advanced through the focal plane in elements as small as 50 nm. Three-dimensional images can be constructed to show the shape of the particles.

11.5.6 Scanning probe microscopy (SPM)

SPM can measure physical, chemical and electrical properties of the sample by scanning the particle surface with a tiny sensor of high resolution. Scanning probe microscopes do not measure a force directly; they measure the deflection of a cantilever which is equipped with a tiny stylus (the tip) functioning as the probe. The deflection of the cantilever is monitored by (i) a tunnelling current; (ii) laser deflection beam from the back side of the cantilever; (iii) optical interferometry; (iv) laser output controlled by the cantilever used as a mirror in the laser cavity; and (v) change in capacitance. SPM generates a three-dimensional image and allows calibrated measurements in three (x, y, z) coordinates. SPB not only produces a highly magnified image, but provides valuable information on sample characteristics. Unlike EM which requires vacuum for its operation, SPM can be operated under ambient conditions and, with some limitation, in liquid media.

11.5.7 Scanning tunnelling microscopy (STM)

STM measures an electric current that flows through a thin insulating layer (vacuum or air) separating two conductive surfaces. The electrons are visualized to "tunnel" through the dielectric and generate a current, I, that depends exponentially on the distance, s, between the tiny tip of the sensor and the electrically conductive surface of the sample. STM tips are usually prepared by etching a tungsten wire in an NaOH solution until the wire forms a conical tip. Pt/Ir wire has also been used. In the contrast current imaging mode, the probe tip is raster-scanned across the surface and a feedback loop adjusts the height of the tip in order to maintain a constant tunnel current. When the energy of the tunnelling current is sufficient to excite luminescence, the tip-surface region emits light and functions as an excitation source of subnanometre dimensions. In situ STM has revealed a two-dimensional molecular lamellar arrangement of long chain alkanes adsorbed on the basal plane of graphite. Thermally induced disordering of adsorbed alkanes was studied by variable temperature STM and atomic scale resolution of the disordered phase was claimed by studying the quenched high-temperature phase.

11.5.8 Atomic force microscopy (AFM)

AFM allows one to scan the topography of a sample using a very small tip made of silicon nitride. The tip is attached to a cantilever that is characterized by its spring constant, resonance frequency and a quality factor. The sample rests on a piezoceramic tube which can move the sample horizontally (x, y motion) and vertically (z motion). The displacement of the cantilever is measured by the position of a laser beam reflected from the mirrored surface on the top side of the cantilever. The reflected laser beam is detected by a photodetector. AFM can be operated in either a contact or a noncontact mode. In the contact mode the tip travels in close contact with the surface, whereas in the noncontact mode the tip hovers 5–10 nm above the surface.

11.5.9 Scattering techniques

These are by far the most useful methods for characterization of suspensions and in principle they can give quantitative information on the particle size distribution, floc size and shape. The only limitation of the methods is the need to use sufficiently dilute samples to avoid interference such as multiple scattering which makes interpretation of the results difficult. However, recently backscattering methods have been designed to allow one to measure the sample without dilution. In principle one can use any electromagnetic radiation such as light, X-ray or neutrons, but in most industrial labs only light scattering is applied (using lasers).

11.5.9.1 Light scattering techniques

These can be conveniently divided into the following classes: (i) time-average light scattering, static or elastic scattering; (ii) turbidity measurements which can be carried out using a simple spectrophotometer; (iii) light diffraction technique; (iv) dynamic (quasi-elastic) light scattering that is usually referred as photon correlation spectroscopy. This is a rapid technique that is very suitable for measuring submicron particles or droplets (nanosize range); (v) backscattering techniques that are suitable for measuring concentrated samples. Application of any of these methods depends on the information required and availability of the instrument.

11.5.9.2 Time-average light scattering

In this method the dispersion that is sufficiently diluted to avoid multiple scattering is illuminated by a collimated light (usually laser) beam and the time-average intensity of scattered light is measured as a function of scattering angle, θ. Static light scattering is termed elastic scattering. Three regimes can be identified:

Rayleigh regime. The particle radius R is smaller than $\lambda/20$ (where λ is the wave length of incident light). The scattering intensity is given by the equation,

$$I(Q) = [\text{Instrument constant}]\,[\text{Material constant}]\,NV_p^2 . \qquad (11.30)$$

Q is the scattering vector that depends on the wavelength of light λ used and is given by,

$$Q = \left(\frac{4\pi n}{\lambda}\right) \sin\left(\frac{\theta}{2}\right), \qquad (11.31)$$

where n is the refractive index of the medium.

The material constant depends on the difference between the refractive index of the particle and that of the medium. N is the number of particles and V_p is the volume of each particle. Assuming that the particles are spherical one can obtain the average size using equation (11.30).

The Rayleigh equation reveals two important relationships: (i) The intensity of scattered light increases with the square of the particle volume and consequently with the sixth power of the radius R. Hence the scattering from larger particles may dominate the scattering from smaller particles. (ii) The intensity of scattering is inversely proportional to λ^4. Hence a decrease in the wavelength will substantially increase the scattering intensity.

Rayleigh–Gans–Debye Regime (RGD) $\lambda/20 < R < \lambda$. The RGD regime is more complicated than the Rayleigh regime and the scattering pattern is no longer symmetrical about the line corresponding to the 90° angle but favours forward scattering ($\theta < 90°$) or backscattering ($180° > \theta > 90°$). Since the preference for forward scattering increases with increasing particle size, the ratio $I_{45°}/I_{135°}$ can indicate the particle size.

Mie Regime $R > \lambda$. The scattering behaviour is more complex than the RGD regime and the intensity exhibits maxima and minima at various scattering angles depending on particle size and refractive index. The Mie theory for light scattering can be used to obtain the particle size distribution using numerical solutions. One can also obtain information on particle shape.

11.5.9.3 Turbidity measurements

Turbidity (total light scattering technique) can be used to measure particle size, flocculation and particle sedimentation. This technique is simple and easy to use; a single or double beam spectrophotometer or a nephelometer can be used.

For nonabsorbing particles the turbidity τ is given by,

$$\tau = \left(\frac{1}{L}\right) \ln\left(\frac{I_0}{I}\right), \qquad (11.32)$$

where L is the path length, I_0 is the intensity of incident beam and I is the intensity of transmitted beam.

The particle size measurement assumes that the light scattering by a particle is singular and independent of other particles. Any multiple scattering complicates the analysis. According to Mie theory, the turbidity is related to the particle number, N, and their cross section, πr^2, (where r is the particle radius) by

$$\tau = Q\pi r^2 N,\tag{11.33}$$

where Q is the total Mie scattering coefficient. Q depends on the particle size parameter α (which depends on particle diameter and wavelength of incident light λ) and the ratio of refractive index of the particles and medium m.

Q depends on α in an oscillatory mode that exhibits a series of maxima and minima whose position depends on m. For particles with $R < (1/20)\lambda$, $\alpha < 1$ and it can be calculated using Rayleigh theory. For $R > \lambda$, Q approaches 2 and between these two extremes, Mie theory is used. If the particles are not monodisperse (as is the case with most practical systems), the particle size distribution must be taken into account. Using this analysis one can establish the particle size distribution using numerical solutions.

11.5.9.4 Light diffraction techniques

This is a rapid and nonintrusive technique for determining particle size distribution in the range 2–300 μm with good accuracy for most practical purposes. Light diffraction gives an average diameter over all particle orientations as randomly oriented particles pass the light beam. A collimated and vertically polarized laser beam illuminates a particle dispersion and generates a diffraction pattern with the undiffracted beam in the centre. The energy distribution of diffracted light is measured by a detector consisting of light sensitive circles separated by isolating circles of equal width. The angle formed by the diffracted light increases with decreasing particle size. The angle-dependent intensity distribution is converted by Fourier optics into a spatial intensity distribution I(r). The spatial intensity distribution is converted into a set of photocurrents and the particle size distribution is calculated using a computer. Several commercial instruments are available, e.g. Malvern Mastersizer (Malvern, UK), Horriba (Japan) and Coulter LS Sizer (USA). A schematic illustration of the set-up is shown in Fig. 11.14.

In accordance with the Fraunhofer theory (which was introduced by Fraunhofer over 100 years ago), the special intensity distribution is given by,

$$I(r) = \int_{X_{min}}^{X_{max}} N_{tot}q_0(x)I(r, x)\,dx,\tag{11.34}$$

where $I(r, x)$ is the radial intensity distribution at radius r for particles of size x, N_{tot} is the total number of particles and $q_0(x)$ describes the particle size distribution.

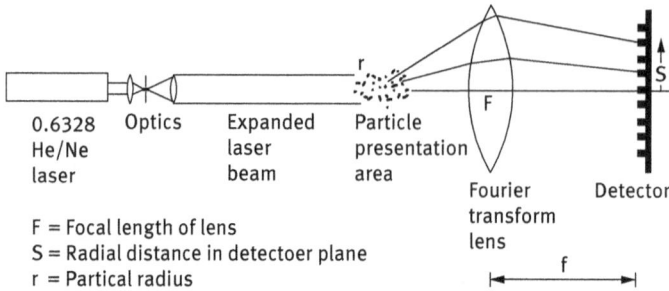

F = Focal length of lens
S = Radial distance in detectoer plane
r = Partical radius

Fig. 11.14: Schematic illustration of light diffraction particle sizing system.

The radial intensity distribution $I(r, x)$ is given by,

$$I(r, x) = I_0 \left(\frac{\pi x^2}{2f} \right)^2 \left(\frac{J_i(k)}{k} \right)^2, \tag{11.35}$$

with $k = (\pi x r)/(\lambda f)$ and where r is the distance to the centre of the disc, λ is the wavelength, f is the focal length, and J_i is the first-order Bessel function.

Fraunhofer diffraction theory applies to particles whose diameter is considerably larger than the wavelength of illumination. As shown in Fig. 11.14, an He/Ne laser is used with $\lambda = 632.8$ nm for particle sizes mainly in the 2–120 μm range. In general, the diameter of the sphere-shaped particle should be at least four times the wavelength of the illumination light. The accuracy of particle size distribution determined by light diffraction is not very good if a large fraction of particles with diameter < 10 μm is present in the suspension. For small particles (diameter < 10 μm), Mie theory is more accurate if the necessary optical parameters, such as refractive index of particles and medium and the light absorptivity of the dispersed particles, are known. Most commercial instruments combine light diffraction with forward light scattering to obtain a full particle size distribution covering a wide range of sizes.

As an illustration, Fig. 11.15 shows the result of particle sizing using a six component mixture of standard polystyrene lattices (using a Mastersizer).

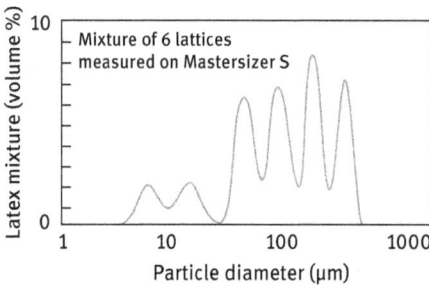

Fig. 11.15: Single measurement of a mixture of six standard lattices using the Mastersizer.

Most practical suspensions are polydisperse and generate a very complex diffraction pattern. The diffraction pattern of each particle size overlaps with the diffraction patterns of other sizes. The particles of different sizes diffract light at different angles and the energy distribution becomes a very complex pattern. However, manufacturers of light diffraction instruments (such as Malvern, Coulters and Horriba) have developed numerical algorithms relating diffraction patterns to particle size distribution.

Several factors can affect the accuracy of Fraunhofer diffraction: (i) particles smaller than the lower limit of Fraunhofer theory; (ii) non-existent "ghost" particles in a particle size distribution obtained by Fraunhofer diffraction that was applied to systems containing particles with edges, or a large fraction of small particles (below 10 μm); (iii) computer algorithms that are unknown to the user and vary with the manufacturer's software version; (iv) the composition-dependent optical properties of the particles and dispersion medium; (v) if the density of all particles is not the same, the result may be inaccurate.

11.5.9.5 Dynamic light scattering – photon correlation spectroscopy (PCS)

Dynamic light scattering (DLS) is a method that measures the time-dependent fluctuation of scattered intensity. It is also referred to as quasi-elastic light scattering (QELS) or photon correlation spectroscopy (PCS). The latter is the most commonly used term for describing the process since most dynamic scattering techniques employ autocorrelation.

PCS is a technique that utilizes Brownian motion to measure particle size. As a result of Brownian motion of dispersed particles, the intensity of scattered light undergoes fluctuations that are related to the velocity of the particles. Since larger particles move less rapidly than smaller ones, the intensity fluctuation (intensity versus time) pattern depends on particle size as is illustrated in Fig. 11.16. The velocity of the scatterer is measured in order to obtain the diffusion coefficient.

In a system where Brownian motion is not interrupted by sedimentation or particle-particle interaction, the movement of particles is random. Hence, the intensity fluctuations observed after a large time interval do not resemble those fluctuations observed initially, but represent a random distribution of particles. Consequently, the fluctuations observed at large time delay are not correlated with the initial fluctuation pattern. However, when the time differential between the observations is very small (a nanosecond or a microsecond) both positions of particles are similar and the scattered intensities are correlated. When the time interval is increased, the correlation decreases. The decay of correlation is particle size-dependent. The smaller the particles are, the faster is the decay.

The fluctuations in scattered light are detected by a photomultiplier and are recorded. The data containing information on particle motion are processed by a digital correlator. The latter compares the intensity of scattered light at time t, $I(t)$, to the intensity at a very small time interval τ later, $I(t + \tau)$, and it constructs the

Large particles

Intensity

Time

Small particles

Intensity

Time

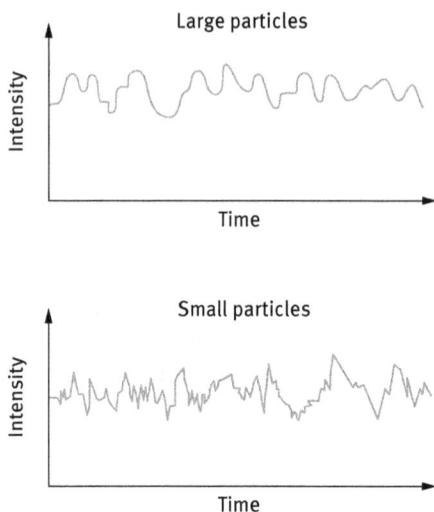

Fig. 11.16: Schematic representation of the intensity fluctuation for large and small particles.

second-order autocorrelation function $G_2(\tau)$ of the scattered intensity,

$$G_2(\tau) = \langle I(t)\, I(t + \tau)\rangle . \tag{11.36}$$

The experimentally measured intensity autocorrelation function $G_2(\tau)$ depends only on the time interval τ, and is independent of t, the time when the measurement started.

PCS can be measured in a homodyne mode where only scattered light is directed to the detector. It can also be measured in heterodyne mode where a reference beam split from the incident beam is superimposed on scattered light. The diverted light beam functions as a reference for the scattered light from each particle.

In the homodyne mode, $G_2(\tau)$ can be related to the normalized field autocorrelation function $g_1(\tau)$ by,

$$G_2(\tau) = A + B g_1^2(\tau) , \tag{11.37}$$

where A is the background term designated as the baseline value and B is an instrument-dependent factor. The ratio B/A is regarded as a quality factor for the measurement or the signal-to-noise ratio and expressed sometimes as the % merit.

The field autocorrelation function $g_1(\tau)$ for a monodisperse suspension decays exponentially with τ,

$$g_1(\tau) = \exp(-\Gamma\tau) , \tag{11.38}$$

where Γ is the decay constant (s^{-1}).

Substituting equation (11.38) into equation (11.37) yields the measured autocorrelation function,

$$G_2(\tau) = A + B \exp(-2\Gamma\tau) . \tag{11.39}$$

The decay constant Γ is linearly related to the translational diffusion coefficient D_T of the particle,

$$\Gamma = D_T q^2 . \tag{11.40}$$

The modulus q of the scattering vector is given by,

$$q = \frac{4\pi n}{\lambda_0} \sin\left(\frac{\theta}{2}\right) , \tag{11.41}$$

where n is the refractive index of the dispersion medium, θ is the scattering angle and λ_0 is the wavelength of the incident light in vacuum.

PCS determines the diffusion coefficient and the particle radius R is obtained using the Stokes–Einstein equation,

$$D = \frac{kT}{6\pi\eta R} , \tag{11.42}$$

where k is the Boltzmann constant, T is the absolute temperature and η is the viscosity of the medium.

The Stokes–Einstein equation is limited to noninteracting, spherical and rigid spheres. The effect of particle interaction at relatively low particle concentration c can be taken into account by expanding the diffusion coefficient into a power series of concentration,

$$D = D_0(1 + k_D c) , \tag{11.43}$$

where D_0 is the diffusion coefficient at infinite dilution and k_D is the virial coefficient that is related to particle interaction. D_0 can be obtained by measuring D at several particle number concentrations and extrapolating to zero concentration.

For polydisperse suspensions, the first-order autocorrelation function is an intensity-weighted sum of autocorrelation functions of particles contributing to the scattering,

$$g_1(\tau) = \int_0^\infty C(\Gamma) \exp(-\Gamma\tau) \, d\Gamma . \tag{11.44}$$

$C(\Gamma)$ represents the distribution of decay rates.

For narrow particle size distribution, the cumulant analysis is usually satisfactory The cumulant method is based on the assumption that for monodisperse suspensions, $g_1(\tau)$ is monoexponential. Hence, $\log g_1(\tau)$ versus τ yields a straight line with a slope equal to Γ,

$$\ln g_1(\tau) = 0.5 \ln(B) - \Gamma\tau , \tag{11.45}$$

where B is the signal-to-noise ratio.

The cumulant method expands the Laplace transform about an average decay rate,

$$\langle \Gamma \rangle = \int_0^\infty \Gamma C(\Gamma) \, d\Gamma . \tag{11.46}$$

The exponential in equation (11.45) is expanded about an average and integrated term,

$$\ln g_1(\tau) = \langle \Gamma \rangle \tau + (\mu_2 \tau^2)/2! - (\mu_3 \tau^3)/3! + \cdots . \tag{11.47}$$

An average diffusion coefficient is calculated from $\langle \Gamma \rangle$ and the polydispersity (termed the polydispersity index) is indicated by the relative second moment, $\mu_2/\langle \Gamma \rangle^2$. A constrained regulation method (CONTIN) yields several numerical solutions to the particle size distribution and this is normally included in the software of the PCS machine.

PCS is a rapid, absolute, nondestructive and rapid method for particle size measurements. It has some limitations. The main disadvantage is the poor resolution of particle size distribution. Also it suffers from the limited size range (absence of any sedimentation) that can be accurately measured. Several instruments are commercially available, e.g. by Malvern, Brookhaven, Coulters, etc. The most recent instrument that is convenient to use is HPPS supplied by Malvern (UK) and this allows one to measure the particle size distribution without the need for too much dilution (which may cause some particle dissolution).

11.5.9.6 Backscattering techniques

This method is based on the use of fibre optics, sometimes referred to as fibre optic dynamic light scattering (FODLS) and it allows one to measure at high particle number concentrations. FODLS employs either one or two optical fibres. Alternatively, fibre bundles may be used. The exit port of the optical fibre (optode) is immersed in the sample and the scattered light in the same fibre is detected at a scattering angle of 180° (i.e. backscattering).

The above technique is suitable for on-line measurements during manufacture of a suspension or emulsion. Several commercial instruments are available, e.g. Lesentech (USA).

11.6 Measurement of rate of flocculation

Two general techniques may be applied for measuring the rate of flocculation of suspensions, both of which can only be applied for dilute systems. The first method is based on measuring the scattering of light by the particles. For monodisperse particles with a radius that is less than $\lambda/20$ (where λ is the wavelength of light), one can apply the Rayleigh equation, whereby the turbidity τ_0 is given by,

$$\tau_0 = A' n_0 V_1^2 , \tag{11.48}$$

where A' is an optical constant (which is related to the refractive index of the particle and medium and the wavelength of light) and n_0 is the number of particles, each with a volume V_1.

By combining Rayleigh theory with the Smoluchowski–Fuchs theory of flocculation kinetics [17], one can obtain the following expression for the variation of turbidity with time,

$$\tau = A'n_0V_1^2(1 + 2n_0kt),\qquad(11.49)$$

where k is the rate constant of flocculation.

The second method for obtaining the rate constant of flocculation is by direct particle counting as a function of time. For this purpose optical microscopy or image analysis may be used, provided the particle size is within the resolution limit of the microscope. Alternatively, the particle number may be determined using electronic devices such as the Coulter counter or the flow ultramicroscope.

The rate constant of flocculation is determined by plotting 1/n versus t, where n is the number of particles after time t, i.e.,

$$\left(\frac{1}{n}\right) = \left(\frac{1}{n_0}\right) + kt.\qquad(11.50)$$

The rate constant k of slow flocculation is usually related to the rapid rate constant k_0 (the Smoluchowski rate) by the stability ratio W,

$$W = \left(\frac{k}{k_0}\right).\qquad(11.51)$$

One usually plots log W versus log C (where C is the electrolyte concentration) to obtain the critical coagulation concentration (ccc), which is the point at which log W = 0.

A very useful method for measuring flocculation is to use the single-particle optical method. The particles of the suspension that are dispersed in a liquid flow through a narrow uniformly illuminated cell. The suspension is made sufficiently dilute (using the continuous medium) so that particles pass through the cell individually. A particle passing through the light beam illuminating the cell generates an optical pulse detected by a sensor. If particle size is greater than the wavelength of light ($> 0.5\,\mu m$), the peak height depends on the projected area of the particle. If particle size is smaller than $0.5\,\mu m$, scattering dominates the response. For particles $> 1\,\mu m$, a light obscuration (also called blockage or extinction) sensor is used. For particles smaller than $1\,\mu m$, a light scattering sensor is more sensitive.

The above method can be used to determine the size distribution of aggregating suspensions. The aggregated particles pass individually through the illuminated zone and generate a pulse which is collected at small angle ($< 3°$). At sufficiently small angles, the pulse height is proportional to the square of the number of monomeric units in an aggregate and independent of the aggregate shape or its orientation.

11.7 Measurement of incipient flocculation

This can be done for sterically stabilized suspensions, when the medium for the chains becomes a θ-solvent. This occurs, for example, on heating an aqueous suspension stabilized with poly(ethylene oxide) (PEO) or poly(vinyl alcohol) chains. Above a certain temperature (the θ-temperature) that depends on electrolyte concentration, flocculation of the suspension occurs. The temperature at which this occurs is defined as the critical flocculation temperature (CFT).

This process of incipient flocculation can be followed by measuring the turbidity of the suspension as a function of temperature. Above the CFT, the turbidity of the suspension rises very sharply.

For the above purpose, the cell in the spectrophotometer that is used to measure the turbidity is placed in a metal block that is connected to a temperature programming unit (which allows one to increase the temperature at a controlled rate).

11.8 Measurement of crystal growth (Ostwald ripening)

Ostwald ripening is the result of the difference in solubility S between small and large particles. The smaller particles have greater solubility than the larger particles. The effect of particle size on solubility is described by the Kelvin equation [19],

$$S(r) = S(\infty) \exp\left(\frac{2\sigma V_m}{rRT}\right), \tag{11.52}$$

where $S(r)$ is the solubility of a particle with radius r, $S(\infty)$ is the solubility of a particle with infinite radius, σ is the solid/liquid interfacial tension, V_m is the molar volume of the disperse phase, R is the gas constant and T is the absolute temperature.

For two particles with radii r_1 and r_2,

$$\frac{RT}{V_m} \ln\left(\frac{S_1}{S_2}\right) = 2\sigma\left(\frac{1}{r_1} - \frac{1}{r_2}\right). \tag{11.53}$$

R is the gas constant, T is the absolute temperature, M is the molecular weight and ρ is the density of the particles.

To obtain a measure of the rate of crystal growth, the particle size distribution of the suspension is followed as a function of time, using either a Coulter counter, a Mastersizer or an optical disc centrifuge. One usually plots the cube of the average radius versus time, which gives a straight line from which the rate of crystal growth can be determined (the slope of the linear curve),

$$r^3 = \frac{8}{9}\left[\frac{S(\infty)\sigma V_m D}{\rho RT}\right]t. \tag{11.54}$$

D is the diffusion coefficient of the disperse phase in the continuous phase and ρ is the density of the particles.

11.9 Bulk properties of suspensions, equilibrium sediment volume (or height) and redispersion

For a "structured" suspension, obtained by "controlled flocculation" or addition of "thickeners" (such polysaccharides, clays or oxides), the "flocs" sediment at a rate dependent on their size and porosity of the aggregated mass. After this initial sedimentation, compaction and rearrangement of the floc structure occurs, a phenomenon referred to as consolidation.

Normally in sediment volume measurements, one compares the initial volume V_0 (or height H_0) with the ultimately reached value V (or H). A colloidally stable suspension gives a "close-packed" structure with relatively small sediment volume (dilatant sediment referred to as clay). A weakly "flocculated" or "structured" suspension gives a more open sediment and hence a higher sediment volume. Thus, by comparing the relative sediment volume V/V_0 or height H/H_0, one can distinguish between a clayed and a flocculated suspension.

11.10 Application of rheological techniques for the assessment and prediction of the physical stability of suspensions [21]

11.10.1 Rheological techniques for prediction of sedimentation and syneresis

As mentioned in Chapter 8, sedimentation is prevented by addition of "thickeners" that form a "three-dimensional elastic" network in the continuous phase. If the viscosity of the elastic network, at shear stresses (or shear rates) comparable to those exerted by the particles or droplets, exceeds a certain value, then sedimentation is completely eliminated.

The shear stress, σ_p, exerted by a particle (force/area) can be simply calculated,

$$\sigma_p = \frac{(4/3)\pi R^3 \Delta\rho g}{4\pi R^2} = \frac{\Delta\rho R g}{3}. \tag{11.55}$$

For a 10 μm radius particle with a density difference $\Delta\rho$ of 0.2 g cm^{-3}, the stress is equal to,

$$\sigma_p = \frac{0.2 \times 10^3 \cdot 10 \times 10^{-6} \cdot 9.8}{3} \approx 6 \times 10^{-3}\,\text{Pa}. \tag{11.56}$$

For smaller particles, smaller stresses are exerted. Thus, to predict sedimentation, one has to measure the viscosity at very low stresses (or shear rates). These measurements can be carried out using a constant stress rheometer (Carrimed, Bohlin, Rheometrics, Haake or Physica). Usually one obtains good correlation between the rate of creaming or sedimentation, v, and the residual viscosity $\eta(0)$. This is illustrated in Fig. 11.17. Above a certain value of $\eta(0)$, v becomes equal to 0. Clearly, to minimize sedimentation one has to increase $\eta(0)$; an acceptable level for the high shear viscosity η_∞ must

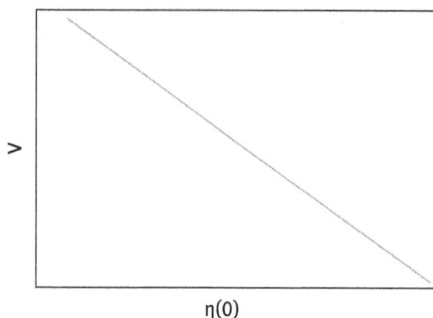

η(0)

Fig. 11.17: Variation of creaming or sedimentation rate with residual viscosity.

be achieved, depending on the application. In some cases, a high $\eta(0)$ may be accompanied by a high η_∞ (which may not be acceptable for the application, for example if spontaneous dispersion on dilution is required). If this is the case, the formulation chemist should look for an alternative thickener.

Another problem encountered with many dispersions is that of "syneresis", i.e. the appearance of a clear liquid film at the top of the container. "Syneresis" occurs with most "flocculated" and/or "structured" (i.e. those containing a thickener in the continuous phase) dispersions. "Syneresis" may be predicted from measurement of the yield value (using steady state measurements of shear stress as a function of shear rate) as a function of time or using oscillatory techniques (whereby the storage and loss modulus are measured as a function of strain amplitude and frequency of oscillation). The oscillatory measurements are perhaps more useful, since to prevent separation the bulk modulus of the system should balance the gravity force that is given by $h\Delta\rho g$ (where h is the height of the disperse phase, $\Delta\rho$ is the density difference and g is the acceleration due to gravity). The bulk modulus is related to the storage modulus G'. A more useful predictive test is to calculate the cohesive energy density of the structure E_c that is given by equation (11.57).

$$E_c = \int_0^{\gamma_{cr}} G'\gamma \, d\gamma = \frac{1}{2}G'\gamma_{cr}^2 . \tag{11.57}$$

The separation of a formulation decreases with increasing E_c. This is illustrated in Fig. 11.18, which schematically shows the reduction in separation (in percent) with increasing E_c. The value of E_c that is required to stop complete separation depends on the particle size distribution, the density difference between the particle and the medium as well as on the volume fraction ϕ of the dispersion.

The correlation of sedimentation with residual (zero shear) viscosity was illustrated in Chapter 8 for model suspensions of aqueous polystyrene latex in the presence of ethyl hydroxyethyl cellulose as a thickener.

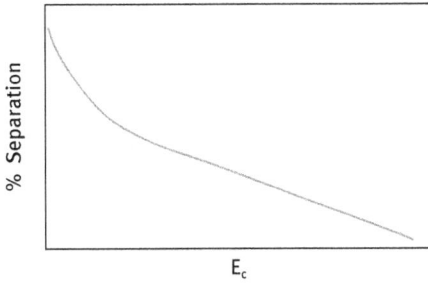

Fig. 11.18: Schematic representation of the variation of percentage separation with E_c.

11.10.2 Role of thickeners

As mentioned above, thickeners reduce creaming or sedimentation by increasing the residual viscosity $\eta(0)$, which must be measured at stresses compared to those exerted by the particles (mostly less than 0.1 Pa). At such low stresses, $\eta(0)$ increases very rapidly with increasing "thickener" concentration. This rapid increase is not observed at high stresses and this illustrates the need for measurement at low stresses (using constant stress or creep measurements). The variation of η with applied stress σ for ethyl hydroxyethyl cellulose (EHEC), a thickener that is applied in some formulations, was shown in Chapter 8. The limiting residual viscosity increases rapidly with increasing EHEC concentration. A plot of sedimentation rate for 1.55 μm PS latex particles versus $\eta(0)$ is shown in Fig. 11.19, which shows an excellent correlation. In this case a value of $\eta(0) > 10$ Pa s is sufficient for reducing the rate of sedimentation to 0.

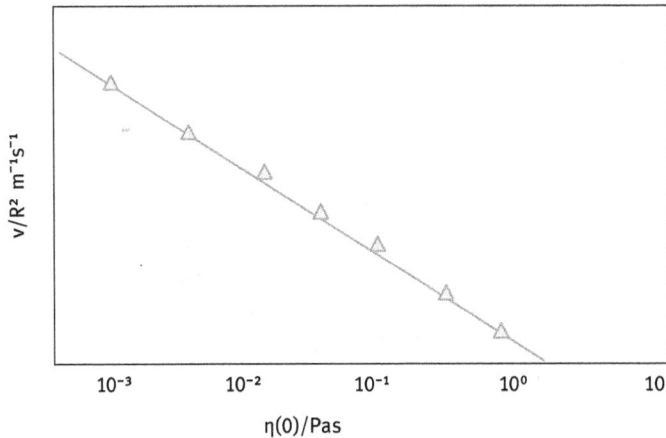

Fig. 11.19: Sedimentation rate versus $\eta(0)$.

11.10.3 Assessment and prediction of flocculation using rheological techniques

Steady state rheological investigations may be used to investigate the state of flocculation of a dispersion. Weakly flocculated dispersions usually show thixotropy and the
change of thixotropy with applied time may be used as an indication of the strength
of this weak flocculation.

The above methods are only qualitative and one cannot use the results in a quantitative manner. This is due to the possible breakdown of the structure on transferring
the formulation to the rheometer and also during the uncontrolled shear experiment.
Better techniques to study flocculation of a formulation are constant stress (creep) or
oscillatory measurements. By careful transfer of the sample to the rheometer (with
minimum) shear the structure of the flocculated system may be maintained.

A very important point that must be considered in any rheological measurement
is the possibility of "slip" during the measurements. This is particularly the case with
highly concentrated dispersions, where the flocculated system may form a "plug" in
the gap of the platens leaving a thin liquid film at the walls of the concentric cylinder
or cone-and-plate geometry. This behaviour is caused by some "syneresis" of the formulation in the gap of the concentric cylinder or cone and plate. To reduce "slip", one
should use roughened walls for the platens. A vane rheometer may also be used.

Steady state shear stress-shear rate measurements are by far the most commonly
used method in many industrial laboratories. Basically, the dispersion is stored at various temperatures and the yield value σ_β and plastic viscosity η_{pl} are measured at
various intervals of time. Any flocculation in the formulation should be accompanied
by an increase in σ_β and η_{pl}. A rapid technique to study the effect of temperature
changes on the flocculation of a formulation is to carry out temperature sweep experiments, running the samples from say 5–50 °C. The trend in the variation of σ_β
and η_{pl} with temperature can quickly give an indication of the temperature range at
which a dispersion remains stable (during that temperature range σ_β and η_{pl} remain
constant).

If Ostwald ripening occurs simultaneously, σ_β and η_{pl} may change in a complex
manner with storage time. Ostwald ripening results in a shift of the particle size distribution to higher diameters. This has the effect of reducing σ_β and η_{pl}. If flocculation
occurs simultaneously (having the effect of increasing these rheological parameters),
the net effect may be an increase or decrease of the rheological parameters. This trend
depends on the extent of flocculation relative to Ostwald ripening and/or coalescence.
Therefore, following σ_β and η_{pl} with storage time requires knowledge of Ostwald
ripening. Only in the absence of these latter breakdown processes can one use rheological measurements as a guide to assessing flocculation.

Constant stress (creep) experiments are more sensitive for following flocculation.
As mentioned before, a constant stress σ is applied on the system and the compliance J (Pa^{-1}) is plotted as a function of time. These experiments are repeated several
times, increasing the stress from the smallest possible value (that can be applied by

the instrument) in small increments. A set of creep curves are produced at various applied stresses. From the slope of the linear portion of the creep curve (after the system reaches a steady state), the viscosity at each applied stress, η_σ, is calculated. A plot of η_σ versus σ allows one to obtain the limiting (or zero shear) viscosity $\eta(0)$ and the critical stress σ_{cr} (which may be identified with the "true" yield stress of the system). The values of $\eta(0)$ and σ_{cr} may be used to assess the flocculation of the dispersion on storage. If flocculation occurs on storage (without any Ostwald ripening), the values of $\eta(0)$ and σ_{cr} may show a gradual increase with increasing storage time. As discussed in the previous section (on steady state measurements), the trend becomes complicated if Ostwald ripening occurs simultaneously (both have the effect of reducing $\eta(0)$ and σ_{cr}).

The above measurements should be supplemented by particle size distribution measurements of the diluted dispersion (making sure that no flocs are present after dilution) to assess the extent of Ostwald ripening. Another complication may arise from the nature of the flocculation. If it occurs in an irregular way (producing strong and tight flocs), $\eta(0)$ may increase, while σ_{cr} may show some decrease and this complicates the analysis of the results. In spite of these complications, constant stress measurements may provide valuable information on the state of the dispersion on storage.

Carrying out creep experiments and ensuring that a steady state is reached can be time consuming. One usually carries out a stress sweep experiment, whereby the stress is gradually increased (within a predetermined time period to ensure that one is not too far from reaching the steady state) and plots of η_σ versus σ are established. These experiments are carried out at various storage times (say every two weeks) and temperatures. From the change of $\eta(0)$ and σ_{cr} with storage time and temperature, one may obtain information on the degree and the rate of flocculation of the system. Clearly, interpretation of the rheological results requires expert knowledge of rheology and measurement of the particle size distribution as a function of time.

One main problem in carrying the above experiments is sample preparation. When a flocculated dispersion is removed from the container, care should be taken not to cause much disturbance to that structure (minimum shear should be applied on transferring the formulation to the rheometer). It is also advisable to use separate containers for assessment of flocculation; a relatively large sample is prepared and this is then transferred to a number of separate containers. Each sample is used separately at a given storage time and temperature. One should be careful in transferring the sample to the rheometer. If any separation occurs in the formulation, the sample is gently mixed by placing it on a roller. It is advisable to use as minimum shear as possible when transferring the sample from the container to the rheometer (the sample is preferably transferred using a "spoon" or by simple pouring from the container). The experiment should be carried out without an initial pre-shear.

Another rheological technique for assessment of flocculation is oscillatory measurement. As mentioned above, one carries out two sets of experiments.

(i) Strain sweep measurements. In this case, the oscillation is fixed (say at 1 Hz) and the viscoelastic parameters are measured as a function of strain amplitude. G^*, G' and G'' remain virtually constant up to a critical strain value, γ_{cr}. This region is the linear viscoelastic region. Above γ_{cr}, G^* and G' starts to fall, whereas G'' starts to increase; this is the nonlinear region. The value of γ_{cr} may be identified with the minimum strain above which the "structure" of the dispersion starts to break down (for example break down of flocs into smaller units and/or break down of a "structuring" agent).

From γ_{cr} and G'', one can obtain the cohesive energy E_c ($J\,m^{-3}$) of the flocculated structure using equation (11.57). E_c may be used in a quantitative manner as a measure of the extent and strength of the flocculated structure in a dispersion. The higher the value of E_c, the more flocculated the structure is. Clearly, E_c depends on the volume fraction of the dispersion as well as the particle size distribution (which determines the number of contact points in a floc). Therefore, for quantitative comparison between various systems, one has to make sure that the volume fraction of the disperse particles is the same and that the dispersions have very similar particle size distributions. E_c also depends on the strength of the flocculated structure, i.e. the energy of attraction between the droplets. This depends on whether the flocculation is in the primary or secondary minimum. Flocculation in the primary minimum is associated with a large attractive energy and this leads to higher values of E_c when compared with the values obtained for secondary minimum flocculation (weak flocculation). For a weakly flocculated dispersion, such as is the case with secondary minimum flocculation of an electrostatically stabilized system, the deeper the secondary minimum, the higher the value of E_c (at any given volume fraction and particle size distribution of the dispersion). With a sterically stabilized dispersion, weak flocculation can also occur when the thickness of the adsorbed layer decreases. Again, the value of E_c can be used as a measure of the flocculation – the higher the value of E_c, the stronger the flocculation. If incipient flocculation occurs (on reducing the solvency of the medium for the change to worse than θ-condition), a much deeper minimum is observed and this is accompanied by a much larger increase in E_c.

To apply the above analysis, one must have an independent method for assessing the nature of the flocculation. Rheology is a bulk property that can give information on the interparticle interaction (whether repulsive or attractive) and to apply it in a quantitative manner one must know the nature of these interaction forces. However, rheology can be used in a qualitative manner to follow the change of the formulation on storage. Providing the system does not undergo any Ostwald ripening, the change of the moduli with time and in particular the change of the linear viscoelastic region may be used as an indication of flocculation. Strong flocculation is usually accompanied by a rapid increase in G' and this may be accompanied by a decrease in the critical strain above which the "structure" breaks down. This may be used as an indication of formation of "irregular" and tight flocs, which become sensitive to the applied strain. The floc structure will entrap a large amount of the continuous phase and this leads to an apparent increase in the volume fraction of the dispersion and hence an increase in G'.

(ii) Oscillatory sweep measurements. In this case, the strain amplitude is kept constant in the linear viscoelastic region (one usually takes a point far from γ_{cr} but not too low, i.e. in the mid-point of the linear viscoelastic region) and measurements are carried out as a function of frequency. Both G^* and G' increase with increasing frequency and ultimately above a certain frequency, they reach a limiting value and show little dependence on frequency. G'' is higher than G' in the low frequency regime; it also increases with increasing frequency and at a certain characteristic frequency ω^* (that depends on the system) it becomes equal to G'' (usually referred to as the crossover point), after which it reaches a maximum and then shows a reduction with a further increase in frequency.

From ω^* one can calculate the relaxation time τ of the system,

$$\tau = \frac{1}{\omega^*}. \tag{11.58}$$

The relaxation time may be used as a guide for the state of the dispersion. For a colloidally stable dispersion (at a given particle size distribution), τ increases with increasing volume fraction of the disperse phase, ϕ. In other words, the crossover point shifts to lower frequency with increasing ϕ. For a given dispersion, τ increases with increasing flocculation, providing the particle size distribution remains the same (i.e. no Ostwald ripening).

The value of G' also increases with increasing flocculation, since aggregation of particles usually results in liquid entrapment and the effective volume fraction of the dispersion shows an apparent increase. With flocculation, the net attraction between the particles also increases and this results in an increase in G'. The latter is determined by the number of contacts between the particles and the strength of each contact (which is determined by the attractive energy).

It should be mentioned that in practice one may not obtain the full curve, due to the frequency limit of the instrument and also measurement at low frequency is time consuming. Usually one obtains part of the frequency dependence of G' and G''. In most cases, one has a more elastic than viscous system.

Most disperse systems used in practice are weakly flocculated and they also contain "thickeners" or "structuring" agents to reduce sedimentation and to acquire the right rheological characteristics for the application, e.g. in hand creams and lotions. The exact values of G' and G'' required depend on the system and its application. In most cases a compromise has to be made between acquiring the right rheological characteristics for the application and the optimum rheological parameters for long-term physical stability. Application of rheological measurements to achieve these conditions requires a great deal of skill and understanding of the factors that affect rheology.

11.10.4 Examples of application of rheology for assessment and prediction of flocculation

11.10.4.1 Flocculation and restabilization of clays using cationic surfactants

Hunter and Nicol [22] studied the flocculation and restabilization of kaolinite suspensions using rheology and zeta potential measurements. Figure 11.20 shows plots of the yield value σ_β and electrophoretic mobility as a function of cetyl trimethyl ammonium bromide (CTAB) concentration at pH = 9. σ_β increases with increasing CTAB concentration, reaching a maximum at the point where the mobility reaches zero (the isoelectric point, IEP, of the clay) and then decreases with a further increase in CTAB concentration. This trend can be explained on the basis of flocculation and restabilization of the clay suspension.

Initial addition of CTAB causes reduction in the negative surface charge of the clay (by adsorption of CTA^+ on the negative sites of the clay). This is accompanied by reduction in the negative mobility of the clay. When complete neutralization of the clay particles occurs (at the IEP), maximum flocculation of the clay suspension occurs and this is accompanied by a maximum in σ_β. On a further increase in CTAB concentration, further adsorption of CTA^+ occurs resulting in charge reversal and restabilization of the clay suspension. This is accompanied by reduction in σ_β.

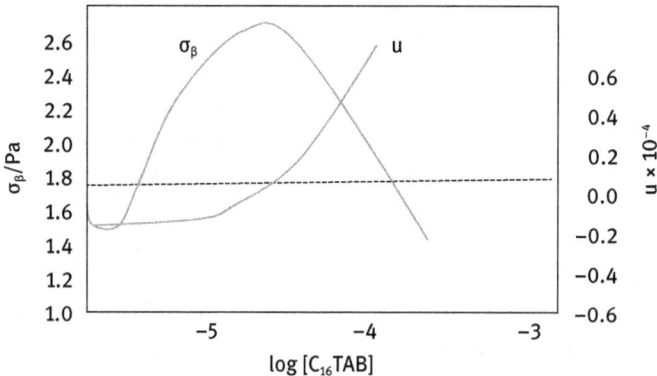

Fig. 11.20: Variation of yield value σ_β and electrophoretic mobility u with C_{16} TAB concentration.

11.10.4.2 Flocculation of sterically stabilized dispersions

Neville and Hunter [23] studied the flocculation of polymethylmethacrylate (PMMA) latex stabilized with poly(ethylene oxide) (PEO). Flocculation was induced by addition of electrolyte and/or increasing temperature. Figure 11.21 shows the variation of σ_β with increasing temperature at constant electrolyte concentration.

It can be seen that σ_β increases with increasing temperature, reaching a maximum at the critical flocculation temperature (CFT) and then decreases with a further increase in temperature. The initial increase is due to the flocculation of the latex with increasing temperature, as a result of a reduction of solvency of the PEO chains with increasing temperature. The reduction in σ_β after the CFT is due to the reduction in the hydrodynamic volume of the dispersion.

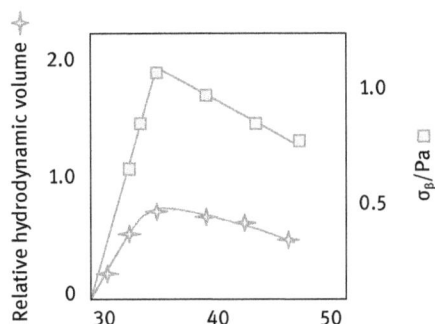

Fig. 11.21: Variation of σ_β and hydrodynamic volume with temperature.

References

[1] Lyklema, J., "Structure of the Solid/Liquid Interface and the Electrical Double Layer", in "Solid/Liquid Dispersions", Th. F. Tadros (ed.), Academic Press, London (1987).
[2] Tadros, Th. F., "Interfacial Phenomena and Colloid Stability, Basic Principles", De Gruyter, Germany (2015).
[3] Lyklema, J., in "Trends in Electrochemistry", J.O'M. Bockris, D. A. J. Rand and B. J. Welch (eds.), Plenum Publishing Corp., New York (1977).
[4] Ven den Hul, H. J. and Lyklema, J., J. Colloid Sci., **23**, 500 (1967).
[5] Penners, N. H. G., "The Preparation and Stability of Monodisperse Colloidal Haematite (α-Fe$_2$O$_3$)", Ph. D. Thesis, Agricultural University, Wageningen, Netherlands.
[6] Tadros, Th. F. and Lyklema, J., J. Electroanal. Chem, **17**, 267 (1968).
[7] Lyklema, J., J. Electroanal. Chem, **18**, 341 (1968).
[8] Hunter, R. J., "Zeta Potential in Colloid Science: Principles and Application", Academic Press, London (1981).
[9] Debye, P., J. Chem. Phys., **1**, 13 (1933).
[10] Bugosh, J., Yeager, E. and Hovarka, F., J. Chem. Phys., **15**, 542 (1947).
[11] Yeager, E., Bugosh, J., Hovarka, F. and McCarthy, J., J. Chem. Phys., **17**, 411 (1949).
[12] Dukhin, A. S. and Goetz, P. J., Colloids and Surfaces, **144**, 49 (1998).
[13] Oja, T., Petersen, G. L. and Cannon, D. C., US Patent 4,497,208 (1985).
[14] O'Brian, R. W., J. Fluid Mech., **190**, 71 (1988).
[15] O'Brian, R. W., J. Fluid Mech., **212**, 81 (1990).
[16] O'Brian, R. W., Garaside, P. and Hunter, R. J., Langmuir, **10**, 931 (1994).
[17] O'Brian, R. W., Cannon, D. W. and Rowlands, W. N., J. Colloid Interface Sci., **173**, 406 (1995).
[18] Rowlands, W. N. and O'Brian, R. W., J. Colloid Interface Sci., **175**, 190 (1995).

[19] Tadros, Th. F., "Dispersion of Powders in Liquids and Stabilisation of Suspensions", Wiley-VCH, Germany (2012).

[20] Kissa, E., "Dispersions, Characterization, Testing and Measurement", Marcel Dekker, New York (1999).

[21] Tadros, Th. F., "Rheology of Dispersions", Wiley-VCH, Germany (2010).

[22] Hunter, R. J. and Nicol, S. K., J. Colloid Interface Sci., **28**, 250 (1968).

[23] Neville, P. C. and Hunter, R. J., J. Colloid Interface Sci., **49**, 204 (1974).

12 Application of suspensions in pharmacy

12.1 Introduction

Pharmaceutical suspensions cover a wide range of sizes, ranging from particles $\leq 1\,\mu m$ (colloidal suspensions) to 50–75 μm particles (coarse suspensions) [1]. The formulation and stabilization of suspensions in pharmacy follow the same basic principles described in Chapters 2–9 and only a brief description of these principles is given in the examples of the various systems.

An aqueous suspension is a useful oral dosage form for administering insoluble or poorly soluble drugs. The large surface area of dispersed drug ensures a high degree of availability for absorption. Unlike tablets or capsules, the dissolution of drug particles in suspension and subsequent in vivo absorption commence upon dilution in the gastrointestinal fluids. Finely divided particles dissolve at a greater rate and have higher relative solubilities than similar macroparticles [2]. When the particle size is greater than 10 μm, the rate of dissolution is proportional to the surface area. However, when the particle size is lower than 10 μm, the rate of dissolution is inversely proportional to the particle radius. Reinhold et al. [3] showed that fine particles of suspended sulfadiazine gave more rapid absorption with higher maximum serum level versus time and greater area under the serum level versus time curve than a similar suspension containing somewhat larger particles of the drug.

The bioavailability of a drug is assumed to decrease in the following order: solutions > suspensions > capsules > compressed tablets > coated tablets. Thus, the parenteral suspension is an ideal dosage for prolonged therapy. Administration of a drug as an aqueous or oleaginous suspension into subcutaneous or muscular tissue results in the formation of a depot at the injection site. The depot acts as a drug reservoir, slowly releasing drug solubility of the insoluble drug and the type of suspending vehicle used, either aqueous or oil for the purpose of maintaining systemic absorption of the drug from the injection site. The suspension form of the drug frequently provides a more prolonged release from the injection site than a comparable solution of the same drug in a suitable injectable oil.

12.2 Particle size reduction

Solids are comparatively easy to grind to sieve size particle [4]. Reduction to a size range of about 50–75 μm produces a powder that is free flowing. Most solids tend to exhibit aggregation or agglomeration in the dry state when individual particles are smaller. Below about 10–50 μm, the increased free surface energy becomes a factor interfering with further particle size reduction (due to cohesion of small particles). The powder may become damp, especially if there is a tendency to attract moisture.

DOI 10.1515/9783110486872-013

The solid "balls up" and this will suggest that the particles are much larger than their actual size. As the pores between the smaller particles become smaller with decreasing particle size, the surface area becomes more accessible to liquid penetration. The aggregates behave like hydrophobic solids, entrap air and are difficult to "wet out".

The most efficient method of producing fine particles is by dry milling prior to suspension preparation. Several methods are used to produce fine drug powders of which the following are worth mentioning. Micropulverization is one of the most rapid, convenient and inexpensive methods of producing fine drug powders. The main advantage of micropulverization is the large distribution of particle sizes produced, normally in the range 10–50 μm. The milling equipment includes hammer mills, micropulverizers, universal mills, end runner mills and ball mills. Micropulverizers are high speed attrition or impact mills especially adapted for fine grinding. Some mills are fitted with classifiers to facilitate particle separation by centrifugation. Since ultrafine particles below 10 μm are infrequently produced, build-up of electrostatic charge on the surface of milled powder is encountered only occasionally. Fluid energy grinding, (jet milling or micronizing) is the most effective method for reducing particles below 10 μm. The ultrafine particles are produced by the shearing action of high velocity streams of compressed air on particles in a confined space. The main disadvantage of fluid energy grinding is the high electrostatic charge built up on the surfaces of the milled powder, which makes powder classification and collection exceedingly difficult. Particles of microcrystalline size can also be obtained by spray drying [5]. This produces a porous, free-flowing, easily wetted and essentially monodispersed powder. By control of the process variables, one can produce spherical particles that may be coated with wetting and suspending agents to aid suspension and/or promote stability of the resulting system.

A special method for producing fine particles of the drug is by controlled crystallization. A solvent that dissolves a solid very readily at room temperature may serve as a crystallizing medium when mixed with another solvent in which the compound is only sparingly soluble. A solution that is nearly saturated at a temperature about 10 °C below the boiling point of the solvent combination is prepared in a temperature range between 60 and 150 °C. Separation of microcrystals from such hot concentrated solutions is commonly induced by cooling and stirring. However, when supersaturation is obtained by agitation and shock cooling of the hot solution and through rapid introduction of another cold miscible solvent in which the drug is only sparingly soluble, formation of minute particles (nucleates) proceeds without appreciable crystal growth and uniform microcrystals of the drug are thus obtained. Microcrystals can also be conveniently produced by bubbling liquid nitrogen through the saturated solution of a solute prior to solvent freezing. In addition, ultrasonic insonation techniques have also been used during shock cooling procedures [6].

12.3 Dispersion and stabilization of the drug suspension

Dispersion equipment, such as bead mills, colloid mills and high pressure homogenizers are used for wet milling of the suspension in order to break up the aggregates and agglomerates. This usually requires the addition of a wetting agent (surfactant) such as dioctyl sulphosuccinate (Aerosol OT) or an alcohol ethoxylate (nonionic surfactant) as described in Chapter 3. The resulting dispersed particles undergo a process of comminution (grinding) to produce the desirable particle size distribution as described in Chapter 3.

The potency stability of a drug in suspension is controlled by the fact that the rate of degradation is related to the concentration of drug in solution, rather than to the total concentration of the drug in the product [7]. Generally, a suspended drug decomposes only in solution as the solid phase gradually dissolves; thus a solution concentration equal to the solubility of the drug is maintained. Drug degradation in a suspension usually follows zero-order kinetics with the rate constant dependent on the solubility rate of the insoluble drug. Improved potency stability may be accomplished either by selecting a pH value or range where the drug is least soluble or by replacing the drug with a more vehicle-insoluble derivative or salt. Decomposition may also be described by a diffusion-controlled process or by catalysis initiated by environmental factors such as oxygen, light and trace metals. The physical stability of a pharmaceutical suspension may be as or more important than the chemical stability. Since the suspension exists in more than one state (liquid and solid), there are several different ways in which the system can undergo either chemical or physical change [8, 9]. This poses problems in predicting the physical and chemical stability of drug suspensions. For example, it is difficult to predict the settling rates of drug suspensions on the basis of Stokes' law. The latter assumes spherical noninteracting particles which are obviously not the case with drug suspensions. In addition, these suspensions exhibit non-Newtonian flow as discussed in Chapter 8. A number of modifications of the basic equations have been suggested to take these factors into account [10].

The chemical stability predictions are sometimes complicated by the difficulty of determining the pH of suspensions, which often changes the liquid junction potential of the measuring electrode. Part of this results from "pH drift" associated with surface coating of the electrodes, and partly from the disparity in pH value between bulk suspension and supernatant vehicle. If two glass electrodes are used, one in the supernate and one in sediment, each electrode will show the same pH value in both portions of the suspension. In addition, slow attainment of saturation often complicates steady state treatment of rate data. Finally, accelerated, elevated temperature stability testing will often have a pronounced and adverse effect on suspension viscosity, particle solubility and size distribution [1].

12.4 Colloid stability of pharmaceutical suspensions

One of the most important criteria in formulating drug suspensions is to prepare a colloidally stable system in a potentially suitable suspension vehicle. This requires the addition of a dispersing agent to produce a homogeneous and stable suspension. As mentioned above, the dispersed system is usually passed through a homogenizer or colloid mill to improve dispersion and reduce particle size. Alternatively, the dispersion is "wet-milled" using ceramic beads (to avoid contamination).

As discussed in Chapters 4 and 5, there are generally two methods for producing colloidally stable suspensions. The first, referred to as electrostatic stabilization, is based on creating electrical double layers around the drug particles, for example by addition of an anionic surfactant such as sodium dodecyl sulphate. The surfactant anions adsorb on the hydrophobic drug particles (by hydrophobic bonding between the alkyl chain and the drug surface) producing a negative charge (sulphate groups) on the particle surface. These negative charges are compensated by unequal distribution of counter and co-ions. An electrical double layer is produced around the particle with an extension (double layer thickness) that depends on electrolyte concentration. This extension is of the order of 100 nm in 10^{-5} 1:1 electrolyte (e.g. NaCl). When two particles with their extended double layers approach to a surface-to-surface distance h that is smaller than twice the double layer extension, strong repulsion occurs as a result of double layer overlap. This electrostatic repulsion, G_{elec}, increases with increasing surface (or zeta potential) and decreasing electrolyte concentration and valency. Combining electrostatic repulsion with van der Waals attraction results in an energy-distance curve that shows a maximum, G_{max}, (repulsive barrier) at intermediate particle separation [11, 12]. G_{max} increases with increasing zeta potential (ζ) and decreasing electrolyte concentration, C, and valency (see Chapter 4). In practical conditions with $\zeta > 60$ mV and $C < 10^{-2}$ mol dm^{-3}, $G_{max} > 25$ kT (where k is the Boltzmann constant and T is the absolute temperature) and this prevents particle aggregation (see Chapter 4).

The second and more effective method of producing colloidally stable drug suspensions is to use nonionic surfactants or block copolymers. A good example of a nonionic surfactant is polysorbate 80 (sorbitan mono-oleate with 20 mol ethylene oxide, EO). An example of an A-B-A block copolymer is poloxamer F127 (consisting of a hydrophobic B chain of polypropylene oxide, about 55 monomer units, and two hydrophilic stabilizing chains A of 100 EO units each). These nonionic surfactants and block copolymers adsorb on the hydrophobic drug particles with the alkyl chain of polysorbate 80 or the PPO chain of poloxamer F127. The strongly hydrated PEO chains provide the stabilizing mechanism, referred to as steric stabilization as discussed in detail in Chapter 5. An adsorbed layer thickness, δ, can be assigned to the stabilizing PEO chain. When two particles, each containing an adsorbed layer of thickness δ, approach to a surface-to-surface distance h that is smaller than 2δ, the stabilizing chains begin to overlap or become compressed resulting in an increase in the PEO segment concentration in the overlap region. This results in two main repulsive energies [13]: (i) G_{mix}, resulting from the unfavourable mixing of the hydrated PEO chains, when these are in good solvent conditions. As a result solvent molecules will diffuse from the bulk to the overlap region thus separating the particles. (ii) G_{el}, resulting from the loss of configurational entropy of the A chains on considerable overlap. Combining G_{mix} and G_{el} with the van der Waals attractive energy G_A gives the total energy of interaction G_T. The G_T–h curve shows a shallow minimum G_{min} at $h \approx 2\delta$ (due to the residual attraction at large h values) of the order of few kT units. When $h < 2\delta$, a sharp increase in the repulsive energy occurs with a further reduction in h. This repulsion prevents any particle aggregation, thus producing a colloidally stable suspension.

12.5 Prevention of Ostwald ripening (crystal growth)

As discussed in Chapter 7, the driving force of Ostwald ripening is the difference in solubility between the smaller and larger particles. The small particles with radius r_1 will have higher solubility than the larger particle with radius r_2. Thus, with time molecular diffusion will occur between the smaller and larger particles, with the ultimate disappearance of most of the small particles. This results in a shift in the particle size distribution to larger values on storage of the suspension. This could lead to the formation of a suspension with average particle size $> 2\,\mu m$. This instability can cause severe problems, such as sedimentation, flocculation and even flocculation of the suspension.

A second driving force for Ostwald ripening in suspensions is due to polymorphic changes. If the drug has two polymorphs A and B, the more soluble polymorph, say A (which may be more amorphous) will have higher solubility than the less soluble (more stable) polymorph B. During storage, polymorph A will dissolve and recrystallize as polymorph B. This can have a detrimental effect on bioefficacy, since the more soluble polymorph may be more active.

The task of the formulation scientist is to reduce crystal growth to an acceptable level depending on the application. This is particularly the case with pharmaceutical suspensions, where crystal growth leads to the shift of the particle size distribution to larger values. Unfortunately, crystal growth inhibition is still an "art", rather than a "science", in view of the lack of adequate fundamental understanding of the process at a molecular level.

Since pharmaceutical suspensions are prepared by using a wetting/dispersing agent, it is important to know how these agents can affect the growth rate. In the first place, the presence of wetting/dispersing agents influences the process of diffusion of the molecules from the surface of the crystal to the bulk solution. The wetting/dispersing agent may affect the rate of dissolution by affecting the rate of transport away from the boundary layer [1], although their addition is not likely to affect the rate of dissolution proper (passage from the solid to the dissolved state in the immediate adjacent layer). If the wetting/dispersing agent forms micelles which can solubilize the solute, the diffusion coefficient of the solute in the micelles is greatly reduced. However, as a result of solubilization, the concentration gradient of the solute is increased to an extent depending on the extent of solubilization. The overall effect may be an increase in crystal growth rate as a result of solubilization. In contrast, if the diffusion rate of the molecules of the wetting/dispersing agent is sufficiently rapid, their presence will lower the flux of the solute molecules compare to that in the absence of the wetting/dispersing agent. In this case, the wetting/dispersing agent will lower the rate of crystal growth.

Secondly, wetting/dispersing agents are expected to influence growth when the rate is controlled by surface nucleation. Adsorption of wetting/dispersing agents on the surface of the crystal can drastically change the specific surface energy and makes it inaccessible to the solute molecules. In addition, if the wetting/dispersing agent is preferentially adsorbed at one or more of the faces of the crystal (for example by electrostatic attraction between a highly negative face of the crystal and cationic surfactant), surface nucleation is no longer possible at this particular face (or faces). Growth will then take place at the remaining faces which are either bare or incompletely covered by the wetting/dispersing agent. This will result in a change in crystal habit.

From the above discussion, it can be seen that surfactants (wetting/dispersing agents), if properly chosen, may be used for crystal growth inhibition and control of habit formation. Inhibition of crystal growth can also be achieved by polymeric surfactants and other additives. For example, Simon Elli et al. [14] found that the crystal growth of the drug sulphathiazole can be inhibited by the addition of poly(vinylpyrrolidone) (PVP). The inhibition effect depends on the concentration and molecular weight of PVP. A minimum concentration (expressed as grams PVP/100 ml) of polymer is required for inhibition, which increases with increasing molecular weight of the polymer. However, if the concentration is expressed in $mol\,dm^{-3}$, the reverse is true, i.e. the higher the molar mass of PVP the lower the number of moles required for inhibition. Carless et al. [15] reported that the crystal growth of cortisone acetate in aqueous suspensions can be inhibited by addition of cortisone alcohol. Crystal growth in this system is mainly inhibited by polymorphic transformation [15]. The authors assumed that cortisone alcohol is adsorbed onto the particles of the stable form and this prevents the arrival of new cortisone acetate molecules which would result in crystal growth. The authors also noticed that the particles change their shape, growing to long needles. This means that the cortisone alcohol fits into the most dense lattice plane of the cortisone acetate crystal, thus preventing preferential growth on that face.

Many block ABA and graft BA_n copolymers (with B being the "anchor" part and A the stabilizing chain) are very effective in inhibiting crystal growth. The B chain adsorbs very strongly on the surface of the crystal and sites become unavailable for deposition. This has the effect of reducing the rate of crystal growth. Apart from their influence on crystal growth, the above copolymers also provide excellent steric stabilization, providing the A chain is chosen to be strongly solvated by the molecules of the medium.

12.6 Prevention of particle settling and suspension separation

Most pharmaceutical suspensions have particle radii $> 1\,\mu m$ and the particle density is $> 1.1\,g\,cm^{-3}$. Thus on storage, the particles will sediment under gravity with the larger particles sedimenting at a faster rate than the smaller ones. A hard sediment (sometimes referred to as "cake" or "clay") is formed at the bottom of the container which is very difficult to redisperse (due to the close packing of the particles and the dilatant nature of the sediment). Several methods are applied to prevent particle sedimentation and formation of hard sediments and these were discussed in detail in Chapter 8. The most commonly used method to prevent particle sedimentation is to use a suspending agent, mostly a high molecular weight polymer such as xanthan gum, carboxymethyl cellulose, alginates, carrageenans, etc. Alternatively, a particulate suspending agent such as bentonite (colloidal aluminium silicate), hectorite (colloidal magnesium aluminium silicate) can be used. As discussed in Chapter 8, all these suspending agents form a "three-dimensional" gel network structure in the continuous medium that is non-Newtonian with a very high residual viscosity, at stresses comparable to those exerted by the particles, thus preventing particle sedimentation.

12.7 Oral suspensions

Drug suspensions are often very effective pharmacologically and are more bioavailable than comparable tablets or capsules. The solids content of an oral suspension may vary considerably. For example, antibiotic preparations may contain 125–500 mg active solid per 5 ml (teaspoonful) dose (2.5–5.0 % solid), while the drop concentrate may provide the same amount of insoluble drug in a 1–2 ml dose. Antacids and radioactive suspensions also contain relatively high amounts of suspended material for oral administration. The pH and particle size of antacid suspensions containing aluminium hydroxycarbonate and magnesium hydroxide must be carefully adjusted. The buffer used for adjusting the pH must be carefully chosen to avoid interaction with the antacid material. For example, phosphate buffers can react with the antacid material forming insoluble phosphates and this reducing the efficacy of the antacid. Suspending agents are used to prevent rapid settling and caking of the antacid particles. Some sweeteners like sucrose, mannitol, sorbitol or saccharin are also added to the antacid suspension.

A very effective antacid suspension is Gaviscon that contains calcium carbonate particles dispersed in a solution of sodium bicarbonate. Sodium alginate is used as a suspending agent. The final formulation appears like a "gel" and its rheology must be controlled to avoid separation on storage.

12.8 Parenteral suspensions

Injectable (parenteral) suspensions must be sterile, pyrogen free and maintain physical and chemical stability over the intended shelf life. These suspensions may be formulated as a ready-to-use injection or require a reconstitution step prior to use. The solids content of parenteral suspensions is usually between 0.5 and 5 %, with the exception of insoluble penicillin, in which the concentration of the antibiotic may exceed 30 %. These sterile suspensions are designed for intramuscular, intradermal, intralesional or subcutaneous administration. The resulting suspension is non-Newtonian and its viscosity must be sufficiently low (at shear rates comparable to that encountered in injection) to facilitate injection. Common vehicles for parenteral suspensions include preserved sodium chloride injection or a parenterally acceptable vegetable oil. Control of the rheology of parenteral suspensions can be achieved by the use of rheology modifiers [16].

12.9 Ophthalmic suspensions

Ophthalmic suspensions (mostly aqueous) that are used for diagnosis and treatment of ocular diseases must be non-irritating to the ocular tissues and homogeneous (particles well dispersed, smooth and free from lumps or aggregates). They should not cause blurred vision, should be sterile and adequately preservable for multiple use. They also should be physically and chemically stable [17]. Physical stability implies absence of sedimentation and caking, absence of flocculation and crystal growth. This requires adequate choice of dispersing and suspending agents. The chemical stability is usually achieved by the use of antioxidants that retard the degradation of the drug.

12.10 Topical suspensions

These are a suitable form of drug delivery for the topical application of dermatologic materials to the skin and sometimes to the mucous membranes. Compared with the solubilized system, suspensions offer better chemical stability of the drug. One of the earliest externally applied "shake lotions" is calamine lotion USP. The protective action and cosmetic properties of topical lotions usually require the use of high concentrations of dispersed phase (> 20 %). A variety of pharmaceutical vehicles have been used, such as diluted oil-in-water emulsion bases, diluted water-in-oil emulsion bases, dermatological pastes and clay suspensions.

12.11 Reconstitutable suspensions

Some drugs have a limited chemical stability when formulated as an aqueous suspension. In this case, the drug is formulated as a dry powder that can be easily dispersed in an aqueous medium before administration [17]. For example, an aqueous suspension of penicillin has a maximum shelf life of 14 days, whereas the manufactured dry powder (reconstitutable suspension) has a shelf life of two years. Another reason for formulating for reconstitution is to avoid problems of physical stability often encountered in conventional suspensions, e.g. particle sedimentation and caking, increased drug solubility due to pH changes from chemical degradation, viscosity changes and crystal growth and polymorphic changes. Several desirable attributes are necessary for a reconstitutable suspension: (i) uniform mixture of the appropriate concentration of each ingredient; (ii) ease of dispersion of the powder on mixing with the vehicle with minimum agitation; (iii) the reconstituted suspension must be easily redispersed and pourable by the patient to provide an accurate and uniform dose; (iv) the final constituted product must have an acceptable appearance, odour and taste. These desirable attributes require adequate selection of wetting/dispersing and suspending agents, good preservative and antioxidant (to ensure good shelf life), acceptable sweetener and flavour, etc.

Three types of dry mixture preparations are possible [18]: (i) Powder blends (powder mixtures) where the product is prepared by mixing the ingredients of the dry mixture in powder form. The selection of the appropriate mixer involves rapid and reliable production of a homogeneous mixture. The main problem with this simple method is the possibility of producing an inhomogeneous mixture. The insoluble drug is usually milled to a smaller size than the sweetener or suspending agent. Too broad a size range of such particles can induce aggregation into layers of different sizes. Poor flow can cause demixing. Another problem in powder blends is the systematic loss of active ingredient during mixing. (ii) Granulated products, where wet granulation is the usual process and the granulating fluid is water or an aqueous binder solution. The drug can be dry blended with the other ingredients or it can be dissolved or suspended in the granulating fluid. The solid ingredients are blended and massed with the granulating fluid in a planetary mixer. The wet mass is formed into granules in a vibratory sieve, oscillating granulator, grater or mill and then the granules are dried. For drugs subject to hydrolysis, nonaqueous granulating fluid can be used. The main disadvantages of the granulated product compared to powder mixing are the requirement of more capital equipment and energy to produce each batch. It is also difficult to remove the last traces of granulating fluid from the interior of the granules. The drug must also be stable to the granulation process. (iii) Combination product, where powdered and granulated ingredients can be combined to overcome the disadvantages of granulated products. The general method is, first, to granulate some of the ingredients and, then, blend the remaining ingredients with the dried granules prior to filling the container. The granules can be made by spray coating in a spray dryer. The main disadvantage of the combination product is the risk of nonuniformity. The mix of granules and nongranual ingredients must not separate into layers of different sizes. To achieve the necessary degree of homogeneity, the particle sizes of the various fractions should be carefully controlled.

Both physical and chemical stability tests must be performed on the reconstitutable suspension. Test of physical stability should evaluate both the dry mixture and reconstituted suspension. Particle segregation, particle fusion and crystallization in the dry mixture can be assessed using electron microscopy. The physical stability of the reconstituted suspension includes measurement of particle size, ease of dispersion and rate of sedimentation. The chemical stability should be determined in both the dry mixture and reconstituted suspension.

References

[1]   Nash, R. A., "Pharmaceutical Suspensions", in "Pharmaceutical Dosage Forms: Disperse Systems", Volume 1, A. Liberman, M. M. Rieger and G. S. Banker (eds.), Marcel Dekker, New York (1988).
[2]   Wagner, J. J., "Biopharmaceutics: Absorption Aspects", J. Pharm. Sci., 50, 359 (1961).
[3]   Reinhold, J. G. et al., "Comparison of the Behaviour of Microcrystalline Sulfadiazine with that of Ordinary Sulfadiazine in Man", Am. J. Med. Sci., 210, 141 (1945).
[4]   Smith, E. A., "Particle Size Reduction to Micron Size", Manuf. Chem. and Aerosol News, July, 31–36 (1967).
[5]   Riegman, S. et al., "Application of Spray Drying Techniques to Pharmaceutical Powders", J. Am. Pharm. Assoc., Sci. Ed., 39, 444 (1950).
[6]   Cohen, R. A. and Skauen, D. M., "Controlled Crystallisation of Hydrocortisone by Ultrasonic Irradiation", J. Pharm. Sci., 53, 1040 (1964).
[7]   Waltersson, J. and Lundgren, P., "Nonthermal Kinetics Applied to Drugs in Pharmaceutical Suspensions", Acta Pharm. Seuc., 20, 145 (1983).
[8]   Higuchi, T., "Some Physical Chemical Aspects of Suspension Formulation", J. Am. Pharm. Assoc., Sci. Ed., 47, 657 (1958).
[9]   Weiner, N., "Strategies for Formulation and Evaluation of Emulsions and Suspensions", Drug Dev. Ind. Pharm., 12, 933 (1986).
[10] Hiestand, E. N., "Theory of Coarse Suspension Formulation", J. Pharm. Sci., 53, 1 (1964).
[11] Deryaguin, B. V. and Landau, L., Acta Physicochem. USSR, 14, 633 (1941).
[12] Verwey, E. J. W. and Overbeek, J. Th. G., "Theory of Stability of Lyophobic Colloids", Elsevier, Amsterdam (1948).
[13] Napper, D. H., "Polymeric Stabilisation of Colloidal Dispersions", Academic Press, London (1983).
[14] Simonelli, P. A., Mehta, S. C. and Higuchi, W. I., J. Pharm. Sci., 59, 633 (1970).
[15] Carless, J. E., Moustafa, M. A. and Rapson, H. D. C., J. Pharm. Pharmac., 20, 630 (1968).

[16] Tadros, Th. F., "Rheology of Dispersions", Wiley-VCH, Germany (2010).

[17] Bapatla, K. M. and Hecht, G., "Ophthalmic Ointment Suspensions", in "Pharmaceutical Dosage Forms: Disperse Systems", Volume 2, H. A. Lieberman, M. M. Rieger and G. S. Banker (eds.), Marcel Dekker, New York (1989).

[18] Ofner, C. M. III, Schnaare, R. L. and Schwartz, J. B., "Reconstitutable Suspensions" in "Pharmaceutical Dosage Forms: Disperse Systems", Volume 2, H. A. Lieberman, M. M. Rieger and G. S. Banker (eds.), Marcel Dekker, New York (1989).

13 Applications of suspensions in cosmetics and personal care

13.1 Introduction

Many cosmetic and personal care formulations contain suspended particles in a liquid vehicle, such as nail polishes, liquid make-ups, eyeliners, mascaras and blushers. In addition, suspended particles such as TiO_2 or ZnO (sunscreens) are incorporated in an oil-in-water (O/W) or water-in-oil (W/O) emulsion. These suspension/emulsion mixtures are referred to as suspoemulsions.

The preparation of suspensions in cosmetic systems and their stabilization against aggregation require fundamental understanding of the colloidal interactions between the various components [1, 2]. Understanding these interactions enables the formulation scientist to arrive at the optimum composition for a particular application. One of the most important aspects is to consider the property of the solid/liquid interface, in particular its interaction between the surfactants and/or polymers that are used for formulating the product and how these can affect the stability/instability of the final product. The fundamental principles involved also help in predicting the long-term physical stability of the formulations.

The incorporation of solid particles in any cosmetic formulation requires the presence of a wetting/dispersing agent as well as a suspending agent (rheology modifier) to prevent any particle sedimentation. Some examples of applications of suspensions in cosmetics are given below, with particular reference to the optimization of the product to achieve the desirable properties.

13.2 Suspensions in sunscreens

Fine particles such as such as titanium dioxide and zinc oxide are used in sunscreen formulations due to their ability to absorb UV radiation. The band gap in these semiconductor materials is such that UV light up to around 405 nm can be absorbed. They can also scatter light due to their particulate nature and their high refractive indices make them particularly effective scatterers. Both scattering and absorption depend critically on particle size [3–5]. Particles of around 250 nm for example are very effective at scattering visible light and TiO_2 of this particle size is the most widely used white pigment. At smaller particle sizes, absorption and scattering maxima shift to the UV region and at 30–50 nm UV attenuation is maximized.

The use of TiO_2 as a UV attenuator in cosmetics was, until recently, largely limited to baby sun protection products due to its poor aesthetic properties (viz; scattering of visible wavelengths results in whitening). Recent advances in particle size control and

DOI 10.1515/9783110486872-014

coatings have enabled formulators to use fine particle titanium dioxide and zinc oxide in daily skin care formulations without compromising the cosmetic elegance [4, 5].

The benefits of a predispersion of inorganic sunscreens are widely acknowledged. However, it requires an understanding of the nature of colloidal stabilization to optimize this predispersion (for both UV attenuation and stability) and exceed the performance of powder-based formulations. Dispersion rheology and its dependence on interparticle interactions is a key factor in this optimization. Optimization of sunscreen actives however does not end there; an appreciation of the end application is crucial to maintaining performance. Formulators need to incorporate the particulate actives into an emulsion, mousse or gel with due regard to aesthetics (skin feel and transparency), stability and rheology.

As mentioned above, TiO_2 and ZnO absorb and scatter UV light. They provide a broad spectrum and are inert and safe to use. Larger particles scatter visible light and they cause whitening. The scattering and absorption depend on the refractive index (which depends on the chemical nature), the wavelength of light and the particle size and shape distribution. The total attenuation is maximized in UVB for 30–50 nm particles. Figure 13.1 shows the effect of particle size on UVA and UVB absorption.

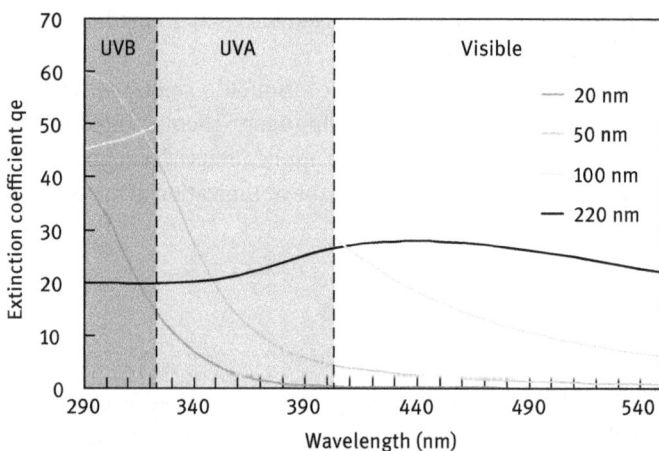

Fig. 13.1: Effect of particle size on UVA and UVB absorption.

The performance of any sunscreen formulation is defined by a number referred to as the sun protection factor (SPF). The basic principle for calculating the SPF [6] is based on the fact that the inverse of the UV transmission through an absorbing layer, $1/T$, is the factor by which the intensity of the UV light is reduced. Thus, at a certain wavelength λ, $1/T(\lambda)$ is regarded as a monochromatic protection factor (MPF). Since the spectral range relevant for the in vivo SPF is between 290 and 400 nm (Fig. 13.1), the monochromatic protections factors have to be averaged over this range.

To keep the particles well dispersed (as single particles) high steric repulsion is required to overcome strong van der Waals attraction. The mechanism of steric stabilization has been described in detail in Chapter 5 and only a summary is given here.

Small particles tend to aggregate as a result of the universal van der Waals attraction unless this attraction is screened by an effective repulsion between the particles. The van der Waals attraction energy, $G_A(h)$, at close approach depends upon the distance, h, between particles of radius, R, and is characterized by the effective Hamaker constant, A,

$$G_A(h) = -\frac{AR}{12h}.$$ (13.1)

The effective Hamaker constant A is given by the following equation,

$$A = (A_{11}^{1/2} - A_{22}^{1/2})^2.$$ (13.2)

A_{11} is the Hamaker constant of the particles and A_{22} is that for the medium. For TiO_2, A_{11} is exceptionally high so that in nonaqueous media with relatively low A_{22} the effective Hamaker constant A is high and despite the small size of the particles a dispersant is always needed to achieve colloidal stabilization. This is usually obtained using adsorbed layers of polymers or surfactants. The most effective molecules are the A-B, ABA block or BA_n graft polymeric surfactants [7] where B refers to the anchor chain. For a hydrophilic particle, this may be a carboxylic acid, an amine or phosphate group or other larger hydrogen bonding type block such as polyethylene oxide. The A chains are referred to as the stabilizing chains, which should be highly soluble in the medium and strongly solvated by its molecules. For nonaqueous dispersions, the A chains could be polypropylene oxide, a long chain alkane, oil soluble polyester or polyhydroxystearic acid (PHS). A schematic representation of the adsorbed layers and the resultant interaction energy-distance curve is shown in Fig. 13.2.

When two particles with an adsorbed layer of hydrodynamic thickness δ approach to a separation distance h that is smaller than 2δ, repulsion occurs as a result of two main effects: (i) unfavourable mixing of the A chains when these are in good solvent

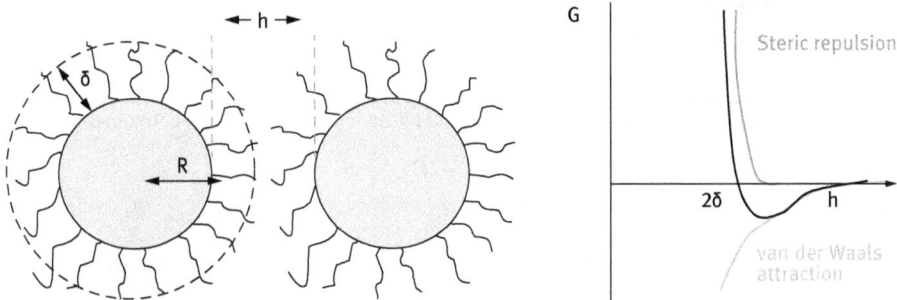

Fig. 13.2: Schematic representation of adsorbed polymer layers and resultant interaction energy G on close approach at distance $h < 2R$.

condition given by a free energy of interaction term G_{mix}; (ii) reduction in configurational entropy on significant overlap given by a free energy of interaction term G_{el}. Combining G_{mix} and G_{el} gives the total steric repulsion G_s as illustrated in Fig. 13.2. Combining G_s with G_A gives the total energy of interaction G. A plot of G versus h is schematically shown in Fig. 13.2. This shows a shallow minimum at $h \approx 2\delta$ and when $h < 2\delta$, G increases very sharply with a further decrease in h. This strong steric repulsion prevents any particle aggregation.

It is useful to consider the parameters responsible for effective steric repulsion: (i) The adsorbed amount Γ; the higher the value the greater the interaction/repulsion. (ii) Solvent conditions as determined by the value of χ; two very distinct cases emerge. Maximum interaction occurs on overlap of the stabilizing layers when the chains are in good solvent conditions, i.e. $\chi < 0.5$. Osmotic forces cause solvent to move into the highly concentrated overlap zone forcing the particles apart. If $\chi = 0.5$, a theta solvent, the steric potential goes to zero and for poor solvent conditions ($\chi > 0.5$) the steric potential becomes negative and the chains will attract, enhancing flocculation. (iii) Adsorbed layer thickness δ. Steric interaction starts at $h = 2\delta$ as the chains begin to overlap and increases as the square of the distance. Here the important factor is not the size of the steric potential but the distance h at which it begins. (iv) The final interaction potential is the superposition of the steric potential and van der Waals attraction as shown in Fig. 13.2.

Most sunscreen formulations consist of an oil-in-water (O/W) emulsion in which the particles are incorporated. These active particles can be in either the oil phase, or the water phase, or both as illustrated in Fig. 13.3. For a sunscreen formulation based on a W/O emulsion, the added nonaqueous sunscreen dispersion mostly stays in the oil continuous phase.

On addition of the sunscreen dispersion to an emulsion to produce the final formulation, one has to consider the competitive adsorption of the dispersant/emulsifier system. In this case the strength of adsorption of the dispersant to the surface modified TiO$_2$ particles must be considered. One should make sure that the adsorption of the dispersant to the TiO$_2$ is very strong and hence it cannot be displaced by the emulsifier, otherwise flocculation of the TiO$_2$ particles may occur resulting in reduction of the sunscreen's protection.

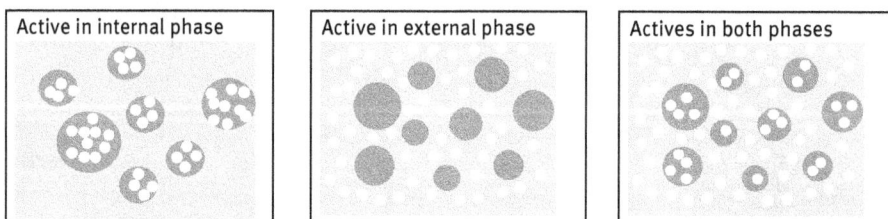

Fig. 13.3: Schematic representation of the location of active particles in sunscreen formulations.

13.3 Suspensions in colour cosmetics

Pigments are the primary ingredient of any colour cosmetic and the way in which these particulate materials are distributed within the product will determine many aspects of product quality including functional activity (colour, opacity, UV protection) but also stability, rheology and skin feel [1, 2]. Several colour pigments are used in cosmetic formulations ranging from inorganic pigments (such as red iron oxide) to organic pigments of various types. The formulation of these pigments in colour cosmetics requires a great deal of skill since the pigment particles are dispersed in an emulsion (oil-in-water or water-in-oil). The pigment particles may be dispersed in the continuous medium in which case one should avoid flocculation with the oil or water droplets. In some cases, the pigment may be dispersed in an oil which is then emulsified in an aqueous medium. Several other ingredients are added such as humectants, thickeners, preservatives, etc. and the interaction between the various components can be very complex.

The particulate distribution depends on many factors such as particle size and shape, surface characteristics, processing and other formulation ingredients but ultimately is determined by interparticle interactions. A thorough understanding of these interactions and how to modify them can help to speed up product design and solve formulation problems.

In this section, I start by briefly describing the fundamental principles of preparation of pigment dispersion. These consist of three main topics, namely wetting of the powder, its dispersion (or wet milling including comminution) and stabilization against aggregation. A schematic representation of this process is shown in Fig. 13.4 [9]. The interaction with other formulation ingredients when these particulates are incorporated in an emulsion (forming a suspoemulsion) will be briefly discussed. Particular attention will be given to the process of competitive adsorption of the dispersant and emulsifier. Optimization of colour cosmetics can be achieved through a fundamental understanding of colloid and interface science. The dispersion stability and rheology of particulate formulations depend on interparticle interactions which in turn depend on the adsorption and conformation of the dispersant at the solid/liquid interface. Dispersants offer the possibility to control the interactions between particles such that consistency is improved. Unfortunately, it is not possible to design a universal dispersant due to specificity of anchor groups and solvent-steric interactions.

Colour formulators should be encouraged to understand the mechanism of stabilizing the pigment particles and how to improve that. In order to optimize performance of the final cosmetic colour formulation, one must consider the interactions between particles, dispersant, emulsifiers and thickeners and strive to reduce the competitive interactions through proper choice of the modified surface as well as the dispersant to optimize adsorption strength.

Wetting of powders of colour cosmetics is an important prerequisite for dispersion of that powder in liquids. It is essential to wet both the external and internal surfaces of the powder aggregates and agglomerates as schematically represented in Fig. 13.4. In all these processes one has to consider both the equilibrium and dynamic aspects of the wetting process [9]. The equilibrium aspects of wetting can be studied at a fundamental level using interfacial thermodynamics. For agglomerates (represented in Fig. 13.4), which are found in all powders, wetting of the internal surface between the particles in the structure requires liquid penetration through the pores.

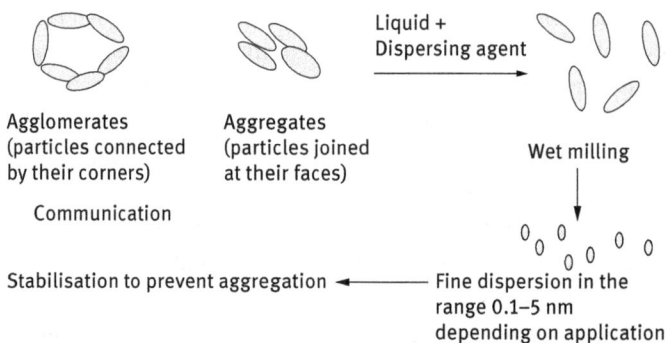

Agglomerates
(particles connected
by their corners)

Aggregates
(particles joined
at their faces)

Liquid +
Dispersing agent

Wet milling

Communication

Stabilisation to prevent aggregation ◄——— Fine dispersion in the
range 0.1–5 nm
depending on application

Fig. 13.4: Schematic representation of the dispersion process.

The dispersion of the powder is achieved by using high speed stirrers such as the Ultra-Turrax or Silverson mixers. This results in dispersion of the wetted powder aggregate or agglomerate into single units [9]. The primary dispersion (sometimes referred to as the mill base) may then be subjected to a bead milling process to produce nanoparticles which are essential for some colour cosmetic applications. Subdivision of the primary particles into much smaller units in the nanosize range (10–100 nm) requires application of intense energy. In some cases, high pressure homogenizers (such as the Microfluidizer, USA) may be sufficient to produce nanoparticles. This is particularly the case with many organic pigments. In some cases, the high pressure homogenizer is combined with application of ultrasound to produce the nanoparticles.

Milling or comminution (the generic term for size reduction) is a complex process and there is little fundamental information on its mechanism. For the breakdown of single crystals or particles into smaller units, mechanical energy is required. This energy in a bead mill is supplied by impaction of the glass or ceramic beads with the particles. As a result, permanent deformation of the particles and crack initiation results. This will eventually lead to the fracture of particles into smaller units. Since the milling conditions are random, some particles receive impacts far in excess of those required for fracture whereas others receive impacts that are insufficient for the fracture process. This makes the milling operation grossly inefficient and only a small fraction of the applied energy is used in comminution. The rest of the energy is dissipated as heat, vibration, sound, interparticulate friction, etc.

The role of surfactants and dispersants on the grinding efficiency is far from being understood. In most cases the choice of surfactants and dispersant is made by trial and error until a system is found that gives the maximum grinding efficiency [9]. As a result of surfactant adsorption at the solid/liquid interface, the surface energy at the boundary is reduced and this facilitates the process of deformation or destruction. The adsorption of surfactants at the solid/liquid interface in cracks facilitates their propagation.

For stabilization of the dispersion against aggregation (flocculation) one needs to create a repulsive barrier that can overcome van der Waals attraction [9]. The process of stabilization of dispersions in cosmetics has been described in detail in Chapters 4 and 5. As discussed in Chapter 4, all particles experience attractive forces on close approach. In order to achieve stability, one must provide a balancing repulsive force to reduce interparticle attraction. This can be done in two main ways by electrostatic or steric repulsion as illustrated in Fig. 13.5 (a) and 13.5 (b) (or a combination of the two, Fig. 13.5 (c)). A polyelectrolyte dispersant such as sodium polyacrylate is required to achieve high solids content. This produces a more uniform charge on the surface and some steric repulsion due to the high molecular weight of the dispersant. Under these conditions the dispersion becomes stable over a wider range of pH at moderate electrolyte concentration. This is electrosteric stabilization. Figure 13.5 (c) shows a shallow minimum at long separation distances, a maximum (of the DLVO type) at intermediate h and a sharp increase in repulsion at shorter separation distances. This combination of electrostatic and steric repulsion can be very effective for stabilization of the suspension [9].

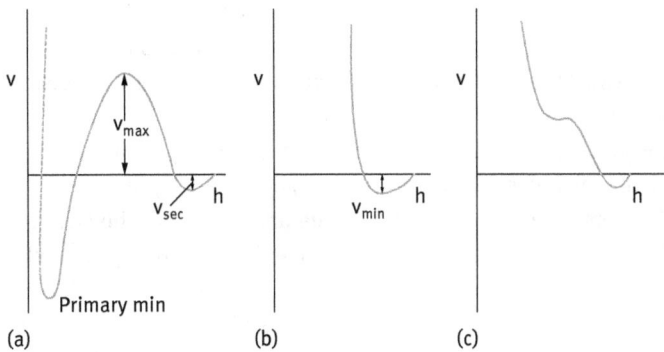

Fig. 13.5: Energy-distance curves for three stabilization mechanisms: (a) electrostatic; (b) steric; (c) electrosteric.

Electrostatic stabilization can be achieved if the particles contain ionizable groups on their surface, such as inorganic oxides, which means that in aqueous media they can therefore develop a surface charge depending upon pH, which affords an electrostatic stabilization to the dispersion. On close approach the particles experience a repulsive potential overcoming the van der Waals attraction which prevents aggregation [9]. This stabilization is due to the interaction between the electric double layers surrounding the particles as discussed in detail in Chapter 4. The double layer repulsion depends upon pH and electrolyte concentration and can be predicted from zeta potential measurements [9]. Surface charge can also be produced by adsorption of ionic surfactants. This balance of electrostatic repulsion with van der Waals attraction is described in the well-known theory of colloid stability by Deryaguin–Landau–Verwey–Overbeek (DLVO theory) [10, 11]. Figure 13.5 (a) shows two attractive minima at long and short separation distances; V_{sec} that is shallow and of few kT units and $V_{primary}$ that is deep and exceeds several 100 kT units. These two minima are separated by an energy maximum V_{max} that can be greater than 25 kT, thus preventing flocculation of the particles into the deep primary minimum.

When the pH of the dispersion is well above or below the isoelectric point or the electrolyte concentration is less than 10^{-2} mol dm^{-3} 1 : 1 electrolyte, the electrostatic repulsion is often sufficient to produce a dispersion without the need for added dispersant. However, in practice this condition often cannot be reached since at high solids content the ionic concentration from the counter- and co-ions of the double layer is high and the surface charge is not uniform. Therefore, a polyelectrolyte dispersant, such as sodium polyacrylate, is required to achieve this high solids content. This produces a more uniform charge on the surface and some steric repulsion due to the high molecular weight of the dispersant. Under these conditions the dispersion becomes stable over a wider range of pH at moderate electrolyte concentration. This is electrosteric stabilization (Fig. 13.5 (c) shows a shallow minimum at long separation distances, a maximum (of the DLVO type) at intermediate h and a sharp increase in repulsion at shorter separation distances). This combination of electrostatic and steric repulsion can be very effective for stabilization of the suspension.

Steric stabilization is usually obtained using adsorbed layers of polymers or surfactants. The most effective molecules are the A-B or ABA block or BA$_n$ graft polymeric surfactants [9] where B refers to the anchor chain. This anchor should be strongly adsorbed to the particle surface. For a hydrophilic particle this may be a carboxylic acid, an amine or phosphate group or other larger hydrogen bonding type block such as polyethylene oxide. The A chains are referred to as the stabilizing chains which should be highly soluble in the medium and strongly solvated by its molecules. When two particles with an adsorbed layer of hydrodynamic thickness δ approach to a separation distance h that is smaller than 2δ, repulsion occurs (Fig. 13.5 (b)) as a result of two main effects: (i) unfavourable mixing of the A chains when these are in good solvent condition; (ii) reduction in configurational entropy on significant overlap.

The efficiency of steric stabilization depends on both the architecture and the physical properties of the stabilizing molecule. Steric stabilizers should have an adsorbing anchor with a high affinity for the particles and/or insoluble in the medium. The stabilizer should be soluble in the medium and highly solvated by its molecules. For aqueous or highly polar oil systems, the stabilizer block can be ionic or hydrophilic such as polyalkylene glycols and for oils it should resemble the oil in character. For silicone oils, silicone stabilizers are best, other oils could use a long chain alkane, fatty ester or polymers such as poly(methylmethacrylate) (PMMA) or polypropylene oxide.

Various types of surface-anchor interactions are responsible for the adsorption of a dispersant to the particle surface: ionic or acid/base interactions; sulphonic acid, carboxylic acid or phosphate with a basic surface e.g. alumina; amine or quat with acidic surface e.g. silica; H bonding; surface esters, ketones, ethers, hydroxyls; multiple anchors – polyamines and polyols (h-bond donor or acceptor) or polyethers (h-bond acceptor). Polarizing groups, e.g. polyurethanes, can also provide sufficient adsorption energies and in nonspecific cases lyophobic bonding (van der Waals) driven by insolubility (e.g. PMMA). It is also possible to use chemical bonding e.g. by reactive silanes.

For relatively reactive surfaces, specific ion pairs may interact giving particularly good adsorption to a powder surface. An ion pair may even be formed in situ, particularly if in low dielectric media. Some surfaces are actually heterogeneous and can have both basic and acidic sites, especially near the IEP. Hydrogen bonding is weak but is particularly important for polymerics which can have multiple anchoring.

The adsorption strength is measured in terms of the segment/surface energy of adsorption χ^s. The total adsorption energy is given by the product of the number of attachment points n by χ^s. For polymers, the total value of $n\chi^s$ can be sufficiently high for strong and irreversible adsorption even though the value of χ^s may be small (less than 1 kT, where k is the Boltzmann constant and T is the absolute temperature). However, this situation may not be adequate, particularly in the presence of an appreciable concentration of wetter and/or in the presence of other surfactants used as adjuvants. If the χ^s of the individual wetter and/or other surfactant molecules is higher than the χ^s of one segment of the B chain of the dispersant, these small molecules can displace the polymeric dispersant particularly at high wetter and/or other surfactant molecules and this could result in flocculation of the suspension. It is, therefore, essential to make sure that the χ^s per segment of the B chain is higher than that of wetter and/or surfactant adsorption and that the wetter concentration is not excessive.

For sterically stabilized dispersions, the resulting energy-distance curve often shows a shallow minimum V_{min} at particle-particle separation distance h comparable to twice the adsorbed layer thickness δ. For a given material, the depth of this minimum depends upon the particle size R, and adsorbed layer thickness δ. So V_{min} decreases with increasing δ/R as illustrated in Fig. 13.6. This is because as we increase the layer thickness, the van der Waals attraction weakens, so the superposition of at-

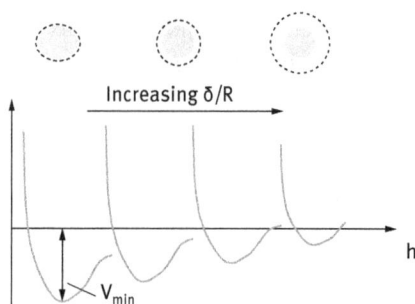

Fig. 13.6: Variation of V_{min} with δ/R.

traction and repulsion will have a smaller minimum. For very small steric layers, V_{min} may become deep enough to cause weak flocculation resulting in a weak attractive gel. So we can see how the interaction energies can also determine the dispersion rheology.

One of the most commonly used dispersants for aqueous media are nonionic surfactants. The most common nonionic surfactants are the alcohol ethoxylates $R-O-(CH_2-CH_2-O)_n-H$, e.g. $C_{13/15}(EO)_n$ with n being 7, 9, 11 or 20. These nonionic surfactants are not the most effective dispersants since the adsorption by the $C_{13/15}$ chain is not very strong. To enhance the adsorption on hydrophobic surfaces a polypropylene oxide (PPO) chain is introduced in the molecule giving $R-O-(PPO)_m-(PEO)_n-H$.

The above nonionic surfactants can also be used for stabilization of polar solids in nonaqueous media. In this case the PEO chain adsorbs on the particle surface leaving the alkyl chains in the nonaqueous solvent. Provided these alkyl chains are sufficiently long and strongly solvated by the molecules of the medium, they can provide sufficient steric repulsion to prevent flocculation.

A better dispersant for polar solids in nonaqueous media is poly(hydroxystearic acid) (PHS) with molecular weight in the region of 1000–2000 Daltons. The carboxylic group adsorbs strongly on the particle surface leaving the extended chain in the nonaqueous solvent. With most hydrocarbon solvents, the PHS chain is strongly solvated by its molecules and an adsorbed layer thickness in the region of 5–10 nm can be produced. This layer thickness prevents any flocculation and the suspension can remain fluid up to high solids content [9].

The most effective dispersants are those of the A-B, A-B-A block and BA_n types. A schematic representation of the architecture of block and graft copolymers is shown in Fig. 13.7.

B, the "anchor chain" (red), is chosen to be highly insoluble in the medium and has a strong affinity to the surface. Examples of B chains for hydrophobic solids are polystyrene (PS), polymethylmethacrylate (PMMA), poly(propylene oxide) (PPO) or alkyl chains provided these have several attachments to the surface. The A stabilizing (blue) chain has to be soluble in the medium and strongly solvated by its molecules. The A chain/solvent interaction should be strong giving a Flory–Huggins parameter

AB end
functionalised

AB / ABA block

B -A$_n$ Graft

Adsorbing polymer B

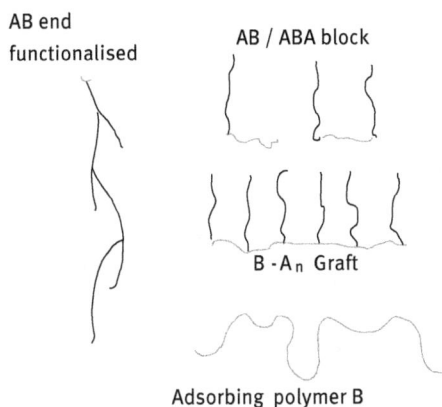

Fig. 13.7: Schematic representation of the architecture of block and graft copolymers.

$\chi < 0.5$ under all conditions. Examples of A chains for aqueous media are polyethylene oxide (PEO), polyvinyl alcohol (PVA) and polysaccharides (e.g. polyfructose). For nonaqueous media, the A chains can be polyhydroxystearic acid (PHS).

One of the most commonly used A-B-A block copolymers for aqueous dispersions are those based on PEO (A) and PPO (B). Several molecules of PEO-PPO-PEO are available with various proportions of PEO and PPO. The commercial name is followed by a letter L (Liquid), P (Paste) and F (Flake). This is followed by two numbers that represent the composition – the first digits represent the PPO molar mass and the last digit represents the % PEO: F68 (PPO molecular mass 1508–1800 + 80 % or 140 mol EO). L62 (PPO molecular mass 1508–1800 + 20 % or 15 mol EO). In many cases two molecules with high and low EO content are used together to enhance the dispersing power.

An example of a BA$_n$ graft copolymer is based on polymethylmethacrylate (PMMA) backbone (with some polymethacrylic acid) on which several PEO chains (with average molecular weight of 750) are grafted. It is a very effective dispersant, particularly for high solids content suspensions. The graft copolymer is strongly adsorbed on hydrophobic surfaces with several attachment points along the PMMA backbone and a strong steric barrier is obtained by the highly hydrated PEO chains in aqueous solutions.

Another effective graft copolymer is hydrophobically modified inulin, a linear polyfructose chain A (with degree of polymerization >23) on which several alkyl chains have been grafted. The polymeric surfactant adsorbs with multipoint attachment with several alkyl chains.

The colour cosmetic pigments are added to oil-in-water (O/W) or water-in-oil (W/O) emulsions. The resulting system is referred to as a suspoemulsion [2]. The particles can be in the internal or external phases or both, as illustrated in Fig. 13.3 for sunscreen formulations. An understanding of competitive interactions is important in optimizing formulation stability and performance. Several possible instabilities might arise in the final formulations: (i) heteroflocuation from particles and droplets of differing charge sign; (ii) electrolyte intolerance of electrostatically stabilized pigments; (iii) competitive adsorption/desorption of a weakly anchored stabilizer which can lead to homoflocculation of the pigment particles and/or emulsion droplet coalescence; (iv) interaction between thickeners and charge stabilized pigments.

Several steps can be made to improve the stability of colour cosmetic formulations which are in fact very similar to those for optimal steric stabilization [2]: (i) use of a strongly adsorbed ("anchored") dispersant, e.g. by multipoint attachment of a block or graft copolymer; (ii) use of a polymeric stabilizer for the emulsion (also with multipoint attachment); (iii) preparation of the suspension and emulsion separately and allowing enough time for complete adsorption (equilibrium); (iv) using low shear when mixing the suspension and emulsion; (v) use of rheology modifiers that reduce the interaction between the pigment particles and emulsion droplets; (vi) increasing dispersant and emulsifier concentrations to ensure that the lifetime of any bare patches produced during collision is very short; (vii) using the same polymeric surfactant molecule for emulsifier and dispersant; (viii) reducing emulsion droplet size.

13.4 Lipsticks and lip balms

These are suspensions of solid oils in a liquid oil or a mixture of liquid oils. They contain a variety of waxes (such as beeswax, carnauba wax, etc.) which give the lipstick its shape and ease of application. Solid oils such as lanolin, palm oil, butter are also incorporated to give the lipstick its tough, shiny film when it dries after application. Liquid oils such as castor oil, olive oil, sunflower oil provide the continuous phase to ensure ease of application. Other ingredients such as moisturizers, vitamin E, collagen, amino acids and sunscreens are sometimes added to help keep lips soft, moist and protected from UV. The pigments give the lipstick its colour, e.g. soluble dyes such as D&C Red No. 21, and insoluble dyes (lakes) such as D&C Red No. 34. Pink shades are made by mixing titanium dioxide with various red dyes. Surfactants are also used in the formulation of lipsticks. The product should show good thermal stability during storage and rheologically it behaves as a viscoelastic solid. In other words, the lipstick should show small deformation at low stresses and this deformation should recover on removal of the stress. Such information could be obtained using creep measurements [9].

13.5 Nail polish

These are pigment suspensions in a volatile nonaqueous solvent. The system should be thixotropic (showing decreasing viscosity with time at a given shear rate and recovery on removal of the shear). On application with the brush it should show proper flow for even coating but should have enough viscosity to avoid "dripping". After application, "gelling" should occur in a controlled time scale. If "gelling" is too fast, the coating may leave "brush marks" (uneven coating). If gelling is too slow, the nail polish may drip. The relaxation time of the thixotropic system should be accurately controlled to ensure good levelling and this requires the use of surfactants.

13.6 Antiperspirants and deodorants

Antiperspirants (commonly used in the US) act both to inhibit sweating and to deodorize. In contrast, deodorants (commonly used in Europe) only inhibit odour. Human skin is almost odourless, but when decomposed by bacteria on the skin, an unpleasant odour develops. There are several possible methods of combating this smell; masking with perfume oils, oxidation of the odoriferous compounds with peroxides, adsorption by finely dispersed ion exchange resins, inhibition of the skin's bacterial flora (the basis of most deodorants), or the action of surfactants, especially appropriate ammonium compounds. Antiperspirants contain astringent substances that precipitate proteins irreversibly and these prohibit perspiration. The general composition of antiperspirant or deodorant is 60–80 % water, 5 % polyol, 5–15 % lipid (stearic acid, mineral oil, beeswax), 2–5 % emulsifiers (polysorbate 40, sorbitan oleate), antiperspirant (aluminium chlorohydrate), 0.1 % antimicrobial, 0.5 % perfume oil. These antiperspirants are thus suspensions of solid actives in a surfactant vehicle. Other ingredients such as polymers that provide good skin feel are added. The rheology of the system should be controlled to avoid particle sedimentation. This is achieved by addition of thickeners. Shear thinning of the final product is essential to ensure good spreadability. In stick application, a "semi-solid" system is produced.

13.7 Foundations

These are complex systems consisting of a suspension-emulsion system (sometimes referred to as suspoemulsions). Pigment particles are usually dispersed in the continuous phase of an O/W or W/O emulsion. Volatile oils such as cyclomethicone are usually used. The system should be thixotropic to ensure uniformity of the film and good levelling. As mentioned above in the section on colour cosmetics, one of the main problems with these suspoemulsions is the competitive adsorption of the dispersant (that is used to disperse the solid particles) and emulsifier used for preparing the emulsion. Strong adsorption of the dispersant on the solid particles is essential to avoid its displacement by the emulsifier molecules. Such displacement may result in flocculation of the pigment particles and this results in the formation of a "rough film" of the pigment. This process can lead to an unacceptable film of the foundation. Any displacement of the emulsifier by the dispersant may result in emulsion coalescence and separation of the foundation formulation.

13.8 Liquid detergents

These are systems used for hand washing (in place of bar soap) and also in washing machines. The formulation consists of an aqueous concentrated surfactant solution (of the anionic/nonionic mixture). At such high surfactant concentrations, liquid crystalline phases of the hexagonal or lamellar type produce a highly viscous liquid. Most liquid detergents contain a builder such as silicates or zeolites. The latter are suspended in the liquid detergent and the high residual viscosity of the liquid detergent may be sufficient to prevent sedimentation of the coarse particles. If this residual viscosity is not high enough to reduce sedimentation, a rheology modifier such as hydroxyethylcellulose or xanthan gum is added to prevent particle sedimentation.

References

[1]   Tadros, Th. F., "Interfacial Phenomena", Vol. 1, De Gruyter, Germany (2015).
[2]   Tadros, Th. F., "Formulation of Cosmetics and Personal Care", De Gruyter, Germany (2016).
[3]   Tadros, Th. F., "Cosmetics", in: "Encyclopedia of Colloid and Interface Science", Th. F. Tadros (ed.), Springer, Germany (2013).
[4]   Kessel, L. M., Naden, B. J., Tooley, I. R. and and Tadros, Th. F., "Application of Colloid and Interface Science Principles for Optimisation of Sunscreen Dispersions", in "Colloids in Cosmetics and Personal Care", Th. F. Tadros (ed.), Wiley-VCH, Germany (2008).
[5]   Robb, J. L., Simpson, L. A. and Tunstall, D. F., Scattering and absorption of UV radiation by sunscreens containing fine particle and pigmentary titanium dioxide, Drug. Cosmet. Ind., **March**, 32–39 (1994).
[6]   Herzog, B., "Models for the Calculation of Sun Protection Factor and Parameters Characterising the UVA Protection Ability of Cosmetic Sunscreens", in "Colloids in Cosmetics and Personal Care", Th. F. Tadros (ed.), Wiley-VCH, Germany (2008).
[7]   Fleer, G. J., Cohen-Stuart, M. A., Scheutjens, J. M. H. M. Cosgrove, T. and Vincent, B., "Polymers at Interfaces", Chapman and Hall, London (1993).
[8]   Napper, D. H., "Polymeric Stabilisation of Colloidal Dispersions", Academic Press, London (1983).
[9]   Tadros, Th. F., "Dispersion of Powders in Liquids and Stabilisation of Suspensions", Wiley-VCH, Germany (2012).
[10] Deryaguin, B. V. and Landau, L., Acta Physicochem. USSR, **14**, 633 (1941).
[11] Verwey, E. J. W. and Overbeek, J. Th. G., "Theory of Stability of Lyophobic Colloids", Elsevier, Amsterdam (1948).

14 Application of suspensions in paints and coatings

14.1 Introduction

Paints or surface coatings are complex multiphase colloidal systems that are applied as a continuous layer to a surface [1]. A paint usually contains pigmented materials to distinguish it from clear films that are described as lacquers or varnishes. The main purpose of a paint or surface coating is to provide aesthetic appeal as well as to protect the surface. For example, a motor car paint can enhance the appearance of the car body by providing colour and gloss and it also protects the car body from corrosion.

When considering a paint formulation, one must know the specific interaction between the paint components and substrates. This subject is of particular importance when one considers the deposition and adhesion of the components to the substrate. The latter can be wood, plastic, metal, glass, etc. The interaction forces between the paint components and the substrate must be considered when formulating any paint. In addition, the method of application can vary from one substrate to another.

For many applications it has been recognized that to achieve the required property, such as durability, strong adhesion to the substrate, opacity, colour, gloss, mechanical properties, chemical resistance, corrosion protection, etc., requires the application of more than one coat. The first two or three coats coats (referred to as the primer and undercoat) are applied to seal the substrate and provide strong adhesion to the substrate. The topcoat provides the aesthetic appeal such as gloss, colour, smoothness, etc. This clearly explains the complexity of paint systems which require fundamental understanding of the processes involved, such as particle-surface adhesion, colloidal interaction between the various components, mechanical strength of each coating, etc.

14.2 The disperse particles

The primary pigment particles (normally in the submicron range) are responsible for the opacity, colour and anti-corrosive properties [2]. The principal pigment in use is titanium dioxide, due to its high refractive index, and the one that is used to produce white paint. To produce maximum scattering, the particle size distribution of titanium dioxide has to be controlled within a narrow limit. Rutile, with a refractive index of 2.76, is preferred over anatase that has a lower refractive index of 2.55. Thus, the primary pigment particles (normally in the submicron range) are responsible for the opacity, colour and anti-corrosive properties. The principal pigment in use is titanium dioxide due to its high refractive index and the one that is used to produce white paint. To produce maximum scattering the particle size distribution of titanium dioxide has to be controlled. Rutile gives the possibility of higher opacity than anatase

DOI 10.1515/9783110486872-015

and it is more resistant to chalking on exterior exposure. To obtain maximum opacity the particle size of rutile should be within 220–140 nm. The surface of rutile is photo-active and it is surface coated with silica and alumina in various proportions to reduce its photo-activity.

Coloured pigments may consist of inorganic or organic particles [2]. For a black pigment one can use carbon black, copper carbonate, manganese dioxide (inorganic) or aniline black (organic). For yellow one can use lead, zinc, chromates, cadmium sulphide, iron oxides (inorganic) or nickel azo yellow (organic). For blue/violet one can use ultramarine, Prussian blue, cobalt blue (inorganic) or phthalocyanin, indanthrone blue, carbazol violet (organic). For red one can use red iron oxide, cadmium selenide, red lead, chrome red (inorganic) or toluidine red, quinacridones (organic).

The colour of a pigment is determined by the selective absorption and reflection of the various wavelengths of visible light (400–700 nm) which impinge on it. For example, a blue pigment appears so because it reflects the blue wavelengths in the incident white light and absorbs the other wavelengths. Black pigments absorb all the wavelengths of incident light almost totally, whereas a white pigment reflects all the visible wavelengths.

The primary shape of pigmented particles is determined by their chemical nature, their crystalline structure (or lack of it) and the way the pigment is created in nature or made synthetically. Pigments as primary particles may be spherical, nodular, needle- or rod-like, plate-like (lamellar).

Pigments are usually supplied in the form of aggregates (where the particles are attached at their faces) or agglomerates (where the particles are attached at their corners). When dispersed in the continuous phase, these aggregates and agglomerates must be dispersed into single units. This requires the use of an effective wetter/dispersant as well as application of mechanical energy. This process of dispersion was discussed in detail in Chapter 3.

In paint formulations, secondary pigments are also used. These are referred to as extenders, fillers and supplementary pigments. They are relatively cheaper than the primary pigments and they are incorporated in conjunction with the primary pigments for a variety of reasons such as cost effectiveness, enhancement of adhesion, reduction of water permeability, enhancement of corrosion resistance, etc. For example, in primer or undercoat (matt latex paint), coarse particle extenders such as calcium carbonate are added in conjunction with TiO_2 to achieve whiteness and opacity in a matt or semi-matt product. The particle size of extenders ranges from submicron to few tens of microns. Their refractive index is very close to that of the binder and hence they do not contribute to opacity from light scattering. Most extenders used in the paint industry are naturally occurring materials such as barytes (barium sulphate), chalk (calcium carbonate), gypsum (calcium sulphate) and silicates (silica, clay, talc or mica). However, more recently synthetic polymeric extenders have been designed to replace some of the TiO_2. A good example is spindrift, which consists of polymer beads that consist of spherical particles (up to 30 μm in diameter) that contain submicron air bubbles and a small proportion of TiO$_2$. The small air bubbles ($< 0.1\,\mu m$) reduce the effective refractive index of the polymer matrix, thus enhancing the light scattering by TiO$_2$.

The refractive index (RI) of any material (primary or secondary pigment) is a key to its performance. As is well known, the larger the difference in refractive index between the pigment and the medium in which it is dispersed, the greater the opacity effect [2].

The refractive index of the medium in which the pigment is dispersed ranges from 1.33 (for water) to 1.4–1.6 (for most film formers). Thus rutile will give the highest opacity, whereas talc and calcium carbonate will be transparent in fully bound surface coatings. Another important fact that affects light scattering is the particle size and hence to obtain the maximum opacity from rutile an optimum particle size of 250 nm is required. This explains the importance of good dispersion of the powder in the liquid that can be achieved by a good wetting/dispersing agent as well as application of sufficient milling efficiency.

For coloured pigments, the refractive index of the pigment in the nonabsorbing or highly reflecting part of the spectrum affects the performance as an opacifying material. For example, Pigment Yellow 1 and Arylamide Yellow G give lower opacity than Pigment Yellow 34 Lead Chromate. Most suppliers of coloured pigments attempt to increase the opacifying effect by controlling particle size.

The nature of the pigment surface plays a very important role in its dispersion in the medium as well as its affinity to the binder. For example, the polarity of the pigment determines its affinity for alkyds, polyesters, acrylic polymers and latexes that are commonly used as film formers (see below). In addition, the nature of the pigment surface determines its wetting characteristics in the medium in which it is dispersed (which can be aqueous or nonaqueous) as well as the dispersion of the aggregates and agglomerates into single particles. It also affects the overall stability of the liquid paint. Most pigments are surface treated by the manufacturer to achieve optimum performance. As mentioned above, the surface of rutile particles is treated with silica and alumina in various proportions to reduce its photo-activity. If the pigment has to be used in a nonaqueous paint, its surface is also treated with fatty acids and amines to make it hydrophobic for incorporation in an organic medium. This surface treatment enhances the dispersibility of the paint, its opacity and tinting strength, its durability (glass retention, resistance to chalking and colour retention). It can also protect the binder in the paint formulation.

The dispersion of the pigment powder in the continuous medium requires several processes, namely wetting of the external and internal surface of the aggregates and agglomerates, separation of the particles from these aggregates and agglomerates by application of mechanical energy, displacement of occluded air and coating of the particles with the dispersion resin. This was discussed in detail in Chapter 3. It is also necessary to stabilize the particles against flocculation either by electrostatic double layer repulsion and/or steric repulsion as discussed in detail in Chapters 4 and 5.

14.3 The dispersion medium and film formers

The dispersion medium can be aqueous or nonaqueous depending on application [2]. It consists of a dispersion of the binder in the liquid (which is sometimes referred to as the diluent). The term solvent is frequently used to include liquids that do not dissolve the polymeric binder. Solvents are used in paints to enable the paint to be made and they enable application of the paint to the surface. In most cases, the solvent is removed after application by simple evaporation and if the solvent is completely removed from the paint film it should not affect the paint film performance. However, in the early life of the film, solvent retention can affect hardness, flexibility and other film properties. In water-based paints, the water may act as a true solvent for some of the components but it should be a nonsolvent for the film former. This is particularly the case with emulsion paints.

With the exception of water, all solvents, diluents and thinners used in surface coatings are organic liquids with low molecular weight. Two types can be distinguished, hydrocarbons (both aliphatic and aromatic) and oxygenated compounds such as ethers, ketones, esters, ether alcohols, etc. Solvents, thinners and diluents control the flow of the wet paint on the substrate to achieve a satisfactory smooth, even thin film, which dries in a predetermined time. In most cases mixtures of solvents are used to obtain the optimum condition for paint application. The main factors that must be considered when choosing solvent mixtures are their solvency, viscosity, boiling point, evaporation rate, flash point, chemical nature, odour and toxicity.

The solvent power or solvency of a given liquid or mixture of liquids determines the miscibility of the polymer binder or resin. It has also a big effect on the attraction between particles in a paint formulation. As mentioned above, the dispersion medium consists of a solvent or diluent and the film former. The latter is also sometimes referred to as a "binder", since it functions by binding the particulate components together and this provides the continuous film-forming portion of the coating. The film former can be a low molecular weight polymer (oleoresinous binder, alkyd, polyurethane, amino resins, epoxide resin, unsaturated polyester), a high molecular weight polymer (nitrocellulose, solution vinyls, solution acrylics), an aqueous latex dispersion (polyvinyl acetate, acrylic or styrene/butadiene) or a nonaqueous polymer dispersion (NAD). Film formers may be based on polymer solutions or polymer latexes. The polymer solution may exist in the form of a fine particle dispersion in nonsolvent. In some cases the system may be a mixed solution/dispersion, implying that the solution contains both single polymer chains and aggregates of these chains (sometimes referred to as micelles). A striking difference between a polymer that is completely soluble in the medium and one which contains aggregates of that polymer is the viscosity reached in both cases. A polymer that is completely soluble in the medium will show a higher viscosity at a given concentration compared to another polymer (at the same concentration) that produces aggregates. Another important difference is the rapid increase in the solution viscosity with increasing molecular weight for a completely soluble polymer. If the polymer makes aggregates in solution, increasing the molecular weight of the polymer does not show a dramatic increase in viscosity.

The earliest film forming polymers used in paints were based on natural oils, gums and resins. Modified natural products are based on cellulose derivatives such as nitrocellulose, which is obtained by nitration of cellulose under carefully specified conditions. Organic esters of cellulose, such as acetate and butyrate, can also be produced. Another class of naturally occurring film formers are those based on vegetable oils and their derived fatty acids (renewable resource materials). Oils used in coatings include linseed oil, soya bean oil, coconut oil and tall oil. When chemically combined into resins, the oil contributes flexibility and with many oils oxidative crosslinking potential. The oil can also be chemically modified, as for example the hydrogenation of castor oil that can be combined with alkyd resins to produce some specific properties of the coating.

Another early binder used in paints are the oleoresinous vehicles that are produced by heating together oils and either natural or certain preformed resins, so that the resin dissolves or disperses in the oil portion of the vehicle. However, these oleoresinous vehicles have been replaced later by alkyd resins, which were probably one of the first applications of synthetic polymers in the coating industry. These alkyd resins are polyesters obtained by reaction of vegetable oil triglycerides, polyols (e.g. glycerol) and dibasic acids or their anhydrides. These alkyd resins enhance the mechanical strength, drying speed and durability over and above those obtained using the oleoresinous vehicles. The alkyds were also modified by replacing part of the dibasic acid with a diisocyanate (such as toluene diisocyanate, TDI) to produce greater toughness and quicker drying characteristics.

Another type of binder is based on polyester resins (both saturated and unsaturated). These are typically composed mainly of co-reacted di- or polyhydric alcohols and di- or tri-basic acid or acid anhydride. They have also been modified using silicone to enhance their durability.

More recently, acrylic polymers have been used in paints due to their excellent properties of clarity, strength and chemical and weather resistance. Acrylic polymers refer to systems containing acrylate and methylacrylate esters in their structure along with other vinyl unsaturated compounds. Both thermoplastic and thermosetting systems can be made; the latter are formulated to include monomers possessing additional functional groups that can further react to give crosslinks following the formation of the initial polymer structure. These acrylic polymers are synthesized by radical polymerization. The main polymer-forming reaction is a chain propagation step which follows an initial initiation process. A variety of chain transfer reactions are possible before chain growth ceases by a termination process.

Radicals produced by transfer, if sufficiently active, can initiate new polymer chains where a monomer is present which is readily polymerized. Radicals produced by chain transfer agents (low molecular weight mercaptans, e.g. primary octyl mercaptan) are designed to initiate new polymer chains [2]. These agents are introduced to control the molecular weight of the polymer.

The monomers used for preparation of acrylic polymers vary in nature and can generally be classified as "'hard" (such as methylmethacrylate, styrene and vinyl acetate) or "soft" (such as ethyl acrylate, butyl acrylate, 2-ethyl hexyl acrylate). Reactive monomers may also have hydroxyl groups (such as hydroxy ethyl acrylate). Acidic monomers such as methacrylic acid are also reactive and may be included in small amounts in order that the acid groups may enhance pigment dispersion. The practical coating systems are usually copolymers of "hard" and "soft" monomers. The vast majority of acrylic polymers consist of random copolymers. By controlling the proportion of "hard" and "soft" monomers and the molecular weight of the final copolymer, one arrives at the right property that is required for a given coating. As mentioned above, two types of acrylic resins can be produced, namely thermoplastic and thermosetting. The former find application in automotive topcoats, although they suffer from some disadvantages like cracking in cold conditions and this may require a process of plasticization. These problems are overcome by using thermosetting acrylics which improve the chemical and alkali resistance. Also it allows one to use higher solid contents in cheaper solvents. Thermosetting resins can be self-crosslinking or may require a co-reacting polymer or hardener.

14.4 Deposition of particles and their adhesion to the substrate

In a paint film the pigment particles need to undergo a process of deposition to the surfaces (that is governed by long-range forces such as van der Waals attraction and electrical double layer repulsion or attraction). This process of deposition is also affected by polymers (nonionic, anionic or cationic) which can enhance or prevent adhesion [3]. Once the particles reach the surface they have to adhere strongly to the substrate. This process of adhesion is governed by short-range forces (chemical or nonchemical). The same applies to latex particles, which also undergo a process of deposition, adhesion and coalescence. The subject of particle deposition and adhesion was discussed in detail before [3].

14.5 Flow characteristics (rheology) of paints

Control of the flow characteristics of paints is essential for their successful application. All paints are complex systems consisting of various components such as pigments, film formers, latexes and rheology modifiers. These components interact with each other and the final formulation becomes non-Newtonian showing complex rheological behaviour [4]. The paint is usually applied in three stages, namely transfer of the paint from the bulk container, transfer of the paint from the applicator (brush or roller) to the surface to form a thin, even film and flow-out of film surface, coalescence of polymer particles (latexes) and loss of the medium by evaporation. During each of these processes the flow characteristics of the paint and its time relaxation produce interesting rheological responses. To understand the rheological behaviour of a paint system, one must start with the basic knowledge of rheology [4] as discussed in detail in Chapter 9. An account of the main rheological characteristics of a paint formulation was given before [1].

References

[1] Tadros, Th. F., "Colloids in Paints", Wiley-VCH, Germany (2010).
[2] Lamboune, R. (ed.), "Paint and Surface Coatings: Theory and Practice", Ellis Horwood, Chichester (1987).
[3] Tadros, Th. F., "Interfacial Phenomena", Vol. 1, De Gruyter, Germany (2015).
[4] Tadros, Th. F., "Rheology of Dispersions", Wiley-VCH, Germany (2010).

15 Application of suspensions in agrochemicals

15.1 Introduction

The formulation of agrochemicals as dispersions of solids in aqueous solution (to be referred to as suspension concentrates or SCs) has attracted considerable attention [1–3]. Such formulations are a natural replacement for wettable powders (WPs). The latter are produced by mixing the active ingredient with a filler (usually a clay material) and a surfactant (dispersing and wetting agent). These powders are dispersed into the spray tank to produce a coarse suspension which is applied to the crop. Although wettable powders are simple to formulate, they are not the most convenient for the farmer. Apart from being dusty (and occupying a large volume due to their low bulk density), they tend to settle fast in the spray tank and they do not provide optimum biological efficiency as a result of the large particle size of the system. In addition, one cannot incorporate the necessary adjuvants (mostly surfactants) in the formulation. These problems are overcome by formulating the agrochemical as an aqueous SC.

Several advantages may be quoted for SCs. Firstly, one may control the particle size by controlling the milling conditions and proper choice of the dispersing agent. Secondly, it is possible to incorporate high concentrations of surfactants in the formulation which is sometimes essential for enhancing wetting, spreading and penetration. Stickers may also be added to enhance adhesion and in some cases to provide slow release.

In recent years there has been considerable research into the factors that govern the stability of suspension concentrates [4]. The theories of colloid stability could be applied to predict the physical states of these systems on storage. In addition, analysis of the problem of sedimentation of SCs at a fundamental level has been undertaken [5]. Since the density of the particles is usually larger than that of the medium (water) SCs tend to separate as a result of sedimentation. The sedimented particles tend to form a compact layer at the bottom of the container (sometimes referred to as clay or cake), which is very difficult to redisperse. It is, therefore, essential to reduce sedimentation and formation of clays by incorporation of an antisettling agent.

15.2 Preparation of suspension concentrates and the role of surfactants/dispersing agents

Suspension concentrates are usually formulated using a wet milling process which requires the addition of a surfactant/dispersing agent [4]. The latter should satisfy the following criteria: (i) A good wetting agent for the agrochemical powder (both external and internal surfaces of the powder aggregates or agglomerates must be spontaneously wetted). (ii) A good dispersing agent to break such aggregates or agglomer-

DOI 10.1515/9783110486872-016

ates into smaller units and subsequently help in the milling process (one usually aims at a dispersion with a volume mean diameter of 1–2 μm). (iii) It should provide good stability in the colloid sense (this is essential for maintaining the particles as individual units once formed). Powerful dispersing agents are particularly important for the preparation of highly concentrated suspensions (sometimes required for seed dressing). Any flocculation will cause a rapid increase in the viscosity of the suspension and this makes the wet milling of the agrochemical a difficult process.

Dry powders of organic compounds usually consist of particles of various degrees of complexity, depending on the isolation stages and the drying process. Generally, the particles in a dry powder form aggregates (in which the particles are joined together with their crystal faces) or agglomerates (in which the particles touch at edges or corners) forming a looser more open structure. It is essential in the dispersion process to wet the external as well as the internal surfaces and displace the air entrapped between the particles. This is usually achieved by the use of surface active agents of the ionic or nonionic type. In some cases, macromolecules or polyelectrolytes may be efficient in this wetting process. This may be the case since these polymers contain a very wide distribution of molecular weights and the low molecular weight fractions may act as efficient wetting agents. For efficient wetting, the molecules should lower the surface tension of water and they should diffuse fast in solution and become quickly adsorbed at the solid/solution interface.

Let us consider an agrochemical powder with surface area A. Before the powder is dispersed in the liquid it has a surface tension γ_{SV} and after immersion in the liquid it has a surface tension γ_{SL}. The work of dispersion W_d is simply given by the difference in adhesion or wetting tension of the SL and SV [4],

$$W_d = A(\gamma_{SL} - \gamma_{SV}) = -A\gamma_{LV} \cos \theta, \tag{15.1}$$

where γ_{LV} is the liquid surface tension and θ is the equilibrium contact angle at the wetting line (see Chapter 3). It is clear from equation (15.1) that if $\theta < 90°$, $\cos \theta$ is positive and W_d is negative, i.e. wetting of the powder is spontaneous. Since surfactants are added in sufficient amounts ($\gamma_{dynamic}$ is lowered sufficiently) spontaneous dispersion is the rule rather than the exception.

Wetting of the internal surface requires penetration of the liquid into channels between and inside the agglomerates. The process is similar to forcing a liquid through fine capillaries as discussed in detail in Chapter 3. To enhance the rate of penetration, γ_{LV} has to be made as high as possible, θ as low as possible and η as low as possible. As discussed in Chapter 3, for dispersion of powders into liquids one should use surfactants that lower θ while not reducing γ_{LV} too much. The viscosity of the liquid should also be kept at a minimum. Thickening agents (such as polymers) should not be added during the dispersion process. It is also necessary to avoid foam formation during the dispersion process.

For the dispersion of aggregates and agglomerates into smaller units one requires high speed mixing, e.g. a Silverson mixer. In some cases the dispersion process is easy and the capillary pressure may be sufficient to break up the aggregates and agglomerates into primary units. The process is aided by the surfactant which becomes adsorbed on the particle surface. However, one should be careful during the mixing process not to entrap air (foam) which causes an increase in the viscosity of the suspension and prevents easy dispersion and subsequent grinding. If foam formation becomes a problem, one should add antifoaming agents such as polysiloxane antifoaming agents [4].

After completion of the dispersion process, the suspension is transferred to a ball or bead mill for size reduction. Milling or comminution (the generic term for size reduction) is a complex process and there is little fundamental information on its mechanism. For the breakdown of single crystals into smaller units, mechanical energy is required [4]. This energy in a bead mill, for example, is supplied by impaction of the glass beads with the particles. As a result, permanent deformation of the crystals and crack initiation result. This will eventually lead to the fracture of the crystals into smaller units. However, since the milling conditions are random, it is inevitable that some particles receive impacts that are far in excess of those required for fracture, whereas others receive impacts that are insufficient to fracture them. This makes the milling operation grossly inefficient and only a small fraction of the applied energy is actually used in comminution. The rest of the energy is dissipated as heat, vibration, sound, interparticulate friction, friction between the particles and beads, and elastic deformation of unfractured particles. For these reasons, milling conditions are usually established by a trial and error procedure. Of particular importance is the effect of various surface active agents and macromolecules on the grinding efficiency. As a result of adsorption of surfactants at the solid/liquid interface, the surface energy at the boundary is reduced and this facilitates the process of deformation or destruction. The adsorption of the surfactant at the solid/solution interface in cracks facilitates their propagation. The surface energy manifests itself in destructive processes on solids, since the generation and growth of cracks and separation of one part of a body from another is directly connected with the development of new free surface. Thus, as a result of adsorption of surface active agents at structural defects in the surface of the crystals, fine grinding is facilitated. In the extreme case where there is a very great reduction in surface energy at the sold/liquid boundary, spontaneous dispersion may take place with the result of the formation of colloidal particles ($< 1 \, \mu m$).

15.3 Control of the physical stability of agrochemical suspension concentrates

When considering the stability of agrochemical suspension concentrates one must distinguish between the colloid stability and the overall physical stability. Colloid stability implies absence of an aggregation between the particles, which requires the presence of an energy barrier that is produced by electrostatic repulsion [6–8], steric repulsion [9, 10] or combination of the two (electrosteric). This is achieved by the use of powerful dispersing agents, e.g. surfactants of the ionic or nonionic type, nonionic polymers or polyelectrolytes. These dispersing agents must be strongly adsorbed onto the particle surfaces and fully cover them. With ionic surfactants, irreversible flocculation is prevented by the repulsive force generated from the presence of an electrical double layer at the particle solution interface (see Chapter 4). Depending on the conditions, this repulsive force can be made sufficiently large to overcome the ubiquitous van der Waals attraction between the particles, at intermediate distances of separation. With nonionic surfactants and macromolecules, repulsion between the particles is ensured by the steric interaction of the adsorbed layers on the particle surfaces (see Chapter 5). With polyelectrolytes, both electrostatic and steric repulsion exist. Physical stability implies absence of sedimentation and/or separation, ease of dispersion on shaking and/or dilution in the spray tanks. As was discussed before [4], to achieve overall physical stability one may apply control and reversible flocculation methods and/or using a rheology modifier.

To distinguish between colloid stability/instability and physical stability one must consider the state of the suspension on standing as schematically illustrated in Fig. 15.1. These states are determined by: (i) magnitude and balance of the various interaction forces, electrostatic repulsion, steric repulsion and van der Waals attraction; (ii) particle size and shape distribution; (iii) density difference between disperse phase and medium which determines the sedimentation characteristics; (iv) conditions and prehistory of the suspension, e.g. agitation which determines the structure of the flocs formed (chain aggregates, compact clusters, etc.); (v) presence of additives, e.g. high molecular weight polymers that may cause bridging or depletion flocculation.

These states may be described in terms of three different energy-distance curves: (a) electrostatic, produced for example by the presence of ionogenic groups on the surface of the particles, or adsorption of ionic surfactants; (b) steric, produced for example by adsorption of nonionic surfactants or polymers; (c) electrostatic + steric (electrosteric), as for example produced by polyelectrolytes. These are illustrated in Fig. 15.2.

States (a)–(c) in Fig. 15.1 correspond to a suspension that is stable in the colloid sense. The stability is obtained as a result of net repulsion due to the presence of extended double layers (i.e. at low electrolyte concentration), the result of steric repulsion produced by adsorption of nonionic surfactants or polymers, or the result of combination of double layer and steric repulsion (electrosteric). State (a) represents the case of a suspension with small particle size (submicron) whereby Brownian diffusion overcomes the gravity force producing uniform distribution of the particles in the suspension, i.e.

$$kT > (4/3)\pi R^3 \Delta\rho gh , \qquad (15.2)$$

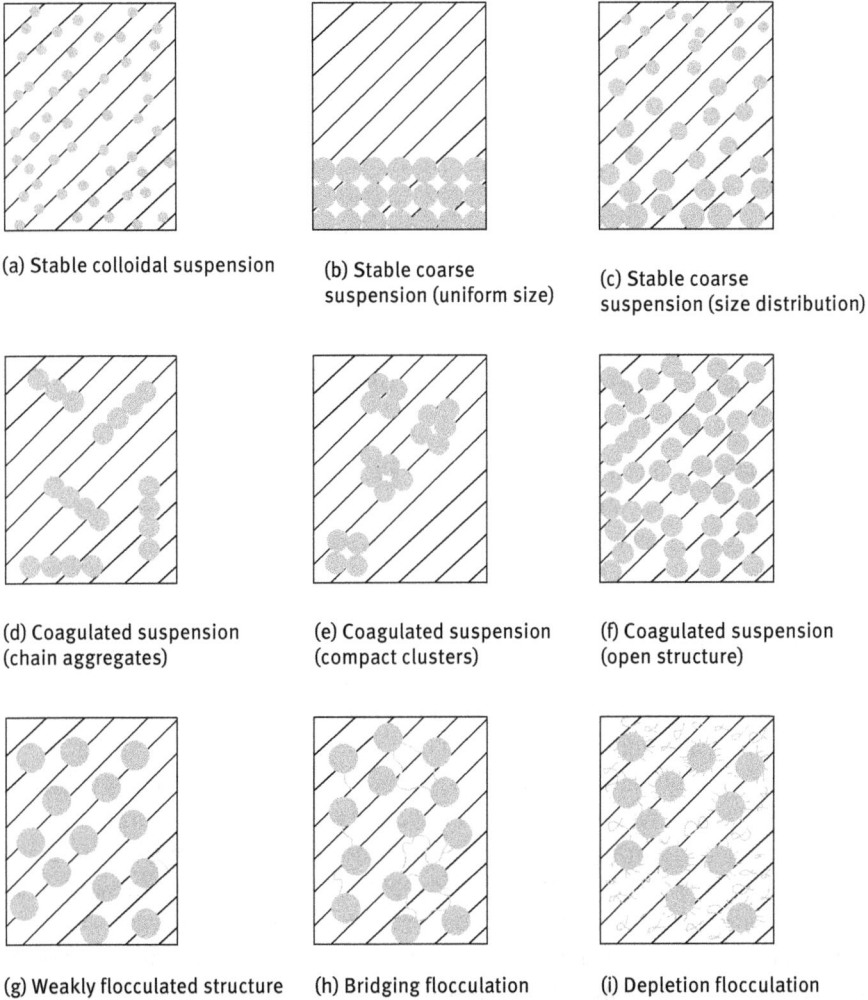

(a) Stable colloidal suspension

(b) Stable coarse suspension (uniform size)

(c) Stable coarse suspension (size distribution)

(d) Coagulated suspension (chain aggregates)

(e) Coagulated suspension (compact clusters)

(f) Coagulated suspension (open structure)

(g) Weakly flocculated structure

(h) Bridging flocculation

(i) Depletion flocculation

Fig. 15.1: States of the suspension.

where k is the Boltzmann constant, T is the absolute temperature, R is the particle radius, $\Delta\rho$ is the buoyancy (difference in density between the particles and the medium), g is the acceleration due to gravity and h is the height of the container.

A good example of the above case is a nanosuspension with particle size well below 1 μm that is stabilized by an ionic surfactant or nonionic surfactant or polymer. This suspension will show no separation on storage for long periods of time [11].

States (b) and (c) represent the case of suspensions where the particle size range is outside the colloid range (> 1 μm). In this case, the gravity force exceeds the Brownian diffusion. With state (b), the particles are uniform and they will settle under gravity forming a hard sediment (technically referred to "clay" or "cake"). The repulsive forces

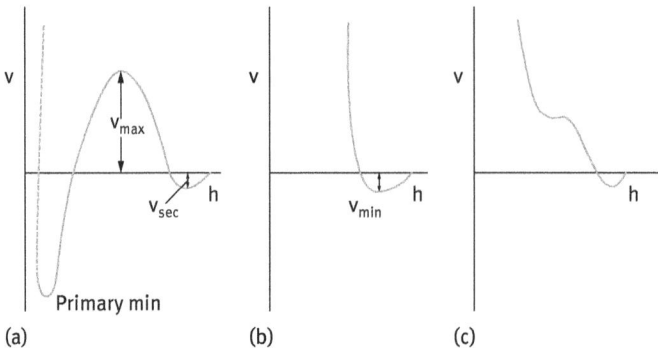

Fig. 15.2: Energy-distance curves for three stabilization mechanisms.

between the particles allow them to move past each other until they reach small distances of separation (that are determined by the location of the repulsive barrier). Due to the small distances between the particles in the sediment, it is very difficult to redisperse the suspension by simple shaking.

With case (c), consisting of a wide distribution of particle sizes, the sediment may contain larger proportions of the larger size particles, but still a hard "clay" is produced. These "clays" are dilatant (i.e. shear thickening) and they can be easily detected by inserting a glass rod in the suspension. Penetration of the glass rod into these hard sediments is very difficult.

States (d)–(f) represent the case of coagulated suspensions which either have a small repulsive energy barrier or its complete absence. State (d) represents the case of coagulation under no stirring conditions in which case chain aggregates are produced that will settle under gravity forming a relatively open structure. State (e) represents the case of coagulation under stirring conditions where compact aggregates are produced that will settle faster than the chain aggregates and the sediment produced is more compact. State (f) represents the case of coagulation at high volume fraction of the particles, ϕ. In this case the whole particles will form a "one-floc" structure that is formed from chains and cross chains that extend from one wall to the other in the container. Such a coagulated structure may undergo some compression (consolidation) under gravity leaving a clear supernatant liquid layer at the top of the container. This phenomenon is referred to as syneresis.

State (g) represents the case of weak and reversible flocculation. This occurs when the secondary minimum in the energy-distance curve (Fig. 15.2 (a)) is deep enough to cause flocculation. This can occur at moderate electrolyte concentrations, in particular with larger particles. The same occurs with sterically and electrosterically stabilized suspensions (Fig. 15.2 (b) and (c)). This occurs when the adsorbed layer thickness is not very large, particularly with large particles. The minimum depth required for causing weak flocculation depends on the volume fraction of the suspension. The

higher the volume fraction, the lower the minimum depth required for weak flocculation.

The above flocculation is weak and reversible, i.e. on shaking the container, redispersion of the suspension occurs. On standing, the dispersed particles aggregate
to form a weak "gel". This process (referred to as sol-gel transformation) leads to reversible time dependency of viscosity (thixotropy). On shearing the suspension, viscosity decreases and when the shear is removed, viscosity is recovered. This phenomenon is applied in paints. On application of the paint (by a brush or roller), the
gel is fluidized, allowing uniform coating of the paint. When shearing is stopped, the
paint film recovers its viscosity and this avoids any dripping.

State (h) represents the case where the particles are not completely covered by
the polymer chains. In this case, simultaneous adsorption of one polymer chain on
more than one particle occurs, leading to bridging flocculation. If the polymer adsorption is weak (low adsorption energy per polymer segment), the flocculation could be
weak and reversible. In contrast, if the adsorption of the polymer is strong, tough flocs
are produced and the flocculation is irreversible. The latter phenomenon is used for
solid/liquid separation, e.g. in water and effluent treatment.

Case (i) represents a phenomenon, referred to as depletion flocculation, produced
by addition of "free" nonadsorbing polymer [12]. In this case, the polymer coils cannot
approach the particles to a distance Δ (that is determined by the radius of gyration of
free polymer R_G), since the reduction of entropy on close approach of the polymer
coils is not compensated by an adsorption energy. The suspension particles will be
surrounded by a depletion zone with thickness Δ. Above a critical volume fraction of
the free polymer, ϕ_p^+, the polymer coils are "squeezed out" from between the particles and the depletion zones begin to interact. The interstices between the particles
are now free from polymer coils and hence an osmotic pressure is exerted outside the
particle surface (the osmotic pressure outside is higher than in between the particles)
resulting in weak flocculation [12].

The magnitude of the depletion attraction free energy, G_{dep}, is proportional to
the osmotic pressure of the polymer solution, which in turn is determined by ϕ_p and
molecular weight M. The range of depletion attraction is proportional to the thickness
of the depletion zone, Δ, which is roughly equal to the radius of gyration, R_G, of the
free polymer [12].

15.4 Ostwald ripening (crystal growth)

This occurs as a result of the difference in solubility between small and large crystals
[3, 4]. Smaller crystals have a higher solubility than larger ones. On storage of the suspension, the smaller crystals dissolve and become deposited on the larger. This leads
to a shift of the particle size distribution to larger values. This results in enhancement

of the sedimentation of the particles, increased flocculation and reduced bioefficacy of the agrochemical.

Another mechanism for crystal growth is related to polymorphic changes in solutions, and again the driving force is the difference in solubility between the two polymorphs. In other words, the less soluble form grows at the expense of the more soluble phase. This is sometimes also accompanied by changes in the crystal habit. Thus, prevention of crystal growth or at least reducing it to an acceptable level is essential in most suspension concentrates. Many surfactants and polymers may act as crystal growth inhibitors if they adsorb strongly on the crystal faces, thus preventing solute deposition. However, the choice of an inhibitor is still an art and there are not many rules that can be used for selection of crystal growth inhibitors.

15.5 Stability against claying or caking

Once a dispersion that is stable in the colloid sense has been prepared, the next task is to eliminate claying or caking. This is the consequence of settling of the colloidally stable suspension particles [5]. The repulsive forces necessary to ensure this colloid stability allow the particles to move past each other forming a dense sediment which is very difficult to redisperse. Such sediments are dilatant (shear thickening, see Chapter 9) and hence the SC becomes unusable.

To reduce settling and prevent the formation of dilatant sediments, one must add an antisettling agent. Most of the antisettling agents used in practice are high molecular weight polymers such as hydroxyethyl cellulose or xanthan gum. These materials show an increase in the viscosity of the medium as their concentration increases [4, 5]. However, at a critical polymer concentration (which depends on the nature of the polymer and its molecular weight) they show a very rapid increase in viscosity with any further increase in their concentration. This critical concentration (sometimes denoted by C^*) represents the situation where the polymer coils or rods begin to overlap. Under these conditions the solutions become significantly non-Newtonian (viscoelastic, see Chapter 9) and they produce stresses that are sufficient to overcome the stress exerted by the particles. The settling of suspensions in these non-Newtonian fluids is not simple since one has to consider the non-Newtonian behaviour of these polymer solutions. In order to adequately describe the settling of particles in non-Newtonian fluids, one need to know how the viscosity of the medium changes with shear rate or shear stress. Most of these viscoelastic fluids show a gradual increase in viscosity with decreasing shear rate or shear stress, but below a critical stress or shear rate they show a Newtonian region with a limiting high viscosity that is denoted as the residual (or zero shear) viscosity.

For more concentrated suspensions, an elastic network is produced in the system which encompasses the suspension particles as well as the polymer chains. In this case, settling of individual particles may be prevented. However, in this case the elastic network may collapse under its own weight and some liquid is squeezed out from between the particles. This is manifested in a clear liquid layer at the top of the suspension, a phenomenon usually referred to as syneresis. If such separation is not significant, it may not cause any problem on application since by shaking the container the whole system redisperses. However, significant separation is not acceptable since it becomes difficult to homogenize the system. In addition, such extensive separation is cosmetically unacceptable and the formulation rheology should be controlled to reduce such separation to a minimum.

Another method that can be applied to reduce settling and formation of hard sediments is to add fine inorganic materials such as swellable clays and finely divided oxides (silica or alumina). These fine inorganic materials form a "three-dimensional" network in the continuous medium which by virtue of its elasticity prevents sedimentation and claying. With swellable clays such as sodium montmorillonite, the gel arises from the interaction of the plate-like particles in the medium [3].

Mixtures of polymers such as hydroxyethyl cellulose or xanthan gum with finely divided solids such as sodium montmorillonite or silica offer one of the most robust antisettling systems. By optimizing the ratio of the polymer to the solid particles, one can arrive at the right viscosity and elasticity to reduce settling and separation. Such systems are more shear thinning than the polymer solutions and hence they are more easily dispersed in water on application. The most likely mechanism by which these mixtures produce a viscoelastic network is probably through bridging or depletion flocculation. The polymer-particulate mixtures also show less temperature dependency for viscosity and elasticity than the polymer solutions and hence they ensure long-term physical stability at high temperatures.

References

[1]   Tadros, Th. F., Advances Colloid and Interface Sci., **12**, 141 (1980).
[2]   Tadros, Th. F., "Surfactants in Agrochemicals", Marcel Dekker, New York (1994).
[3]   Tadros, Th. F., "Colloids in Agrochemicals", Wiley-VCH, Germany (2009).
[4]   Tadros, Th. F., "Dispersion of Powders in Liquids and Stabilisation of Suspensions", Wiley-VCH, Germany (2012).
[5]   Tadros, Th. F. (ed.), "Solid/Liquid Dispersions", Academic Press, London (1987).
[6]   Bijesterbosch, B. H., "Stability of Solid/Liquid Dispersions", in "Solid/Liquid Dispersions", Th. F. Tadros (ed.), Academic Press, London (1987).
[7]   Kruyt, H. R. (ed.), "Colloid Science", Vol. I, Elsevier, Amsterdam (1952).
[8]   Verwey, E. J. W. and Overbeek, J. Th. G., "Theory of Stability of Lyophobic Colloids", Elsevier, Amsterdam (1948).
[9]   Tadros, Th. F., "Polymer Adsorption and Dispersion Stability", in "The Effect of Polymers on Dispersion Properties", Th. F. Tadros (ed.), Academic Press, London (1981).
[10] Napper, D. H., "Polymeric Stabilisation of Colloidal Dispersions", Academic Press, London (1983).
[11] Tadros, Th. F., "Nanodispersions", De Gruyter, Germany (2016).
[12] Asakura, S. and Oosawa, F., J. Chem. Phys., **22**, 1235 (1954); J. Polymer Sci., **93**, 183 (1958).

Index

www.ingramcontent.com/pod-product-compliance
Lightning Source LLC
Chambersburg PA
CBHW080904220326
41598CB00034B/5471